SHUTTLE-MIR
МИР-ШАТТЛ

The United States and Russia Share History's Highest Stage

By Clay Morgan

The NASA History Series
NASA SP-2001-4225

National Aeronautics and
Space Administration
Lyndon B. Johnson Space Center
Houston, Texas 77058

How to Use This Book

This book is an example of what may become the norm for history texts—an illustrated narrative accompanied by a CD-ROM.

The text tells the story from the human side. It is based on reflections and quotes from the astronauts, cosmonauts, and team members who participated in the historic partnership. The main chapters give the perspectives of the seven U.S. astronauts living on Mir. The "STS boxes" share the Space Shuttle crews' experiences from the 11 Shuttle-Mir missions. The "Meanwhile on Earth" sections provide details of what was happening on Earth while the attention of the Program focused on the situations in space. The text reflects conventional usage; that is, temperature is given in degrees Fahrenheit, and metrics are used as appropriate.

While this publication provides an accurate overview of the Shuttle-Mir Program, the reader is encouraged to explore the companion CD-ROM. It contains a complete, searchable text of the book itself plus source publications, mission data, status reports, drawings and illustrations, videos and hundreds of images, and even a virtual Shuttle-Mir children's book. Of particular interest are the Shuttle-Mir oral history transcripts and the letters written by American astronauts while they were in residence on the Russian space station. Many of the quotes in *Shuttle-Mir: The U.S. and Russia Share History's Highest Stage* came from sources located on the CD-ROM.

The reader is invited to explore the Shuttle-Mir story through the words, images, and insights of those who took part in it.

Library of Congress Cataloguing-in-Publication Data

Morgan, Clay 1950-
Shuttle-Mir = [Mir-shattl] : the U.S. and Russia share history's highest stage / by Clay Morgan
p. cm. — (NASA SP ; 4225) (The NASA history series)
In English.
Parallel title romanized.
Includes bibliographical references and index.
1. Astronautics— United States—International cooperation.
2. Astronautics—Russia—International cooperation. 3. Space shuttles. 4. Mir (Space station)
I. Title: Mir-shattl. II. Title. III. Series. IV. Series: The NASA history series.

TL788.4.M67 2001
629.44—dc21

2001026605

Foreword

Phase 1 - Shuttle-Mir is an amazing story, superbly presented here in text, image and sound, by Clay Morgan, Rebecca Wright, Sandra Johnson, and Paula Vargas, and by the many people who lived the story itself. It was an honor to play a small part in helping make that story real.

One of the many parts of this story that continues to amaze me is the perseverance and undying dedication of those who actually made it happen, especially despite the adversities that arose, seemingly at every turn. These came from internal critics and external doubters, and from the technical challenges, including fire, depressurization, and even "floods" of condensing liquids inside the station. The dedicated team members included the U.S. and Russian space crews who lived and trained far from home; the U.S. and Russian Flight Surgeons who learned every nuance of each other's flight program and medical system, and who gained fluency in each other's language in a few months; the engineers and specialists who lived and worked away from friends and families for months, and who made sense of totally new situations and systems; and the payload experts who worked tirelessly in trying physical conditions, and who integrated the U.S. hardware into Russian space modules and still maintained safety. These are the same caliber of people who put Gagarin into space, who put men on the Moon, who kept the Mir flying beyond its predicted service life, and who brought the Shuttle to a level of capability, reliability, and predictability that as a total package surpasses any launch vehicle in the world. These are people I would trust with my life, and indeed I will do so in the very near future as I begin my deployment onboard the International Space Station (ISS) with my Russian crewmates.

An interesting irony of our continuing partnership with the Russians is that, on the Russian side, the ISS team is virtually the same one that executed Phase 1, as it did many earlier phases of the Soviet and Russian space programs. On the U.S. side, although some Shuttle-Mir veterans are now working other details of the current ISS Program, these people are only a few amongst the dozens of U.S. team members who were not here for Phase 1, or who were so immersed in the separate Space Shuttle and Station Programs that Phase 1 is now barely a memory. To some of them, Shuttle-Mir may be barely a factor in the necessary change from short-duration "Space Shuttle thinking" to long-duration "Space Station thinking." It may bring sadness to some, but this much is true: We will never do things the "old way" again.

Of course, the current ISS team members will write their own story. They will also achieve great things as they overcome many of the same adversities that faced Phase 1. Meanwhile, they will build an incredible piece of hardware in space, using pieces from all over the globe.

But there will never be the same groundbreaking, the same pathfinding, the same cultural breakthroughs that we saw in Phase 1. It had its high points and low points, its high drama and political circus, but on the whole it must be seen as a success achieved by humans of diverse technical and cultural backgrounds, performing on a very public stage.

With the world and the politicians constantly looking over their shoulders, men and women worked through their problems face-to-face, building trust in each other and in each other's goals. This was often done at very personal levels and with high stakes, both physical and emotional, to achieve exactly what the participants set out to do and more: to execute a joint program of scientific achievement and space exploration by partners who had been archenemies less than ten years before. During Phase 1, Russians worked side by side with U.S. specialists at NASA facilities, and Americans lived not only on the Mir space station but also on a formerly secret military base. More importantly, they solved, successfully and safely, every problem they faced together. Learning to solve problems jointly is the skill we must not lose, or we will have to start over.

Shuttle-Mir was a unique challenge at a unique time in history. It may not be fully appreciated for quite a while, possibly not until after we finish the highest risk portions of the ISS and we have time to reflect on what made this all possible, and perhaps not until after we have sorted out some of the relational growing pains we still see in the new operational relationship. The most important message in the story presented here, however, is the story of the people, from the highly visible ones to the ones hidden behind their stacks of documents and boxes of experiment hardware. It is a story of cultural and linguistic misunderstandings as well as technical "mind-melds" and operational tugs-of-war.

Phase 1 - Shuttle-Mir succeeded for three reasons. First, it succeeded because we had the unwavering support and guidance of key leaders such as George Abbey, Dan Goldin, and Yuri Koptev despite the most intense political pressure from outside the two space agencies. Second, it succeeded because we had the initial program structure, set up by Tommy Holloway, Valery Ryumin, Jim Nise, and others, which worked superbly. And finally, it succeeded because the Russians and Americans always found a way to meet each other, sometimes halfway, sometimes on totally different paths, but always striving to find that common place, always trying to learn and to teach at the same time.

We still have so much to learn and so much to teach each other, and we must now include the rest of the world in our story. The story of international space exploration only begins in low Earth orbit. It should end in the stars.

By Frank L. Culbertson, Jr.
Phase 1 Program Manager

Shuttle-Mir
The U.S. and Russia Share History's Highest Stage

Contents

CD-ROM Contents

As in no other time in history, the Shuttle-Mir Program forged a whole new kind of discovery in space—one of cooperation and shared exploration. The story of this partnership continues in the searchable companion CD-ROM.

Text Features

Complete text from the book, plus:
- "Around the Clock, Around the Globe"
- "Before Mir: Early Soviet Space Stations"
- "Getting Back to Gravity"
- "Lifting of Secrecy"
- "NASA and Human Spaceflight"
- "NASA Public Affairs Office"
- "Russian Space Agency and Energia"
- "Safety and Risk"
- "Skylab: America's First Space Station"
- "Working Groups"

Profiles of Astronauts, Cosmonauts, U.S. Mir Residents, Team Members. Shuttle crew biographies listed by missions and increments.

Oral Histories

Stories collected from many Shuttle-Mir participants tell about the day-to-day operations and explain aspects not included in the book. The participants' names serve as links to the transcripts.

Source Materials

The following publications and reports are some of the source materials used for the Shuttle-Mir illustrated history.

- *Brief History of NASA*, by Stephen J. Garber and Roger D. Launius
- *Competition vs. Cooperation 1959 – 1962*, from a NASA report, *The Partnership: A History of the Apollo-Soyuz Test Project*, by Edward Clinton Ezell and Linda Neuman Ezell
- Congressional Mir Safety Hearings, September 18, 1997
- Inspector General's Letter to Congress Concerning Shuttle-Mir Program, August 29, 1997
- *Mir Hardware Heritage*, by D.S.F. Portree
- *Mir Mission Chronicle*, by Sue McDonald
- NASA Shuttle-Mir Press Briefings
- NASA STS Mission Summaries (listed by mission)
- NASA STS Mission Highlights
- NASA Mir Increment Summaries (listed by increment)
- "Six Months on Mir" by Shannon Lucid
- "Part I, Early Space Station Activities – 1923 – July 1965," by Roland W. Newkirk and Ivan D. Ertel with Courtney G. Brooks
- *Phase 1 Research Program Overview*, by John Uri
- *Phase 1 Program Joint Report*, edited by George C. Nield and Pavel Mikhailovich Vorobiev
- The Outer Space Treaty – 1967
- *U.S. Human Spaceflight (1961 – 1998)*, by Judy Rumerman
- "What's in a Name?" by Frank L. Culbertson, Jr.

Timelines

Two timelines present a chronological view of the Shuttle-Mir Program. The Text Timeline lists events with brief explanations; the Graphic Timeline depicts American and Russian spaceflight operations.

Children's Book

This virtual storybook for young space enthusiasts takes a look at living in space.

Video Overview

This 90-second video, *Discovery in Space*, provides a brief overview of Shuttle-Mir.

Science

Research Program Overview, by John J. Uri, Phase 1 Mission Scientist, NASA/Johnson Space Center, Houston, TX; and Oleg N. Lebedev, RSC-Energia, Korolev, Russia.

Phase 1 Science Report featuring information on:
- Advanced Technologies
- Earth Sciences
- Fundamental Biology
- Human Life Sciences
- ISS Risk Mitigation
- Life Support Risk Mitigation
- Microgravity
- Space Sciences

"Combustion Experiments on Mir," submitted by Paul Ferkul, et al.
"Fluid Physics Experiments on Mir," submitted by Jeffrey S. Allen, National Center for Microgravity Research and Suzanne Saavedra, NASA Lewis Research Center
"Space Acceleration Measurement System on Mir," submitted by R. DeLombard, NASA Glenn Research Center, Cleveland, Ohio; and S. Ryaboukha, RSC Energia, Kaliningrad, Russia

Personal Reflections

Letters Home. While onboard the Russian Space Station, the NASA Mir Astronauts shared their thoughts in letters sent to friends and family on Earth. The CD-ROM includes a sample of these.

Personal Mementos. Each of the U.S. Mir residents provided a special memento reflecting their long-duration assignment. This section features those items and a brief description of each.

Galleries

The various galleries offer a visual exploration of Shuttle-Mir. Photos and videos are listed by mission or increment. Animations feature computer-generated images of the Mir deorbit, the collision of the Progress Resupply Vehicle and Mir, and the STS-86 fly-around of Mir. Also featured are the Shuttle-Mir STS mission and increment patches, diagrams of U.S. and Russian spacecraft, a video tour of spaceflight history, and Mir's fiery plunge into the Pacific Ocean as captured by CNN.

History's Highest Stage

Two hundred and more miles above Earth's surface, at the top of a technological ladder humankind had climbed for millennia, the space station Mir set history's highest stage for the world's two most powerful nations.

As the 20th century was nearing its end, the United States and Russia were alike and unalike in many ways. They were alike in their mighty military powers and near equals in the importance of their previous explorations of space. And, for much of the century, they had been alike in the amount of influence that their opposing ideologies had held over "the hearts and minds"—and lives—of people all around the world. The United States had championed free markets, free speech, and free minds. The former Soviet Union had worked for planned economies and planned lives.

On the other hand, the two nations were dramatically unalike in the ways their fortunes had turned. The United States was embarking on a robust economic expansion, partly fueled by the freedom of information in a new Internet world. One of its main exports, democracy, was ascendant in much of the world. In contrast, Russia was emerging from the collapse of the Soviet Union. The Russians themselves were now attempting democracy, but their economy was in shambles and the future of their legendary space program looked worse than uncertain.

It was within this context that the United States and Russia shared the orbital stage of Mir. They wrote, rewrote, revised, and improvised their desired roles for the fast-approaching 21st century—as the leading actors in the exploration of space.

The Shuttle-Mir Program, also known as Phase 1 of the International Space Station

The Russian Space Station Mir and the U.S. Space Shuttle *Atlantis* docked June 29, 1995, for the first time. Mir-19 crewmembers Anatoly Solovyev and Nikolai Budarin photographed the docked Shuttle-Mir while performing a brief fly-around in the Soyuz spacecraft, based at the Mir.

Program, combined spacecraft and statecraft and mirrored the whole history of spaceflight. This joint effort provided the opportunity for Americans and Russians to share expertise and knowledge while residing in space as well as while working together on the Earth. But, the cooperation that was Shuttle-Mir grew out of a fierce ideological and technological competition between America and the old Soviet Union.

Although allied during World War II, the United States and the Soviet Union quickly became competitors in many arenas after 1945. Militarily, both began working on ways to deliver nuclear weapons to be used at great distances. The Americans preferred crewed bombers. The Soviets favored guided missiles, and they quickly led the U.S. in rocket technology. Both nations based their early rocket designs on Nazi Germany's V-2 missile, and both incorporated German rocket scientists into their programs. Both developed guided missiles; and, by 1955, the U.S. Air Force had successfully launched its first Atlas guided missile.

But, on October 4, 1957, a Soviet R-7 rocket launched the Sputnik-1 satellite into Earth orbit. This "October Surprise" shocked and frightened many Americans. Senator Lyndon B. Johnson, the future U.S. President, was at his ranch in Texas on that day. He later wrote, "In the Open West you learn to live closely with the sky. It is part of your life. But now, somehow, in some new way, the sky seemed almost alien."

Less than a month after Sputnik-1, the Soviets launched Sputnik-2. This satellite carried the first Earth creature into orbit, a dog named Laika. The animal died during the mission, but Sputnik-2 demonstrated that the Soviets could launch large payloads and that they might soon put a human into space.

Americans also feared that the Soviets might be capable of mounting a surprise attack against the United States. Several U.S. rocket failures during this time increased apprehension. The Space Race had begun before the United States could even get into orbit. However, America finally succeeded when it launched its Explorer 1 satellite in January 1958. Later that year, in large part to counter the Soviet threat in space, the U.S. Congress established the National Aeronautics and Space Administration, and NASA began operating on October 1, 1958.

During the early years of the Space Race, success was defined by highly visible "firsts." Uncomfortable as it was for America, the Soviet Union achieved nearly all of these firsts. Presidential candidate John Kennedy wrote in 1960, "The first man-made satellite to orbit the Earth was named Sputnik. The first living creature in space was Laika. The first rocket to the Moon carried a Red flag. The first photograph of the far side of the [Moon] was made with a Soviet camera. If a man orbits Earth this year his name will be Ivan."

The man's name, incidentally, was Yuri.

Yuri Gagarin's one-orbit flight, on April 12, 1961, kicked the Space Race into high gear. Indeed, it turned the Space Race into a race to the Moon. On May 25, 1961—barely six weeks after Gagarin's flight and only 20 days after American Alan Shepard's 15-minute suborbital flight in a Mercury capsule—President Kennedy spoke to Congress on "Urgent National Needs." Along with other defense spending, Kennedy called for five Polaris nuclear submarines. Then he turned to the Space Race, saying, "[If] we are to win the battle that is now going on around the world between freedom and tyranny, the dramatic achievements in space which occurred in recent weeks should have made clear to us all

... the impact of this adventure [is] on the minds of men everywhere, who are attempting to make a determination of which road they should take."

Kennedy announced that the United States should "commit itself to achieving the goal, before this decade is out, of landing a man on the Moon and returning him safely to the Earth. No single space project in this period will be more impressive to mankind, or more important for the long-range exploration of space; and none will be so difficult or expensive to accomplish." Kennedy added that, "in a very real sense, it will not be one man going to the Moon ... it will be an entire nation. For all of us must work to put him there." Thus, Project Apollo was born, and the American

nation embraced it. NASA was handed its greatest mission, along with the money with which to accomplish it.

The Soviets never publicly stated their commitment to landing their own cosmonauts on the Moon, but they spent billions on their own lunar program. They built a lunar lander and the giant N-1 Moon rocket. Although the United States had to recover from its tragic Apollo capsule fire, which took the lives of three astronauts, the Soviet Union was unable to keep up with the pace and the successes of the Apollo program. When Neil Armstrong stepped onto the Moon's surface on July 20, 1969, Moscow sent Washington its congratulations.

America eventually landed six Apollo spacecraft on the Moon, with the last one launching on December 7, 1972. Perhaps understandably, after the Moon prize had been won, Americans' interest in space exploration waned and Congress' willingness to spend money on space diminished. Although some have criticized Project Apollo for being an exercise in political one-upmanship, Apollo's engineering payoff was immense. Project

Apollo and its precursor, Gemini, pioneered many technologies and techniques that would be crucial to the Shuttle-Mir Program. These included the rendezvous of two spacecraft in orbit, the docking and undocking of those two spacecraft, and the use of spacewalks to build and maintain structures in orbit.

While competition was the rule in the early years of space exploration, cooperation was always considered and, at times, attempted. In November 1959, NASA's Deputy Administrator Hugh L. Dryden met privately with Soviet space scientists. They agreed that their countries should cooperate more closely in space science, and Dryden made it clear that NASA was ready to talk about issues of mutual interest. However, the Soviets were not prepared to proceed; and the Soviets' downing of an American U-2 spyplane over the Ural Mountains damaged the case for cooperation. In September 1960, President Dwight D. Eisenhower called for making space a nonmilitarized zone, much as the Antarctic continent had been declared earlier. President Kennedy in his 1961 inaugural address proclaimed, "Let both sides seek to invoke the wonders of science instead of its terrors. Together let us explore the stars."

Soon after that, in his State of the Union speech, Kennedy invited the Soviet Union and all nations "to join with us in developing a weather prediction program, in a new communications satellite program and in preparation for probing the distant planets of Mars and Venus, probes which may someday unlock the deepest secrets of the universe." Later in 1961, at a Vienna summit meeting with Soviet Premier Nikita Khrushchev, Kennedy suggested that America and the Soviet Union "go to the Moon together." Khrushchev determined that such cooperation would not be practical, because any rocket boosters used would have military uses.

The remainder of 1961 saw heightened Cold War tensions, including the Soviets' building of the Berlin Wall. However, in spring 1962, following John Glenn's three-orbit spaceflight in a Mercury capsule, it was Khrushchev who called for closer cooperation in space activities. Kennedy wrote him a letter recommending cooperation in weather satellites, tracking services, and other space matters. Khrushchev responded favorably, but stated that as long as the Cold War situation remained, cooperation in space would be limited.

The next year, the United States and the Soviet Union found more common ground in space. In December, after much work by NASA's Dryden and Soviet Academician A. A. Blagonravov (and after being delayed by the Cuban missile crisis), the two countries signed a bilateral space agreement that increased cooperation in space. In 1963, President Mstislav Keldysh of the Soviet Academy of Sciences intimated to a British astronomer that the Soviets might have given up on their own manned lunar project. Partly in response to this, President Kennedy made a speech at the United Nations in which he raised the possibility of a "joint expedition to the Moon." Surprisingly, the Soviet government and media ignored the proposal. In the United States, reaction was mixed. Congress tacked on to NASA's appropriations bill the words, "No part of any appropriation made available to the National Aeronautics and Space Administration by this Act shall be used for expenses of participating in a manned lunar landing to be carried out jointly by the United States and any other country without consent of the Congress." The same language was used in NASA's funding bills for fiscal years 1964 to 1966. America would go to the Moon alone.

But, America would not have space to itself. The United States and the Soviet Union agreed on this by signing the 1967 Outer Space Treaty, which proclaimed, "The exploration and use of outer space, including the Moon and other celestial bodies, shall be carried out for the benefit and in the interests of all countries, irrespective of their degree of economic or scientific develop-

ment, and shall be the province of all mankind."

With the 1969 landing on the Moon of Apollo 11, there was not much left to compete for in space. Although cooperation remained tied to politics, the path toward Shuttle-Mir became easier.

That path actually began just before America landed on the Moon, in early July 1969. At that time, NASA astronaut Frank Borman, who had recently orbited the Moon with his Apollo 8 crewmates, James Lovell and William Anders, undertook a nine-day goodwill tour of Soviet space facilities as U.S. President Richard Nixon's

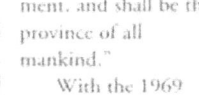

representative. Borman suggested to Soviet space leaders that Americans and Soviets might work together onboard a space laboratory in low Earth orbit. A few days after Apollo 11 returned to Earth, NASA Administrator Thomas O. Paine wrote to Keldysh and suggested cooperative space ventures.

In March 1970, Nixon declared that international cooperation would be an object of America's post-Apollo space program, saying, "I believe that both the adventures and the applications of space missions should be shared.... Our progress will be faster and our accomplishments will be greater if nations will join together in this effort." In September, NASA Administrator Paine wrote to his Soviet colleagues to propose that a Soviet Soyuz spacecraft dock with NASA's planned Skylab orbital workshop.

In October 1970, at Keldysh's invitation, five NASA representatives flew to Moscow for talks on providing compatability to the rendezvous and docking systems of manned spacecraft and space stations. The Soviet delegation included among its members docking system engineer Vladimir Syromiatnikov, who ultimately worked on every Soviet/Russian docking system including the one for Shuttle-Mir. His U.S. counterpart, Caldwell Johnson, brought pictures of NASA's "neuter" docking system. Working to ensure docking compatibility in future generations of spacecraft, these two delegations established three "joint working groups." The successful template of these working groups would continue through the Apollo-Soyuz Test Project, and into the Shuttle-Mir and the International Space Station Programs.

The Nixon Administration remained interested in space cooperation. In January 1971, a delegation led by NASA Acting Administrator George M. Low traveled to Moscow for wide-ranging discussions. U.S. Secretary of State Henry Kissinger had instructed Low to explore the possibility of an early U.S.-Soviet docking mission. Upon his return, Low reported to President Nixon that "apart from our formal negotiations, I did have one private conversation with Keldysh {along lines discussed earlier with Dr. Kissinger} ... I explained to Keldysh that we believe it technically possible to modify Apollo spacecraft and Soyuz modules so as to permit them to dock in the 1973-74 period, several years before entirely new systems like the Space Shuttle would become available."

Keldysh was warm to the idea. Nixon and Kissinger wanted an early international space mission as a high-flying demonstration of their policy of détente (easing of tensions) between the United States and the Soviet Union. From NASA's point of view, an early international docking mission made programmatic sense. Nixon was scaling back and redirecting the Agency, and no space missions were planned between the last Skylab visit in 1974 and the planned first Space Shuttle test flight in 1978 (which actually occurred in 1981). An international docking mission would help fill the gap.

An international mission would also replace competition with cooperation while keeping politics—foreign relations—as an important reason to explore space. After Apollo, spacecraft would remain statecraft, but they would often be flying more than one nation's flag.

In 1975, the Apollo-Soyuz Test Project became the first international human spaceflight. It epitomized the Nixon-Ford Admininstrations' emphasis on détente with Communist nations. It also had ambitious technological goals, especially the compatibility of the U.S. and Soviet rendezvous and docking systems.

On July 15, 1975, Soyuz-19 launched from the Baikonur Cosmodrome in the then-Soviet Republic of Kazakhstan, carrying cosmonauts Aleksei A. Leonov and Valeri N. Kubasov. Less than eight hours later, a Saturn IB rocket carrying a modified Apollo capsule launched from Kennedy Space Center in Florida— 10,000 miles

from Baikonur. Onboard the Apollo were astronauts Thomas P. Stafford, Vance D. Brand, and Donald K. "Deke" Slayton. At the Kennedy Space Center, the Soviet Ambassador to the United States, Anatoly F. Dobrynin, told the firing-room team, "The best of wishes to all of you, and, of course, to both our crews. My heart is with you." NASA Administrator James C. Fletcher announced, "You're making history today. This is the first step on a long mission and a first step on a long program with the Soviet Union."

Once in orbit, the U.S. crew flipped around the cone-shaped Apollo capsule and mated it with a 10-foot-long docking module, which would later attach directly to the Soyuz. On July 17, 1975, the two spacecraft rendezvoused over South America and docked over the Atlantic Ocean. Command Module Pilot Brand later recalled that the docking occurred as if in slow motion. "We came together at a slow rate and we felt a little gentle bump when we docked," he said. "We looked through an optical sight to line it up. There was some relative movement between the spacecraft and then it quickly stopped."

After three hours of tests and preparation, the hatches were opened and spacecraft commanders Stafford and Leonov performed the ceremonial handshake that would be repeated many times during the Shuttle-Mir Program. "Glad to see you," Stafford said in Russian to Leonov. "Very, very happy to see you," Leonov replied in English.

NASA tradition includes the creation of "patches" to symbolize and reflect the missions and programs of its Agency. The Shuttle-Mir patch contains the rising Sun that signifies the dawn of this new era of human spaceflight—the first phase of the U.S.-Russian space partnership. The Shuttle-Mir combination, docked to acknowledge the union of the two space programs, orbits over an Earth devoid of any definable features or political borders to emphasize that the Earth is the home planet for all humanity.

On Earth, U.S. President Gerald Ford and Soviet Communist Party General Secretary Leonid Brezhnev expressed congratulations, and the two crews exchanged flags and other commemorative items. Over the next three days, the crews performed joint activities, such as sharing meals, carrying out scientific experiments, and giving television viewers tours of the joined spacecraft. On July 18, they closed the hatches; and the next day, the Apollo and Soyuz undocked.

Brand said, "It was a positive experience from beginning to end. It was the middle of the Cold War, and [the Soviets] probably saw us as monsters and vice versa before we started training. We opened a crack in the door regarding communication between the two superpowers. Apollo-Soyuz was the first for this kind of cooperation."

Apollo-Soyuz proved to be enormously successful, both symbolically—in those very tense times—and technically for the future. Figuratively, where Apollo-Soyuz could be seen as an introductory handshake, Shuttle-Mir would become a confirming embrace, albeit two decades later.

Looking back on his Apollo-Soyuz experience, Stafford said that, although some Americans were worried about technology transfer to the Soviets, the United States may have learned more than the Soviets did. "Russia was such a closed society. The amount of wheat they produced in a year was a state secret…. We saw the control center and the launch site," although Stafford had to, in his words, "pry it open." But, according to Stafford, "we learned more about where they were than they learned about where we were." In Stafford's opinion, "What they could have learned from us was management," because their management structures were so "vertically separated."

After Apollo-Soyuz, NASA officials began seriously discussing the possibility of the Space Shuttle docking with the Soviets' Salyut space station. Much thought was also given to what was called "air-to-air extravehicular activity capability"; that is, transferring astronauts between the vehicles using spacesuits instead of directly docking the two spacecraft. In August 1976, NASA's Glynn S. Lunney, who had managed the Apollo-Soyuz Test Project, suggested flying a cosmonaut onboard a Shuttle. In late 1976, NASA and Soviet officials agreed to a Shuttle-Soyuz docking, but nothing was signed because of the upcoming U.S. presidential elections. In May 1977, both sides agreed to renew the 1972 Space Cooperation Agreement for a second five years. The Agreement called for the study of a Space Shuttle docking with a Salyut space station in 1981 and for setting up two working groups, one for science and one for operations.

However, the new U.S. President Jimmy Carter and his administration were also worried about technology transferring to the Soviets and, by late 1978, had discontinued all docking discussions. When the Soviet Union invaded Afghanistan in December 1979, any chance for a Shuttle-Salyut docking collapsed as the U.S. administration did what it could to isolate the Soviet regime internationally, including boycotting the 1980 Olympics.

Throughout the 1980s, events served to buoy and then batter the prospects of U.S.-Soviet space cooperation. For the first time after Apollo-Soyuz, America returned to human spaceflight on April 12, 1981, when NASA astronauts John Young and Robert Crippen launched aboard the first Space Shuttle, *Columbia*. In December, Polish government authorities cracked down on the democratic Solidarity movement. In the spring of 1982, President

Ronald Reagan allowed the Space Cooperation Agreement to expire; and in January 1984 in his State of the Union speech, Reagan called on NASA to build its own space station. He echoed President Kennedy's "Urgent National Needs" Moon speech by saying he wanted the station built within a decade.

However, in June 1984, Reagan proposed "a joint simulated space rescue mission" with the Soviets, and Congress and the media renewed attention to space cooperation. Regardless, the Soviets did not respond. In March 1985, Mikhail Gorbachev came to power in Moscow. There began a great period of uncertainty and reform, which would lead ultimately to the collapse of the Soviet government.

Tragedy struck the U.S. space program again in January 1986 with the explosion of the Space Shuttle *Challenger* and the deaths of its crew of seven. The accident left the American public shocked and unsure about the future of human spaceflight. NASA's Shuttle fleet would remain grounded for the next two years. Meanwhile, the Base Block of the Soviets' new space station Mir was being built.

In 1986, the first Mir crew activated the new space station. They then transferred via a Soyuz capsule to the Salyut-7 station to collect equipment, and returned to Mir before returning to Earth. Two years later, in 1988, the Space Shuttle returned to flight with the launch of STS-26, *Discovery*. The Soviets tried out their own Space Shuttle in November 1988, when they launched the *Buran* orbiter. *Buran* flew successfully—without a crew. However, it soon was scrapped, partly for economic reasons, and the Soviets still had no reusable spacecraft with which to ferry crews and resupply Mir.

The 1990s loomed full of uncertainty, but this would be the decade that brought the two space powers together.

In April 1989, President George H.W. Bush had reestablished, by executive order, the National Space Council, led by Vice President Dan Quayle. Part of its job was to find a direction for America's space initiatives in a time when the nation would no longer be engaged in a technology race with the Soviet Union. NASA Administrator Daniel S. Goldin worked closely with Vice President Quayle and important Space Council members, including Executive Secretary Mark Albrecht, Apollo-Soyuz Commander General Stafford, and the Council's Senior

Director for Civil Space Policy, George W. S. Abbey. The Council began to see several unique opportunities for engaging the former Soviet Union in a space station program.

The Space Council saw a space station as a logical next step—back to the Moon and on to Mars. They were interested in reducing the cost of the space station. They also wanted to take advantage of the enormous space assets that the Soviets fielded; for example, their unique heavy-lift boosters and the trustworthy Soyuz crew return vehicle. The Soviets also had an existing space station in their Mir, onboard which American astronauts might gain valuable experience in long-duration spaceflight even before an American station was launched.

Moreover, several on the Space Council, as well as others in the Bush Administration, saw another reason to engage the post-Soviets in a cooperative space venture: as a way to help hold the Russian nation together at a time when the Russian economy was faltering and its society was reeling. In the words of Brian Dailey, Albrecht's sucessor, "If we did not do something in this time of social chaos … in Russia, … then there would be potentially a hemorrhaging of technology … 'away from Russia' … to countries who may not have a more peaceful intention behind the use of those technologies."

Bringing Russia into America's future space plans would help get America to Mars and preserve what President Bush was calling "a new world order." George Abbey, with the help of Dailey and others, began working on a plan that would partner the American Space Shuttle with the Russian Mir program.

Events moved quickly. In May 1990, U.S. Vice President Quayle discussed space cooperation with Soviet President Gorbachev. In 1991, the Soviet Union collapsed. In June 1992, the U.S. and the new Russian Federation issued a "Joint Statement on Cooperation in Space." It called for Russian cosmonauts aboard Space Shuttles, U.S. astronauts onboard Mir, and a Shuttle-Mir docking mission in 1994 or 1995. The agreement also opened the door to U.S. commercial purchases of Russian space services.

In July 1992, at Quayle's urging and Russian President Boris Yeltsin's concurrence, NASA Administrator Dan Goldin met with the Russian Space Agency's General Director Yuri Koptev in

Russia to visit Russian space facilities and to work toward implementation of a new space agreement. Tom Stafford accompanied Goldin as his advisor. In August, NASA's Bryan O'Connor met in Moscow with the Russian Space Agency's Boris Ostroumov and with Valery Ryumin of RSC-Energia to discuss implementation of the cooperative human spaceflight programs. In October,

Goldin and Koptev met again in Moscow to sign the "Implementing Agreement on Human Space Flight Cooperation." This detailed plan included a cosmonaut flight on Shuttle mission STS-60, an astronaut's Soyuz flight and long-duration stay onboard Mir, and a Shuttle-Mir docking that would include a Russian Mir crew exchange and the pick up of the U.S. long-duration astronaut.

The Goals of Shuttle-Mir

Goal 1 Learn How to Work with International Partners

International cooperation is both a challenge and a benefit of shared space ventures. Working with Russia in the Shuttle-Mir Program provided the U.S. with experience for the International Space Station, with its 16 participating nations. In Shuttle-Mir, astronauts and cosmonauts trained in each other's languages and took part in each other's operations. NASA and the Russian Space Agency put together a truly "joint program." NASA's Program Director Frank L. Culbertson said, "It was important that people realize we were not only teaching each other, but observing each other and learning from each other, and that both sides had a lot to offer. It wasn't just a one-way street by any means, and I think we've proven that over and over."

Goal 2 Reduce Risks of Developing and Assembling a Space Station

The International Space Station is a much larger, more complex program than was Shuttle-Mir. Lessons learned during Shuttle-Mir helped reduce risks. The Russian-built, American-integrated docking system made dockings simpler and safer. The astronaut experiences aboard Mir led to many modifications, including single-command shutdown of ventilation systems to prevent the spread of fire, quick-disconnect cables in case of depressurization, and additional lighting for the International Space Station to help navigation during rendezvous. "We were always looking for ways," Culbertson said, "to try out hardware that was going to be used on the station; [to] develop procedures that were going to be used; to react to the situations that arose; and [to] develop processes to respond to them." Sometimes, he said, "we got more than we bargained for." Regardless, Shuttle-Mir "certainly taught us a lot about what could happen during the International Space Station."

Goal 3 Gain Experience for NASA on Long-Duration Missions

Before the Shuttle-Mir Program, NASA's longest-duration experiences had been the Skylab missions of 1973-74. Space Shuttle flights are limited to about two weeks. Shuttle-Mir enabled U.S. astronauts to spend 975 days—more than 27 consecutive months—on a space station. That was more time in orbit than had been accumulated since the Shuttle Program began in 1981.

Goal 4 Conduct Life Science, Microgravity, and Environmental Research Programs

Science conducted by astronauts aboard Mir included over 100 investigations in eight disciplines, including Advanced Technologies, Earth Sciences, Fundamental Biology, Human Life Sciences, International Space Station Risk Mitigation, Life Support Risk Mitigation, Microgravity, and Space Sciences. According to Culbertson, it is important to remember "that anytime we have the opportunity to conduct research on any platform—whether it's at sea, in Antarctica, in space—we do it."

In September 1992, Johnson Space Center (JSC) Director Aaron Cohen established a U.S.-Russian Programs Office to accommodate the evolving relationship between NASA and the Russian space community. Donald Puddy became Special Assistant for U.S.-Russian Programs, responsible for coordinating "all JSC activities that support joint U.S. and Russian Federation programs."

It was at this point that what would become the Shuttle-Mir Program got rolling, in all practicality, with the inclusion of a Russian cosmonaut in the STS-60 Space Shuttle crew. Shuttle training for the cosmonaut had to begin immediately. The Russians selected Cosmonaut Sergei K. Krikalev, who would become the "prime," and Vladimir Titov, who would be backup and who would later fly in 1995 onboard STS-63. According to Tommy E. Capps, who was responsible for spaceflight training, "We had from November [1992] to February [1993] to get them up to speed on the Shuttle." This meant, "Basic training for Shuttle. Understanding what Shuttle was. Understanding a little bit about their roles and responsibilities. So, it was a very difficult job for all of us. We had a lot of initial contacts with the Russians on training plans, trying to learn to communicate with each other, even though through an interpreter it was [still] a different language [even] after we communicated."

Capps and others at the Johnson Space Center worried how these two non-English speakers would adjust to American society. Capps later said, "We started trying to do things like take them out for driving lessons. We were really apprehensive about them driving in Houston, in all of our aggressive freeway traffic and so forth." But, a few months later, Capps visited Russia. About this he said, "After a few hours in Moscow, I realized that we were totally silly for wasting our time trying to teach these guys how to drive in Texas. If they could survive Moscow traffic, then they had no problem whatsoever driving." Capps didn't mention that Krikalev was a champion Russian aerobatics pilot.

In April 1993, newly elected U.S. President Bill Clinton and Russian President Yeltsin met at a summit in Vancouver, Canada, where both sides agreed to an "enhanced" Shuttle-Mir Program. Late in May, NASA and Russian representatives met to consider such an expansion. Among other subjects, they agreed to consider two 3-month and four 6-month U.S. flights aboard Mir through 1997, with astronauts delivered to the station by Russian Soyuz-TMs and U.S. Space Shuttles; using the new Russian Mir modules, Spektr and Priroda, for U.S. equipment; and using joint spacewalks to extend Mir's useful lifetime through the end of 1997.

In September 1993, U.S. Vice President Al Gore and Russian Prime Minister Viktor Chernomyrdin chaired the first meeting of the

Frank Culbertson

Astronaut Frank Culbertson served as NASA's Shuttle-Mir Program Manager, leading the Program from offices at the Johnson Space Center. He was responsible for the Program's overall management as well as for executing many crucial decisions, such as the one to continue the Program after a fire onboard Mir. Culbertson often represented Phase 1 to Congress and to the media and public, and he led many negotiations with the Russians. Culbertson also served as Technical Director for the Program's Management Working Group with his Russian counterpart, Valery Ryumin.

During Shuttle-Mir, Culbertson functioned under a tremendous workload, but he recognized his teammates were similarly burdened. NASA psychiatrist Chris Flynn, who worked in Russia as a Flight Surgeon during Shuttle-Mir, talked about Culbertson's awareness of human stress factors.

Flynn said, "One of the things that I appreciated Frank Culbertson for ... is that maybe more than any other manager that I've met at NASA, he appreciates that it's not just wiring. It's not just fuses that have limits. It's not just metal [with] a limit of how much it can be stressed. But, ... human beings have limits, [too]. And, that ... gets lost without people like Frank Culbertson."

Before the Shuttle-Mir Program, Culbertson piloted Space Shuttle mission STS-38 and commanded STS-51, logging over 344 hours in space. He became an astronaut in 1985, after earning a Bachelor of Science degree in aerospace engineering from the U.S. Naval Academy and serving as an aviator in the U.S. Navy.

After Shuttle-Mir, Culbertson served as the International Space Station's Manager for Operations until he was chosen to command Expedition 3 to the International Space Station. Selected as his two Russian crewmates were Flight Engineer Cosmonaut Mikhail Turin and Soyuz Vehicle Commander Lt. Col. Vladimir Dezhurov, who was a commander on the Russian space station during the Shuttle-Mir Program.

first U.S.-Russian Joint Commission on Energy and Space (Gore-Chernomyrdin Commission). The United States and Russia agreed to begin "Phase 1" of International Space Station cooperation immediately with Shuttle-Mir expanded to include up to two years of total U.S. time on Mir.

On November 1, 1993, the "Addendum to the Program Implementation Plan" merged the previously planned U.S. space station *Alpha* with Russia's planned Mir-2, firmly establishing the International Space Station and its three "phases." Phase 1 would be the Shuttle-Mir Program. Phase 2 would be the construction of the International Space Station. Phase 3 would be the station's full operation.

In December 1993, the Gore-Chernomyrdin Commission formally agreed to expand Shuttle-Mir to include up to 10 Shuttle flights to Mir and four or more long-duration stays for a total of 24 months by U.S. astronauts on Mir. On

December 16, 1993, NASA's Dan Goldin and the Russian Space Agency's Yuri Koptev signed the "Contract for Human Space Flight Activities," and the Shuttle-Mir Program was ready to do real space business.

During the formative days of the Shuttle-Mir Program, NASA Associate Administrator Mike Mott played an important role in defining what the overall program would be like. Associate Administrator Arnauld Nicogossian represented the interests of science. Besides the earlier-mentioned Dan Goldin, George Abbey, Mark Albrecht, Brian Dailey, and Bryan O'Connor, key NASA negotiators and players at this time included Guy Gardner, the Deputy Associate Administrator for Russia, as well as new Johnson Space Center Director Carolyn Huntoon and technical negotiator James Nise, who would later take charge of contracts for the Shuttle-Mir Program. William Shepherd, Robert Clark, and Lee Evey also contributed greatly.

Russian negotiators included the Russian Space Agency Chief Yuri Koptev, Cosmonaut and Shuttle-Mir Project Manager Valery Ryumin, and Technical Director Boris Ostroumov.

About negotiating with the Russians, James Nise later said, "The Russians don't delegate much stuff 'downhill'; that is, to lower than top-level managers. All the power [resided] in one person at the time. Now, they will work stuff 'down below,' but it all gets signed at the very top. So, it's kind of difficult in that regard."

Of course, much was difficult in many regards; but on February 3, 1994, NASA announced that astronauts Norman Thagard and Bonnie Dunbar would train for missions on Mir. On the same day, Space Shuttle *Discovery* (STS-60) launched with cosmonaut Sergei Krikalev onboard.

The U.S. and Russia, the world's leading actors in the drama of space, stepped onto the highest stage together again.

Valery Ryumin

Russian Shuttle-Mir Manager Valery Ryumin knew space and spacecraft, and he had a personal knowledge of life in cramped quarters—both in space and on Earth. He once served as a tank commander in the Soviet Army and he set space endurance records onboard one of Mir's predecessors, the Salyut-6 space station.

Born in 1939 in the city of Komsomolsk-na-Amur in the Russian Far East, Ryumin graduated in 1958 from the Kaliningrad Mechanical Engineering Technical College with the specialty "Cold Working of Metal." After time in the Army, he moved on to electronics and in 1966 graduated from the Department of Electronics and Computing Technology of the Moscow Forestry Engineering Institute with the specialty "Spacecraft Control Systems." He worked in the positions of Ground Electrical Test Engineer, Deputy Lead Designer for Orbital Stations, and Deputy General Designer for Testing. He also helped develop and prepare all Soviet orbital stations, beginning with Salyut-1.

In 1973, he joined the cosmonaut corps and logged a total of 362 days in space, spending two days in 1977 onboard Soyuz-25, 175 days in 1979 onboard Soyuz vehicles and the Salyut-6 space station, and 185 days in 1980 onboard Soyuz vehicles and Salyut-6. From 1981 to 1989, Ryumin served as Flight Director for Salyut-7 and the new Mir space station. In 1992, he became Manager of the Russian portion of the Shuttle-Mir Program. In 1998, he became the ninth and final cosmonaut of Shuttle-Mir, flying onboard STS-91, the last Shuttle flight to Mir.

Twice honored as a "Hero of the Soviet Union," Ryumin is a complex and outspoken man who had to lose weight and improve his English in preparation to fly as a Shuttle crewmember. Commander Charlie Precourt said Ryumin had such a firm grounding in space systems that he learned the Shuttle's systems quickly. And, 18 years after his last space mission, Ryumin was able to visit the space station he had been working so hard to make a success.

Space Shuttle *Discovery*

Launched: February 3, 1994, 10:05 a.m. EST
Kennedy Space Center, Pad 39-A

Orbit: 191 nautical miles

Inclination: 57 degrees

Landed: February 11, 1994, 2:18 p.m. EST
Kennedy Space Center

Mission: 8 days, 7 hours, 9 minutes

STS-60 CREW
Commander Charles F. Bolden
Fourth Shuttle flight
Pilot Kenneth S. Reightler, Jr.
Second Shuttle flight
Mission Specialist N. Jan Davis, Ph.D.
Second Shuttle flight
Mission Specialist Ronald M. Sega, Ph.D.
First Shuttle flight
Mission Specialist Franklin R. Chang-Díaz, Ph.D.
Fourth Shuttle flight
Cosmonaut Sergei K. Krikalev
Russian Space Agency
First Shuttle flight and third space mission

PAYLOADS
Wake Shield Facility-1
SPACEHAB-2 (Space Habitation Module-2)
Capillary Pumped Loop Experiment
Three-Dimensional Microgravity Accelerometer
Astroculture Experiment
Bioserve Pilot Lab
Commercial Generic Bioprocessing
Commercial Protein Crystal Growth
Equipment for Controlled Liquid Phase Sintering
Immune Response Studies
Organic Separations Experiment
Space Experiment Facility
Penn State Biomodule
Space Acceleration Measurement System
Orbital Debris Radar Calibration Spheres
University of Bremen Satellite
Ball Bearing Experiment
Orbiter Stability Experiment
Medicines in Microgravity
Heat Flux Experiment

Joint U.S.-Russian Investigations
Radiological Effects
Sensory Motor Investigation
Metabolic Investigation
Visual Observations from Space

STS-60
A Cosmonaut Flies on the Shuttle
February 3 – 11, 1994

The five astronauts and one cosmonaut of STS-60 gather in the tunnel that connects *Discovery* and the SPACEHAB module. They are (upper right) Commander Bolden, (clockwise) Mission Specialists Sega, Davis, Chang-Díaz, and Krikalev, Pilot Reightler.

Sergei Krikalev became the first Russian cosmonaut to fly on the U.S. Space Shuttle when he launched with his five NASA crewmates onboard *Discovery*. Krikalev and his backup, Vladimir Titov, joined the STS-60 mission after the U.S. crew had already been assigned; however, Krikalev was able to take full part in the mission. His roles included manipulating the Shuttle's payload bay "arm" and operating the Space Acceleration Measurement System experiment, as well as participating in the joint science experiments. Krikalev's backup Titov would go on to fly on the STS-63 "near Mir" mission. Besides gaining practical experience on an American Space Shuttle, Krikalev helped further diplomatic and public relations in ways that hearkened back to the Apollo-Soyuz Test Project of 1975 and pointed forward to the Shuttle-Mir flights.

On February 8, 1994, the ABC-TV program *Good Morning America* telecast a live hookup between the *Discovery* crew and the three cosmonauts aboard the space station, while the Orbiter flew over the Pacific Ocean and Mir was above the southern United States. The next day, Krikalev and STS-60 Commander Charles Bolden received a call from Russian Prime Minister Viktor Chernomyrdin. During the conversation, Chernomyrdin said that he wanted to hear Krikalev speak some Russian because the day before on ABC-TV he thought he had heard him speaking to the Mir cosmonauts in English. But, Krikalev said that was "not exactly correct…. I was speaking Russian, but the interpreter in New York was speaking in English." He added, politely, that the American astronauts could speak some Russian words "without an accent."

During STS-60, Krikalev observed and participated in a typical, busy Space Shuttle science

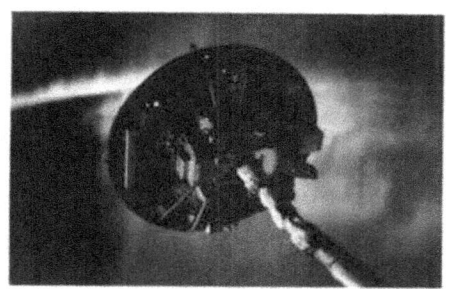

The Southern Lights, between New Zealand and Australia, serve as the backdrop for the Wake Shield Facility, extended by the Remote Manipulator System on *Discovery* during STS-60.

mission that included many varied experiments and investigations. In the payload bay was the commercially developed SPACEHAB-2 facility, providing an additional 1,100 cubic feet of working and storage space. Experiments being carried in SPACEHAB-2 involved materials processing, biotechnology, and hardware and technology development payloads. Also in the bay was the Wake Shield Facility (WSF), a 2-foot-diameter, stainless-steel disk designed to generate an "ultra-vacuum" environment in space within which to grow thin semiconductor films for next-generation advanced electronics.

STS-60 also carried the 100th Get Away Special (GAS) payload. This program provides an opportunity for a variety of smaller experiments to be conducted in space. Among STS-60's GAS payloads were efforts to create a new kind of ball bearing, measure the vibration level during normal Orbiter and crew operations, and understand the boiling process in microgravity.

One disappointment for the STS-60 crew and for scientists was the failure to deploy the WSF to free-fly away from *Discovery*. Several factors postponed the deployment until controllers were concerned that there wasn't enough time remaining to deal with problems that might occur during WSF's free flight. WSF remained at the end of the Shuttle's arm and, from that position, was able to conduct part of its mission.

Sergei K. Krikalev

Sergei Krikalev endured a tough training schedule to prepare for STS-60. One of his more difficult challenges was the English language. He told NASA manager Travis Brice, "By the time I get through with all my studies for the classes … I normally have … from about one o'clock to two o'clock in the morning to do my English language training."

Sergei Konstantinovich Krikalev was born in 1958 in Leningrad and joined NPO Energia in 1981. An aerobatics pilot, he became Champion of Moscow in 1983 and Champion of the Soviet Union in 1986. For Energia, he tested spaceflight equipment, developed space operations methods, and participated in ground control operations. When the Salyut-7 space station failed in 1985, he worked on the rescue mission team, developing procedures for docking with the uncontrolled station and for repairing the station's onboard system.

Selected as a cosmonaut in 1985, Krikalev was first assigned to the *Buran*-Shuttle Program. In early 1988, he began training for his first long-duration flight onboard the Mir space station. He launched to Mir for a five-month mission on November 26, 1988. He flew to Mir again in 1991, and some called him "the cosmonaut without a country" after the Soviet Union broke apart during his time on the space station. He returned to Earth as a Russian citizen on March 25, 1992.

In October 1992, NASA announced that an experienced cosmonaut would fly onboard a

future Space Shuttle mission. Krikalev was one of two candidates the Russian Space Agency named for mission specialist training with the crew of STS-60. In April 1993, Krikalev was assigned as the crew's prime mission specialist. He launched with his American crewmates on February 3, 1994, to inaugurate the operational phase of the Shuttle-Mir Program and become the first cosmonaut to fly on an American Space Shuttle.

After his STS-60 Shuttle mission, Krikalev actively supported U.S.-Russian joint operations throughout the Shuttle-Mir Program, working in the Mission Control Center at NASA's Johnson Space Center and at the Mission Control Center-Moscow and training as backup cosmonaut for STS-63.

Also, Krikalev flew on *Endeavour* on STS-88 (December 4-15, 1998), the first International Space Station assembly mission. Recently, he served as flight engineer on the first International Space Station Expedition crew with American Commander Bill Shepherd and Soyuz Commander Yuri Gidzenko.

Meanwhile on Earth
February 1994 – March 1995

The launch of *Discovery* (STS-60) on February 3, 1994, shifted Shuttle-Mir into a higher gear. The United States and Russia began the operational phase of the Program, with cosmonaut Sergei Krikalev flying as a crewmember of a Space Shuttle. A future Shuttle mission, STS-63, would be reconfigured to include a rendezvous—but no docking—with the Mir space station.

On Earth, events were moving quickly, too. On the same day that Krikalev launched on STS-60, astronauts Norm Thagard and Bonnie Dunbar were announced as the first NASA astronauts to go to Russia to train for a flight on Mir. Thagard was the prime candidate, with Dunbar as backup. They would train, often side by side, at the Gagarin Cosmonaut Training Center in Star City, Russia.

The next day, February 4, began the first meetings of "Team Zero"—the first of ten working groups staffed by U.S. and Russian counterparts-experts from NASA, the Russian Space Agency, Energia, the Institute for Biomedical Problems, the Gagarin Cosmonaut Training Center, and other organizations and companies. These working groups were organized on the model of the Apollo-Soyuz Test Project in 1975. Their structure divided mission planning and execution into different functions, including management, cargo and scheduling, public affairs, safety, operations, science, training, integration, extra-vehicular activities, and medicine. Each nation designated for each group a cochair, who facilitated joint meetings (usually weekly via teleconferences, and occasionally face-to-face). Cochairs were authorized to sign protocols, documenting agreements made within their purview. Working group members generally came to trust each other and work well together, finding common ground in shared interests and humor. The Russian Shuttle-Mir Manager, Valery Ryumin, was heard to joke that the management group was called Working Group Zero because "Managers don't do anything."

On February 23, NASA's Office of Life Sciences and Microgravity Science Applications began soliciting the scientific community for investigations to be carried out by astronauts

onboard Mir. On the same day, astronaut Ken Cameron, who would later command *Atlantis* (STS-74) on a mission to Mir, was announced as NASA's first Director of Operations-Russia. Cameron, his fellow astronauts Thagard and Dunbar, along with Flight Surgeons Dave Ward and Mike Barratt left the same day for Star City, Russia.

In May, NASA and Russian Space Agency negotiating teams met in both Houston and Moscow. Also in May, the first meeting of the Task Force on the Shuttle-Mir Rendezvous and Docking Missions was held at the Johnson Space Center. The NASA Advisory Council created this task force—often called the Stafford Committee for its chairman, Apollo-Soyuz Commander Tom Stafford—to review Shuttle-Mir planning, management, training, operations, and rendezvous and docking and to make recommendations.

In June, Randy Brinkley, NASA's Space Station Program Manager, delegated Shuttle-Mir management to the Space Shuttle Program. Director of Space Shuttle Operations Brewster Shaw announced the creation of a Phase 1 Management Group, with Tommy Holloway of the Space Shuttle Program Office as its chairperson. According to Tom Cremins, who would become NASA's Deputy Assistant to the Director in Russia during Phase 1, "Tommy had a real small staff. I remember going to some of his meetings, and there were probably five or six direct staff." The group put together a "roles and responsibilities" document, organizing Phase 1 of the International Space Station Program. This would also be known as "Shuttle-Mir."

On June 23, 1994, Vice President Gore and Russian Foreign Minister Chernomyrdin met again in Washington, D.C. There, the two countries' space agency chiefs—Dan Goldin and Yuri Koptev—signed a "Definitized Contract" between the Russian Space Agency and NASA, which Gore and Chernomyrdin jointly announced. The contract provided for: $400 million in U.S. expenditures; up to 21 additional months of flight time aboard Mir for U.S. astronauts; up to nine more Shuttle dockings with Mir after STS-71; a docking module to permit repeated Shuttle dockings with Mir; the use of

the Russian modules Spektr and Priroda for U.S. experiments; a joint U.S.-Russian research program onboard Mir; joint technology development; and the extension of Mir's lifetime beyond 1995—to allow time for U.S.-Mir operations.

Cremins later commented on some of the non-space-related motivations behind America's financial stake in the new Russia. "A lot of it was directly related to trying to discourage Russian scientists and engineers from leaving the country and working for folks who didn't necessarily have a … good global interest … as well as to discourage some sales that were going on at that time; for example, the sale of rocket technology to India." According to Cremins, the United States was "trying to create a positive venue for Russia and us, to be working in a nonthreatening manner. So, that was really what generated—at the top level—a lot of the support for doing the flights to Mir."

On October 3, 1994, a Soyuz rocket launched from the Baikonur space complex, carrying Mir Principal Expedition-17 Cosmonauts Aleksandr Viktorenko, Elena Kondakova, and Valeri Polyakov. These cosmonauts would greet NASA-1 Astronaut Thagard and his Mir-18 crewmates when they docked with the space station in March 1995.

NASA's Shuttle-Mir Program became a formal stand-alone program on October 6, 1994, when the NASA Associate Administrator for Spaceflight, Jeremiah Pearson III, signed a letter establishing the program plan and officially appointing Tommy Holloway as Manager.

Despite the injection of U.S. funds into the Russian economy, the Russian government was in danger of financial ruin. On October 11, 1994, the ruble collapsed. On November 2, Ryumin informed Holloway that launch targets for the Priroda and Spektr modules would not be met. (Spektr would eventually be launched on May 20, 1995, while Thagard was onboard Mir. Priroda would launch nearly a year later, on April 23, 1996, when Shannon Lucid was a Mir resident.)

Between November 3 and 14, 1994, tools, devices, and procedures planned for use on Shuttle-Mir missions were tested on STS-66. In December, Gore and Chernomyrdin agreed to the formation of a Joint Medical Policy Board

Tommy Holloway

In October 1994, Tommy W. Holloway became the first Manager of the Phase 1 Program when this historic partnership became a formal stand-alone NASA program. He remained in this position until August 1995, when he was assigned as the Manager for the Space Shuttle Program. In 1999, Holloway became Manager of the International Space Station Program.

Early in the Phase 1 Program, there was some uncertainty over whether Phase 1 was more part of the Space Shuttle Program, because of all the rendezvous and dockings, or more part of the Space Station Program, because of the long-duration astronaut residencies onboard Mir. Holloway provided definition and leadership.

Frank Culbertson, who followed Holloway as Phase 1 Manager, said, "Tommy Holloway did a really good job of pulling together a compromise ... that established a program office—

a framework of working groups with which to negotiate and work the issues.... Some of it was based on the Apollo-Soyuz experience ... Some of it was just good common sense."

According to Culbertson, Holloway got everyone to agree that "this is the way we're going to manage this program. This is the way it's going to exist. And, he was able to establish a budget for it, a schedule, and everything. And, I thought that that was the real foundation of our success—was that early work Tommy did."

Holloway began his career with NASA in 1963 at the Johnson Space Center, planning activities for Gemini and Apollo flights in the Mission Control Center. He was a flight director in Mission Control for early Space Shuttle flights and became Chief of the Flight Director Office in 1985. In 1989, he was named Assistant Director for the Space Shuttle Program for the Mission Operations Directorate. He became the Deputy Manager for Program Integration with the Space Shuttle Program in 1992.

In September 1999, he received the prestigious Robert R. Gilruth Award, which is given to recognize outstanding managers of operations, engineering, and science programs.

to coordinate development of a common system of medical support for Shuttle-Mir. They also signed a customs agreement providing for duty-free clearance of NASA's Shuttle-Mir supplies shipped to Russia.

Regardless of the agreements, early in the Shuttle-Mir Program almost everything was difficult. This included getting funds to managers in Russia where society remained in a deep governmental and economic crisis, according to NASA manager Travis Brice, who worked on Shuttle-Mir training, communications, and coordination. When Johnson Space Center (JSC) managers needed to get funds to their Star City office, they first had to transfer money to NASA Headquarters, in Washington, D.C., which then transferred the money to the U.S. State Department. According to Brice, "Then the State Department transferred [the money] over to the Paris Embassy.... The Paris Embassy then transferred it to the Moscow Embassy. The Moscow Embassy then transferred it to our Moscow NASA Office, who then gave the money—in cash, because that was the way the Russian society worked—to our Director of Operations-Russia in Star City.... It typically had a two-month lag in it."

In January 1995, astronauts Thagard and Dunbar were joined by astronauts Shannon Lucid and John Blaha, who reported to Star City to begin their Mir training.

The early days in Star City have been compared to living in a monastery and included problems with communications, language, supply, and support. (See "Training and Operations," p. 38.) Besides the operational growing pains, there were also culture deprivation and loneliness. Astronaut Bonnie Dunbar later spoke of her "little support group" of Russian women at Star City, including "Olga ... [who] taught technical English at an institute in downtown Moscow. A tremendously warm, nice woman.... She took me on tours of Star City, before we ever had maps, and showed me where everything was, all the little nooks and crannies to buy food. And, Galena," an engineer in Building 2, who, "particularly if I was by myself, would open the door to the hall and pull me inside. It was like disappearing into the bowels of the building. And, we'd have tea with her colleagues there."

Another Russian especially helpful to the Americans in Star City was Natasha Dorishenko,

hired as an administrative assistant. According to Bill Readdy, Director of Operations-Russia, Dorishenko was "absolutely spectacular" at bridging the gap between the different cultures and agencies, and in helping physically to set up the NASA offices in Russia.

Also "spectacular" in her dedication and contributions to the Shuttle-Mir Program was American Jessie Gilmore, executive secretary to Program Manager Culbertson. Gilmore coordinated the Phase 1 Office in Houston and, according to Culbertson, made the whole Russian-American relationship in Houston seem like a family.

"Everybody here kind of worked together as a family," Gilmore said. "It wasn't an eight-to-four job. If [Culbertson] needed something—nine, ten o'clock at night—he could call anybody, and they came in and got it done."

Susan Anderson, Lindy Fortenberry, and other program support specialists worked to make the NASA side of the Shuttle-Mir Program succeed. They escorted visiting Russians, scheduled and rescheduled meetings, and even worked to make it possible for astronauts to vote from orbit.

Space Shuttles

NASA's four Space Shuttle Orbiters launch into orbit like rockets and return to Earth as gliders. During Shuttle-Mir, this capability with Mir's long-duration spaceflight qualities created a new and effective combination. Since Shuttle-Mir, the Shuttles remain the main element of NASA's Space Transportation System. They are used to construct and service the International Space Station as well as for scientific research and space applications,

such as deploying and repairing satellites.

A Space Shuttle can reach orbits of 115 to 400 miles and can launch with payloads of up to 63,500 pounds. Typical missions have crews of five to seven astronauts and last from five to 16 days. The longest a Shuttle has stayed in orbit is 17½ days, on mission STS-80 in November 1996.

The Space Shuttle system is composed of several large components: the Orbiter itself, the three main engines, the external

tank, and the two solid rocket boosters. During the Shuttle-Mir Program, NASA used three Space Shuttles: *Discovery* for the first two and the last missions, STS-60, STS-63, and STS-91; *Atlantis* for the middle seven missions, STS-71, STS-74, STS-76, STS-79, STS-81, STS-84, and STS-86; and *Endeavour* for the next-to-final mission, STS-89. (For additional information on the Space Shuttle, refer to the CD-ROM.)

Space Shuttle Life

A Space Shuttle normally supports crews for up to two weeks at a time. The living area of a Shuttle consists of the flight deck—where the main controls and most of the windows are—and the mid-deck—where lockers hold most of the equipment and supplies. Frequently, the useable area of a Shuttle is expanded greatly by the addition of a Spacelab or SPACEHAB module in the payload bay.

To get the most out of every mission, a Shuttle crew's workday is intense, with activities scripted to a greater detail than on Mir. Beyond this aspect, life on a Shuttle is much like that on Mir. Shuttle crewmembers wear ordinary clothing—for example, rugby shirts and shorts. The atmosphere is kept at about 79 percent nitrogen and 21 percent oxygen with the temperature between 61°F and 90°F. This air, which is cleaner than Earth's air and is pollen-free, is maintained at a pressure equal to Earth's at sea level. Filters remove carbon dioxide and other impurities. Excess moisture is also removed, keeping the humidity comfortable.

Meals can be eaten anywhere in the crew areas, although most astronauts prefer to dine "family style" in the mid-deck area. A galley oven provides heating for some foods. Hot and cold water is available, but a Shuttle does not have a refrigerator. Most foods are dehydrated to save weight and storage. Some foods are thermostabilized, heat-sterilized, then sealed in cans or plastic pouches. Some foods, such as cookies and nuts, are ready-to-eat. One crewmember can "ready"

(Opposite page) STS-86 *Atlantis* rolls out to the launch pad on crawler transport. (Above) This composite image consists of *Atlantis* with its Orbiter payload bay open, revealing the single SPACEHAB module and the Orbiter docking system. The cloudy Earth background is from an image taken by STS-71 on its way to dock with the Mir space station.

meals for four people in about five minutes. Special trays separate the different foods and keep them from floating off. Despite microgravity, most foods stick to spoons, forks, and fingers.

Menus provide about 2,700 calories daily because astronauts need as many calories in space as they do on Earth. Before flights, crewmembers choose among more than 70 foods and 20 beverages. A typical day's meals might include a breakfast of orange drink, peaches, scrambled eggs, sausage, cocoa, and a sweet roll; a lunch of cream of mushroom soup, ham and cheese sandwich, stewed tomatoes, banana, and cookies; and a dinner of shrimp cocktail, beefsteak, broccoli au gratin, strawberries, pudding, and cocoa.

Sanitation is crucial. Microbes can multiply quickly in the confined microgravity environment and could potentially infect the entire crew. As a result, all living areas are regularly cleaned. Eating utensils are cleaned with wet wipes containing a strong disinfectant. Crewmembers' pants are changed weekly; socks, shirts, and underwear are changed every two days; and all used clothing is sealed in airtight plastic bags. Garbage and trash are also sealed in plastic bags and brought back to Earth.

Shuttle crewmembers take sponge baths. An airflow system directs any loose wastewater into the Orbiter's waste collection system,

where it is sealed in plastic watertight bags. For shaving, crewmembers use shaving cream and wipe their faces with a towel, or they use a wind-up shaver with a vacuum that sucks up cut whiskers. Crewmembers use a toilet much like one on an airliner. In place of gravity, airflow directs waste to the bottom of the toilet, where it goes directly into a sealed container to be processed and stored.

Sleep and recreation are important to good health in space. Crewmembers can sleep in "free float," or they can use special berths with good air circulation to eliminate pockets of carbon-dioxide buildup. Crewmembers follow scientifically planned exercise programs to counter microgravity's effects on the heart, veins, and muscles. For recreation, they have games, books, writing materials, and tape and CD players.

Mir – Russia's Space Station

On February 20, 1986, a Soviet rocket launched from Baikonur, Kazakhstan, and carried the first module for the Russian Space Station Mir—beginning a journey in space that would last for 15 years. Through those years, Mir was expanded with add-on modules, served as the residence for numerous space travelers, and provided the opportunity for research, exploration, and discovery. For detailed information about the Mir space station, refer to page 163 and the CD-ROM.

Space Shuttle *Discovery*

Launched: February 3, 1995, 12:22 a.m. EST
Kennedy Space Center, Pad 39-B

Orbit: 213 nautical miles

Inclination: 51.6 degrees

Landed: February 11, 1995, 6:51 a.m. EST
Kennedy Space Center

Mission: 8 days, 6 hours, 28 minutes

STS-63 CREW

Commander James D. Wetherbee
Third Shuttle flight

Pilot Eileen M. Collins
First Shuttle flight

Mission Specialist Bernard A. Harris, M.D.
Second Shuttle flight

Mission Specialist C. Michael Foale, Ph.D.
Third Shuttle flight

Mission Specialist Janice E. Voss, Ph.D.
Second Shuttle flight

Cosmonaut Vladimir G. Titov
Russian Space Agency
First Shuttle flight and fourth space mission

PAYLOADS

Space Habitation Module-03 Experiments

Spartan-204

Cryo Systems Experiment

Orbital Debris Radar Calibration System

IMAX Cargo Bay Camera

Air Force Maui Optical Site

Solid-Surface Combustion Experiment

STS-63
First Rendezvous
February 3 – 11, 1995

STS-63 crewmembers traveled within 35 feet of the
Russian Space Station Mir. Pictured aboard the flight
deck of *Discovery* are (bottom row, left to right)
Pilot Collins, Commander Wetherbee; Mission Specialist
Harris; (top row) Mission Specialists Titov, Foale, Voss.

STS-63 would have been an historic mission even without the added assignment of the first-ever Shuttle-Mir rendezvous. This was the first Space Shuttle mission to be piloted by a woman— Eileen Collins. The crew performed 20 experiments in the SPACEHAB science module, and they deployed, rendezvoused with, and then retrieved the Spartan-204 free-flying astronomy package. Two astronauts— Mike Foale and the first African-American spacewalker, Bernard Harris—conducted an extravehicular activity (EVA) that tested new equipment and demonstrated how well an astronaut could manipulate a large object in space. These accomplishments, plus the crew's many other duties, would have made STS-63 one of the busiest Shuttle missions ever.

But, the rendezvous with the Russian space station became STS-63's primary mission, and on the success of this rendezvous hinged the future of the Shuttle-Mir Program. Although the orbital physics of rendezvous were well understood, many techniques were undemonstrated and the stakes

were high. *Discovery* had a mass of 87 tons; Mir weighed 103 tons; and each measured more than 100 feet long. Even a small human error or mechanical glitch could be magnified by the mass and momentum of the spacecraft, jeopardizing the nine lives aboard *Discovery* and Mir as well as the future of human spaceflight.

Lifting off just past midnight on the morning of February 3, 1995, *Discovery*'s launch lit up the night sky as it roared up America's Atlantic coast so that it could match Mir's orbital inclination of 51.6 degrees. Pilot Collins has said that a night launch is like being "in a room that's on fire." Light from the rockets poured through the windows as the spacecraft roared and shook.

Observers on the ground watched the light from the Orbiter's main engines for the full 8½ minutes until the engines shut down.

Regardless of spectacle, a problem occurred almost immediately. As NASA Flight Director William Reeves said, "We launched, and, lo and behold … we had a leaking jet. One of the thrusters on the Shuttle was leaking propellant … and the Russians didn't know what to think of it. They were concerned about fuel contamination on their vehicle; and if we couldn't arrest the leak, they didn't want the Shuttle coming too close to the Mir." Among other worries, if contamination got onto some of the Soyuz capsule's surfaces, the Mir crew could not use it as an escape vehicle.

Early in the flight the propellant spewed in a conical pattern, "like a snowstorm for five miles up into space," according to Commander Jim Wetherbee. The Russians didn't want *Discovery* to come within 1,000 feet of Mir. But, NASA flight controllers and the *Discovery* crew "worked the problem," at times rolling the Orbiter to warm the thrusters in the Sun. As Reeves tells the story, the Russian engineers were "very sharp and astute … and asked all the right questions." They changed the minimum separation to 400 feet, still not close enough for meaningful data.

According to Flight Director Phil Engelauf, "[It] wasn't until the morning of the rendezvous that we had finally gotten an agreement from the Russians that we were going to be able to go ahead and make the close approach." Onboard *Discovery*, Wetherbee told veteran cosmonaut

Vladimir Titov that if the leak was still a problem, he would not bring the Orbiter close to Mir—no matter what the flight controllers said. But, when the Shuttle crew woke up on the morning of the rendezvous, the leak had diminished.

Another concern was whether *Discovery* might actually bump Mir. The Russians remembered the Apollo-Soyuz docking mission in 1975 when the Apollo capsule docked with the Soyuz with a much bigger bump than the Russians had expected, said Wetherbee. But, he added, the Space Shuttle is "like a big ocean liner coming in, and things are done more slowly. It's a very good system designed by really good engineers, and the handling qualities are perfect…. I think a lot of people are surprised at how stable the vehicle looks, how motionless it looks as it's coming in, and it's very controllable. So, that was another thing that gave the Russians a lot of confidence on [STS]-63 …"

The *Discovery* crew brought the Orbiter up to within 35 feet of Mir. "Oh, beautiful, beautiful, beautiful!" Titov reported. According to Wetherbee, the Orbiter crew was "just blown away by the sight … of that huge, giant space station out the window … Mir looked so brilliant and white and bright." They saw the Mir crew at the windows. Valeri Polyakov and Aleksandr Viktorenko waved, and Elena Kondakova held up a little cosmonaut doll.

STS-63 validated several precision-control techniques; performed an inspection of Mir by eye, photo, and video; and tested the joint

Mir space station as viewed from *Discovery* during STS-63.

operations capabilities of the Mission Control Centers in Houston and Moscow. *Discovery* then left the vicinity of Mir.

Titov used the Shuttle's arm to release the Spartan-204 package. It flew free for two days to study far ultraviolet radiation. The crew conducted science in the SPACEHAB module and shot footage for an IMAX movie. Harris and Foale conducted their EVA to test the spacesuits' thermal properties. Harris also manipulated the Spartan-204 package to test a spacewalker's ability to work with large objects in space.

The Snowball in the Fax

As the "go/no-go" decision for the STS-63 rendezvous approached, NASA Flight Director William Reeves was in Moscow and had to explain the thruster propellant leak situation "one more time." According to Reeves, "All of [the Russian] systems experts and safety people were there…. They all sat there and they looked at me, and they said, 'We understand what you're saying, but what about the 180-gram snowball?'

"And, I'm sitting there with this blank look on my face, and I said, 'What are you talking about?'

"And, they said, 'Well, the 180-gram snowball in your fax.'

"I said, 'What fax?'"

It turned out that some NASA engineers had worked out an extremely remote, worst-case scenario of a thruster jet freezing and getting packed with ice. Communications were poor at this time between Moscow and Houston; when the engineers sent this scenario in a fax, it went straight to the Russians. Reeves had never seen it.

The Russian space officials said to Reeves, "Well, you know, this piece of ice could form in the jet and it could come loose and all of a sudden this big chunk of ice could hit the Mir."

Reeves said, "Oh, no, it's not going to

happen…. You've got to give me time to go figure out what's going on here and explain this."

Reeves got to the bottom of the story, and he explained it to the Russians. Reeves then consulted with Houston and waited. Time was getting short.

Finally, Russian operations chief Victor Blagov came out of his office. According to Reeves, "He grabbed me…. And, between my broken Russian and his broken English, we [could] carry on a conversation. And, he says, 'You know, we're getting very close to making a decision, and we're going to allow you all to come to 30 feet.'

"And then he just grinned at me."

The rendezvous was approved.

Norm Thagard

NASA-1

March 14 – July 7, 1995

An End and a Beginning

It was the end of the old way in space and the beginning of the new. Norm Thagard's NASA-1 mission was all about learning. Thagard symbolized the fledgling Shuttle-Mir Program as he launched from Kazakhstan on a Soyuz rocket with his Commander Vladimir Dezhurov and Flight Engineer Gennady Strekalov. They were on their way to spend 115 days in orbit and begin America's experience on Mir. Thagard's personal objectives were to learn how the Russians did long-duration spaceflight and "to be a cosmonaut and fly as a crewmember on a Russian crew." Shuttle-Mir Program Manager Frank Culbertson would later say that Thagard's stay onboard Mir was "the hardest" of the seven American flights. This was largely because Thagard was the first, and almost everything was new for everyone involved.

Thagard was well-qualified for his own Herculean labors of learning. Yet, in several ways he launched under-prepared for other aspects of his mission. He had had only one year of intensive training, and that training took place under a Russian pedagogical system, within the Russian culture, and in the Russian language. Also, his onboard scientific investigations had to be quickly designed and assembled, and Thagard was often learning the Russian protocols as they were being worked out. Furthermore, the important Spektr science module arrived late in his flight. And so, as Thagard faced his many challenges, he also met one problem that few had expected: He did not have enough meaningful work to do. This created a kind of slow torture for a perpetual-motion astronaut. Nonetheless, Thagard's overall success opened the door wide for the next six American Mir residents.

Norman E. Thagard was born in Marianna, Florida, in 1943. He came to his Shuttle-Mir experience as an example of what many American parents preach to their children: "You can be whatever you want to be—if you work for it." Thagard had told his high school classmates in Jacksonville, Florida, that he wanted to be a medical doctor, a fighter pilot, an engineer, and an astronaut. He became all four. He went to Florida State University to earn bachelor's and master's degrees in engineering. He joined the U.S. Marines in 1967, achieved the rank of Captain, and flew 163 combat missions while serving in Vietnam. After returning to the United States, he worked on a Ph.D. in engineering, then went to the University of Texas Southwestern Medical School, earning his M.D. in 1977. The next year, while he was interning in

Two days before the scheduled launch, the Soyuz-TM spacecraft destined to carry the Mir-18 crew to orbit rolled out from the processing and integration facility to the launch pad on a special flatcar pushed by a locomotive (top photo). On arrival at the pad, the Soyuz rocket stack was quickly hoisted to its vertical position via hydraulic lifts mounted to the flatcar (bottom photo).

(Photo, page 16) NASA-1 Astronaut Norm Thagard training in Mir mockups, Star City, Russia.

(Photo, page 19) Russian suit technicians prepare Mir-18 crewmembers for launch; (left to right) Flight Engineer Gennady Strekalov, Commander Vladimir Dezhurov, and NASA-1 Astronaut Norm Thagard.

South Carolina, NASA selected him for astronaut training. Thagard flew as a Mission Specialist on STS-7, STS-51B, and STS-30, and as the Payload Commander on STS-42. After his fourth Shuttle flight, he was considering what to do "after NASA"—and thinking about teaching at his alma mater. Then, the opportunity arose to fly with the Russians. In February 1994, Thagard went to Russia with a small cadre of NASA people, including his backup astronaut Bonnie Dunbar and Flight Surgeons Dave Ward and Mike Barratt. They were among the few NASA pioneers of training and flying with the Russians. Others—including Frank Culbertson, Ken Cameron, and Peggy Whitson—helped develop management, operations, and the science program.

Thagard's wife, Kirby, and their youngest son, Danny, joined him in Star City. They lived in a three-room apartment in a rundown building. According to Thagard, "It wasn't a luxury apartment … but by Russian standards it certainly was." Kirby Thagard taught half-time at a local school, and Danny attended a Russian high school. Thagard trained in Russian, studying all the systems of the Soyuz capsule and the Mir space station. Fortunately, he found some similarities between the NASA and the Russian methods of training, including the use of single-system training leading to highly sophisticated full-systems simulations. For each system, he and Dunbar took final oral examinations. At these, Thagard said, members of the training department "would array themselves at a table … and they would fire questions at you, in Russian, on their system, and you had to answer the questions.… You had to pass all exams at the end if you were going to be certified to fly."

Two weeks before launch, Russian space officials announced that Thagard and his crewmates had passed all their tests and were ready for their mission. The mission commander was Lieutenant Colonel Vladimir Dezhurov, who would be making his first spaceflight. The flight engineer was Gennady Strekalov. Like Thagard, Strekalov had been in space four times.

The crew was flown to the launch facilities at Baikonur, Kazakhstan, where Thagard was again impressed by the similarities between the Russian and the American space programs. These included the Russian counterpart of NASA's terminal countdown dry test, in which the crew gets into the launch vehicle and goes through all the procedures in a rehearsal for launch.

Afterwards, the crew flew back to Star City, from where they were sent to a lodge in the woods, 100 miles west of Moscow. There, they spent two days resting and relaxing with their families. "It was nice," Thagard said. "Get-togethers in the evenings … big fires in the fireplace." They played ping pong, took saunas, and went cross-country skiing. On one slope, Thagard skied with such abandon that he worried about breaking a leg and ruining his opportunity for a flight.

The Mir-18 crew then went into quarantine, moving into Star City's equivalent of Johnson Space Center's isolation facility. Each crewmember had his own room. As in Houston, their spouses could visit, but according to Thagard, "You're in the place by yourself, basically." In a way, quarantine could be seen as the beginning of the actual mission. Three days before liftoff, the crew flew back to Baikonur, where Thagard enjoyed the Russian preflight traditions, including a party that the flight crew hosted for the flight support people.

On launch day, March 14, 1995, the crew was taken into a "suit-up" room, where a glass window separated the still-quarantined crew from the next room, full of media representatives and Russian and American officials. Thagard's family was there, too. Usually, Russian space-flight crews' families remain in Star City, but Mir "guest researchers" from other nations had often brought family members to Baikonur. Thagard and his family were able to converse through the glass, using a microphone and a speaker. "It was not exactly a private conversation," said Thagard. There were cameras "and media people, sort of lurking around, recording and looking." After a few last words, the curtain was closed. The crew "suited up," and the suit pressure tests were run. Then, the crew went back to the window, this time for a final ceremonial conversation with officials of NASA, the Russian Space Agency, and the Russian government.

Launch time approached. The crew walked out onto the Kazakhstan steppe into the bitter

March winds. On such a cold morning, a Space Shuttle would not have been cleared for launch. But, the American and Russian programs had always differed in that Soviet and Russian space vehicles could launch and land in a wide range of weather conditions, while American space-flights had narrower tolerances. One reason for this was that Kennedy Space Center usually enjoyed Florida's subtropical weather, while Baikonur often suffered an extreme continental climate. A tragic illustration of weather's effects on launches was the Space Shuttle *Challenger* accident in Florida in 1986, partly caused by the cold January temperatures.

Still, on the day of launch, Thagard said to Strekalov, "Gennady, we can't launch today. It's too cold."

Strekalov answered, "Oh, the colder the better."

"Well, all right." Thagard said, "But, it's still too windy." The gusts were almost gale-force. A Shuttle couldn't launch in that much wind.

Baikonur Cosmodrome

Norm Thagard and his Mir-18 crewmates launched in a Soyuz capsule from Baikonur Cosmodrome on the desolate, windswept steppes of Kazakhstan, about 1,300 miles southeast of Moscow.

Baikonur had launched all Soviet and Russian human spaceflight missions, as well as many important unmanned rockets. Its construction began in secrecy in 1955. To mislead Western militaries, the Soviets misrepresented its name and coordinates as those of the actual town of Baikonur, about 200 miles to the northeast. Sputnik-1, the world's first artificial satellite, launched from Baikonur in 1957, as did cosmonaut Yuri Gagarin, the first human in orbit, in 1961. Baikonur was the Proton rocket launch site for all of Mir's modules, including Spektr and Priroda, which docked with Mir during the Shuttle-Mir Program. After the Soviet Union broke up, Russia began leasing the site from the new Republic of Kazakhstan.

Baikonur's 45-degrees-north latitude helps spacecraft attain orbits that pass over most of the inhabited Earth. Its severe continental climate—with temperatures from 40°F below zero to 113°F above zero—obligated the Soviets to develop rockets that could launch in nearly any weather.

During Shuttle-Mir, NASA officials were impressed by Baikonur—by its remoteness, certainly, but also by how its launch capabilities were maintained in spite of Russia's extreme financial difficulties. Former NASA astronaut Joe Engle described Baikonur as "an enormous, vast, desolate" area, and he said that many structures there had "not been kept up ... and yet the particular facilities that they need to launch vehicles from are operational.

The Russians launched the Mir crews from the desolate, windswept steppes of Kazakhstan, 1,300 miles southeast of Moscow.

"I was impressed," Engle said. "It reminded me a little bit of Edwards Air Force Base," in California, "a remote, desert-type environment that is remote for a reason.

"They're launching vehicles that sometimes are not successful, so they need a big area where there are not any people around ... so they can drop them in the desert and not hurt anybody. They also need an area that is secure, where they can perform tests and things that they don't want to be public domain at the time—just like we do at our remote test facilities."

Nonetheless, Engle said, "It was sad to see the facilities that I know at one time were really first class—to see them run down and deteriorating. I guess it's a sign of the times."

NASA's Charles Stegemoeller, who was Project Manager for the U.S. research experiments in the Spektr module, said he had a lot of respect for what the Russians had accomplished. Stegemoeller said, "Never have so many done so much, with so little, for so long. There's just an intense sense of pride. I mean, the little old lady who sews ... thermal insulation ... on the outside of the spacecraft was the same one who was doing it for the last 30 years. She was hunched over this table ... doing this hand-stitch work, while we were unpacking our gear and putting it in the module."

The Shuttle-Mir Program and other international investments in Russia's space launch capabilities led to renovations and new construction at Baikonur. After Shuttle-Mir, Baikonur remained a busy facility. Today, it is Russia's launch site for the International Space Station Program.

As part of the prelaunch operations, (left to right) Thagard, Dezhurov, and Strekalov, carrying their portable life support systems, participate in preflight ceremonies and receive permission to fly. A successful liftoff (pictured below right) for the Mir-18 crew occurred March 14, 1995.

Strekalov just said, "As long as it's not a hurricane." Coincidentally, the Mir-18 mission's code name was "Uragan," meaning "hurricane." The preparations for launch continued. In another way, Thagard found Baikonur's frigid conditions ideal. "It was the only time in my life when I was actually glad I had a pressure suit on, because those things are usually hot and uncomfortable—especially if you start moving around in them. And yet, it was just perfect for that day at Baikonur."

In a final preflight ceremony, the Mir crewmembers took their marks in front of the base commander. They stood to attention, and Mir-18 Commander Dezhurov saluted. He told the General that they were ready to go fly, and the base commander gave his permission.

At this point, Thagard noted two dissimilarities with NASA's launch operations. First, after enduring the strict quarantine and even having to wipe their fingers with alcohol, the crew now rubbed shoulders with a crowd of support people, officials, and well-wishers. Second, after riding in a bus out to the launch pad, they passed through another crowd effectively milling in the shadow of the dangerous Soyuz rocket. "You could see the vapor coming off the liquid oxygen tanks," Thagard said, "telling you that the rocket's fully fueled and ready to go." At Baikonur, everyone assumed his own risk in spite of the fact that in 1960 an R-16 ICBM missile exploded on the launch pad, killing more than 100 people. In contrast, at NASA's Shuttle launch facility all but the most necessary personnel are moved back three miles, even before the Space Shuttle is fueled.

The Mir-18 crew next rode the elevator to the top of the rocket. Thagard was first to crawl into the Soyuz's upper module, which held the crew's food, water, and supplies. Strekalov then climbed down into the descent module and began turning on the systems. Once all three of the crew were inside, the hatch was shut. At this point, Thagard noted that the crew on the rocket was actually safer than the crowd outside on the ground below. The Soyuz capsule had an escape rocket system—similar to the systems for the Mercury and Apollo capsules—which could rocket the crew capsule up and to safety in the event of an emergency. Thagard's crewmate Strekalov had, in fact, been saved by this system when a Soyuz rocket that he and future Shuttle-Mir cosmonaut Vladimir Titov were riding exploded on the pad in 1983. Although NASA's Space Shuttle Orbiter has "return to launch site" procedures in case of emergencies after the Shuttle is launched, there is no automatic escape device in the event of an explosion on the pad.

The crew waited about two hours before launch. Their experience at this point was similar to that of an astronaut on the Space Shuttle's mid-deck. They could not see outside. The Soyuz windows were covered by an aerodynamic shroud. This would fall away later, after the craft had made its way through the denser parts of the atmosphere. Thagard's duties were to switch transmissions between two television cameras—one pointed at himself and the other aimed toward Dezhurov and Strekalov.

Finally, the time arrived. In Thagard's words, "We just lifted off." He said the experience was

> Because his flight was the first, Thagard's flight was "the hardest" of the seven.

Norman E. Thagard

Norm Thagard was born in Marianna, Florida. He holds bachelor's and master's degrees in engineering science from Florida State University and a Doctor of Medicine degree from the University of Texas Southwestern Medical School. He served in Vietnam as a U.S. Marine fighter pilot and flew four Shuttle missions between 1978 and 1992, logging more than 25 days in space.

In regard to his Russian crewmates, Thagard said: "I thought it was extremely ironic, because when I was flying missions in Vietnam in 1969 as an F-4 pilot, I thought

that there was an excellent chance that at some point in time I'd have interactions with the Russians, but I thought they would be of a somewhat different nature than they turned out [to be].

"If anyone in 1969 had ever told me that I would wind up having a Lieutenant Colonel in the Russian force as a commander, I would have said, 'You're crazy.'"

After his Mir experience, Thagard left NASA to become a professor at Florida State University.

First U.S. Mir astronaut to train in Russia • First American to launch onboard a Soyuz • First to complete a residency onboard Mir
Set American space endurance record of 115 days • Launched on Soyuz, March 14, 1995 • Returned with STS-71, July 7, 1995

similar to a Space Shuttle launch, except with "not as much noise, not as much vibration." On the Shuttle, maximum g-forces occur during the last minute of powered flight. On the Soyuz, these occur towards the end of the first stage of the three-stage process. Thagard did not feel the same "sense of power" that he had felt onboard Shuttles. The Soyuz was never quite as noisy, nor did it have as much of those "popcorn-like" vibrations he had felt on his Shuttle launches when the solid rocket boosters were firing. Conversely, he said that the Soyuz was "never as smooth as the Shuttle in second stage, because once the Shuttle's solid rocket boosters separate, it's like you're being propelled by a giant electric motor. It's very smooth and fairly quiet."

Yet another difference during ascent between the Shuttle and the Soyuz was the sensation when the main engines cut off. "When the Shuttle main engines cut off, they just cut off," Thagard said. "It's not a huge emphatic thing. But, when the main engines cut off on the Soyuz, it was very emphatic—almost like a clank or a clang." He attributed this difference to the fact that a Space Shuttle has already throttled back to 65 percent of its thrust by the time its main engines shut down, whereas the Soyuz's third-stage engine is still at full speed when shut down.

The Soyuz's flight path took the vehicle over northeast Russia to a point about 250 miles ahead of Mir. Over the next two days, the Soyuz would slowly expand the distance between the two spacecraft until it caught up with Mir from behind. This rendezvous technique differed from the Shuttle's method to get close to Mir. The Soyuz used fewer but longer maneuvering burns and natural orbital dynamics. A Shuttle would have used more and smaller burns, all targeted to bring the Orbiter to a point about eight miles behind the space station. Then, the Shuttle's crew would have driven the vehicle slowly toward Mir with a series of manually controlled jet firings.

When the Mir-18 crew had launched, the Mir space station was flying over central Africa. Coincidentally, NASA's STS-67 Space Shuttle *Endeavour* was high above Indonesia. This brought to 13 the number of men and women in space at the same time—a new world record. Once in orbit, Thagard was able to look outside, and the first thing he saw was the Soyuz solar panels. The crew was on the radio now, talking to the person who had been their trainer in Star City for the past year. Thagard called this continuity "one of the nicest features" of the Russian space program. It fostered trust, consistency, and better communications. In NASA's Space Shuttle

Program, trainers did not keep working with the crews once the crews had been trained. Furthermore, immediately after a Shuttle lifted off, NASA ground control switched from Kennedy Space Center in Florida to Johnson Space Center in Texas. In Russia, ground control didn't revert to Mission Control-Moscow until the Soyuz had "been around the world a few times." According to Thagard, the trainer on the radio was "a young Captain in the Russian Air Force— a really good guy." He had accompanied the crew down to Baikonur, and he had been helping Thagard in many ways. This underscored another difference between the Russian and American programs. In Russia, emphasis was often on the person. At NASA—as in America in general— emphasis was often on the job.

While this situation reflected the two different societies, the situation also illustrated the different emphases of the two space programs. The Soviet and Russian space agencies had been working mainly at long-term residence in orbit, with its associated need for human constancy. NASA had been developing frequent, short-term access to space, with its need for interchangeable people.

Once the Mir-18 crew received a "go for orbit" from ground controllers, they were able to take off their Sokol pressure suits, open the hatch between the two modules, and make all of the Soyuz's limited volume accessible to them. There wasn't much space and, according to Thagard, there was also absolutely no privacy. Indeed, to avoid having to use the facilities, some Soyuz crewmembers took enemas shortly before launch. However, other than being cramped, Thagard said, the Soyuz was not "a bad place to be for a couple of days." During their journey, the crew checked systems and collected biomedical data on the effects of microgravity on the human body. They rode mainly in the Soyuz's living module, returning to the descent vehicle for course adjustments and communications.

> ... when the main engines cut off on the Soyuz, it was very emphatic—almost like ... a clang.

In another way, the Soyuz's small volume might have had its good side. Thagard later commented on its possible effect on the phenomenon known as space motion sickness. As a military jet pilot, a medical doctor, and a veteran of four Space Shuttle flights, Thagard had experience and training to draw from. To varying degrees, he had felt some space adaptation symptoms on all of his Shuttle flights. Yet, on the Soyuz, he "never got beyond stomach awareness." He attributed the difference to the small volume of the Soyuz. "You just don't move around," he said. "And, you certainly don't have as many head movements." On a Space Shuttle, "you basically hit the deck running ... You've got so much to do, and so little time in which to do it, that immediately upon clearance and a 'go for orbit,' you're up—just darting ... and throwing big head movements." In Thagard's opinion, "There is absolutely no question that that's what causes and exacerbates space motion sickness." (See Life in Microgravity, p. 80.)

Thagard, the first American to reside on the Mir, arrived at his destination aboard the Soyuz, a Soviet spacecraft, with his Russian crewmates two days after launch. His first view of the Mir resembled the image, pictured at right, that was taken by the crew of STS-63 as part of their "near Mir" mission. (For detailed information on the Russian space station, refer to page 163, Mir Space Station, and to the CD-ROM.)

Norm Thagard's responsibilities during rendezvous and docking with the space station were to control the radios and television cameras and to help monitor Soyuz systems. As the Soyuz neared Mir on March 16, he took a brief look at his future home through Dezhurov's periscope, but his main view of Mir was through the television. The docking was "perfectly normal," according to Thagard. Dezhurov could have taken over manually if necessary, but the automatic system worked well. Thagard compared the docking experience to that of backing a car slowly into a cushioned loading dock. "It's a definite contact—no question about it—as though you'd just bumped into something, but not a violent sort of collision."

The Russians insisted that Thagard be the first one to go onboard Mir. There, they were met by the Mir-17 crew of Commander Aleksandr Viktorenko, Valeri Polyakov, who was about to set a world record of 438 straight days in orbit, and Elena Kondakova, who was on her way to setting a women's space record of 169 days. Kondakova was holding a little tray to which she had attached some bread and salt—the traditional Russian greeting to visitors. Everybody hugged and, according to Thagard, "Good times were had by all. It was a nice time. It was a fun time." Thagard was now one of six residents on Mir.

Dr. Polyakov gave Thagard good reason for his cheerfulness. Thagard said, "He didn't look like a person—either from a physical or a psychological standpoint—who had been on a space station for over 14½ months. His legs were just as big as tree trunks, and he was in a great mood. Of course, I'm sure knowing that he was going home in a few days would probably put him in a great mood. Nonetheless, I got the feeling that

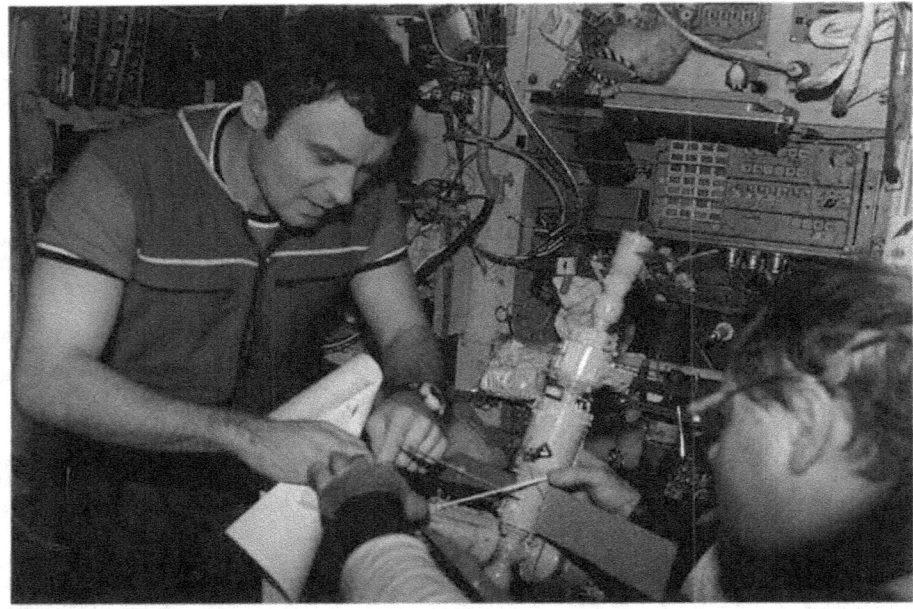

Mir-18 Commander Dezhurov (left) and Flight Engineer Strekalov use a model of the Mir space station while reviewing plans for their extravehicular activity.

he had done perfectly fine." Thagard was very interested in the physical and psychological aspects of being in space for months at a time. Seeing Polyakov reassured him that "indeed, at least some humans can do it without much of a problem, and he was clearly one of those who did…. I figured, gee, if he did that well after 14½ months, I probably didn't have much to worry about for just three [months]." Thagard was determined to follow Polyakov's example of frequently wearing the "Penguin-3" overalls with elastic straps that provided resistance to body movements and compensated somewhat for the microgravity. He was determined to walk off the Space Shuttle at the end of his mission.

The two crews then spent a six-day handover period on Mir in which the departing crew briefed the arriving crew on the state of the space station. The handover information was invaluable to Thagard. For example, the stowage and inventory control onboard Mir had gotten quite poor over the years. Thagard said, "When something's been up there for years and years, the ground never really knows the full state of everything.

They just don't. The only way the new crew can get all of the up-to-date information is by talking with the old crew. There were things I probably never would have found if they hadn't physically led me by the hand and said, 'Okay, this is here and that's there.'"

Command structure was another thing that Thagard noted early in the flight. Although he was still a rookie at spaceflight, Mir-18's Commander Dezhurov became more authoritarian than he had been before the flight. As the flight progressed, veteran cosmonaut Strekalov would sometimes chafe under Dezhurov, then "level a blast" at his commander, who would "back off a little." Thagard himself tried Strekalov's oratorical tactic to a good result. Interpersonal relations warmed during the mission. According to Thagard, "Over the course of time, things just got better and better and better. By the end of the mission, any time Velodya [Dezhurov] would address me, it was always, 'My friend.' It was great, but it didn't start out that way." Other U.S. Mir astronauts would similarly comment that Russian Mir commanders were more autocratic than their U.S. Space Shuttle counterparts, who generally added "please" and "thank you" to their orders.

There were, of course, other factors involved in the style of the command structure onboard Mir. The United States and Russia had recently

been military antagonists, and Dezhurov might have felt some national need to keep a strict command. Indeed, Thagard may have been the first American ever to serve directly under command of a Russian officer. Thagard later said, "I thought it was extremely ironic, because when I was flying missions in Vietnam in 1969 as an F-4 pilot, I thought that there was an excellent chance that at some point in time I'd have interactions with the Russians, but I thought they would be of a somewhat different nature than they turned out [to be]. If anyone in 1969 had ever told me that I would wind up having a Lieutenant Colonel in the Russian force as a commander, I would have said, 'You're crazy. Maybe if I get captured as a POW.'"

On March 21, the Mir-17 crew was ready to leave. As much as he enjoyed them, Thagard was ready to see them go. A contingent of six crewmembers stretched the resources of the space station, affecting many systems, including air quality and toilet facilities. The leave-taking was emotional. When Mir-17's Soyuz backed away from Mir and did its fly-around of the station, Thagard heard on the radio Polyakov's glee at leaving. Mir-17 Commander Viktorenko was saying, "'Shhh,' just trying to calm him down—to get him to be a little bit less rambunctious. It was just fascinating to listen to that kind of stuff."

Thagard and his crewmates were now on their own. They settled into their daily routine, which typically began at 8 a.m. Moscow time. The first two hours were spent washing up, eating breakfast, and preparing for the day's tasks. The workday ran from 10 a.m. until 7 p.m., with breaks for lunch and exercise. After dinner, the crew prepared reports on the day's activities and reviewed their plans for the next day. From 10 to 11 p.m., the crewmembers had personal time, followed by their nine-hour sleep period.

During the first week, Dezhurov and Strekalov replaced a condenser in the air-conditioning system. The entire crew collected body fluid samples for metabolic experiments. They took air and water samples for hygiene, sanitation, and radiation experiments. And, each man spent time in the Chibis suit, which measured cardiovascular system responses to lower body negative pressure.

Thagard's complement of 28 science investigations encompassed seven disciplines, including fundamental biology and microgravity studies; human metabolic, neurosensory, motor performance, and cardiovascular responses to long-duration

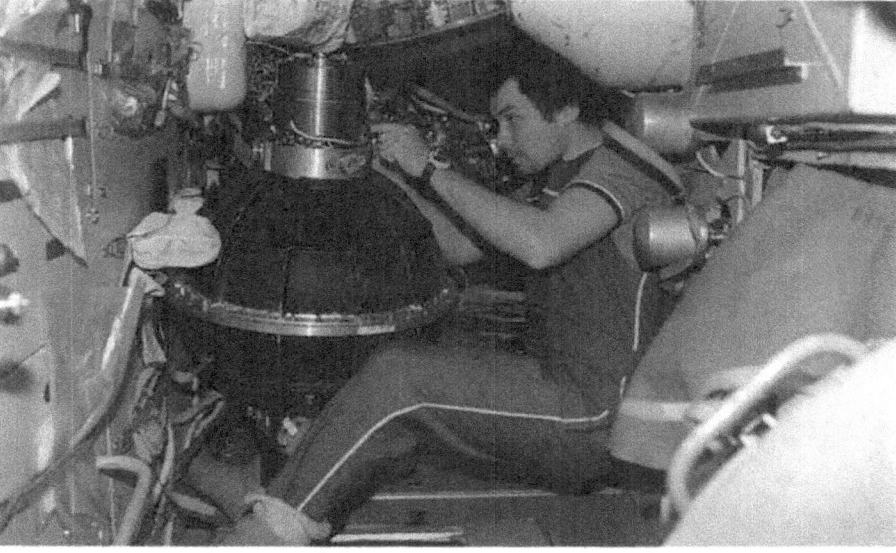

spaceflight; and the scientific characterization of the Mir environment.

In a press conference on March 24, Thagard compared Mir to a utility room that had been lived in for nine years. However, he found the air quality to be excellent; in fact, the air seemed cleaner than the air on a Space Shuttle. The space station's air continually recirculated through filters for years, while a Shuttle launched full of ambient Florida air. Thagard said he had had problems finding equipment and getting things started, and he offered this advice for future Mir astronauts: "Train. Train. Train."

An unmanned Progress resupply spacecraft arrived from Baikonur on April 11. The vehicle carried over two tons of supplies, including water, food, fuel, and equipment, as well as a personal "care package" for each crewmember. When the Progress docked, it was already past midnight, but the crew stayed up to open the hatches. Thagard said, "One of the things you notice is that the air smells different inside the vehicle, but it's not any special air supply or anything. I guess it's just the Baikonur air that was in there." He would notice the same sort of odor when the Spektr module arrived later in the mission. The crew began unloading the Progress on April 13. Among the biological experiments were 48 fertilized Japanese quail eggs, which the crew put into an incubator. The crew planned to stop the development of each egg at a different point in the flight so researchers could later study how embryo development is affected by microgravity.

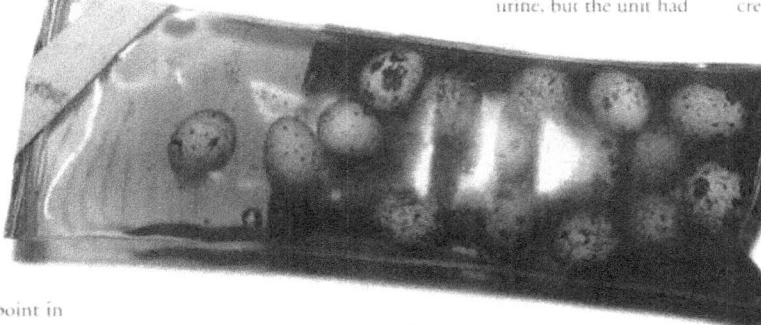

Later in April, the crew learned that scheduled spacewalks for work on solar arrays had been postponed due to a delay in the launch of Spektr.

The crew continued their routine experiment work. They replaced a humidity control fan with one from the Progress, installed a battery unit in the Kristall module, and began removing an unused shower in the Kvant-2 module. They used a machete to help cut the shower into small enough pieces to fit into the Progress. Where the shower had been, they installed a new set of gyrodynes that, when spinning, helped keep spacecraft "locked" into specific attitudes—or tilts—in space.

The crew also worked on defrosting a troublesome freezer left over from a previous European Space Agency mission. Thagard needed the freezer to store biological samples, such as blood and urine, but the unit had been causing problems since early April and much effort went into keeping the freezer from frosting up. At about six weeks into Thagard's mission, the freezer failed entirely.

During the middle of Thagard's stay onboard Mir, before the arrival of the Spektr module, one of the more negative—and unexpected—aspects of his flight became a problem. Thagard's crewmates, Dezhurov and Strekalov, had plenty of work, keeping Mir systems running and preparing for spacewalks. However, Thagard—the everachiever—now found himself without enough to do. Some of his science was not going well. Further, the Russians were not ready to have an American take a more active part in running the station. It was a bad situation. According to Thagard, "The most important thing from a psychological standpoint is to be reasonably busy with meaningful work." In contrast, "My Russian crewmates, from just before the first spacewalk ... almost to the end of the mission, were chronically overworked. Underwork. Boredom. Overwork. Tension. So, you don't want to be at either extreme; you want to be somewhere in the middle."

One item Thagard's wife had sent up for him on the Progress was a New York Times crossword puzzle book. However, said Thagard, "Velodya and Gennady were very busy, extremely so, and there was no way I could sit there and work a crossword puzzle—even if I were bored—while Velodya and Gennady [were] running around working." Instead Thagard found ways to busy himself, but not with what he considered meaningful work.

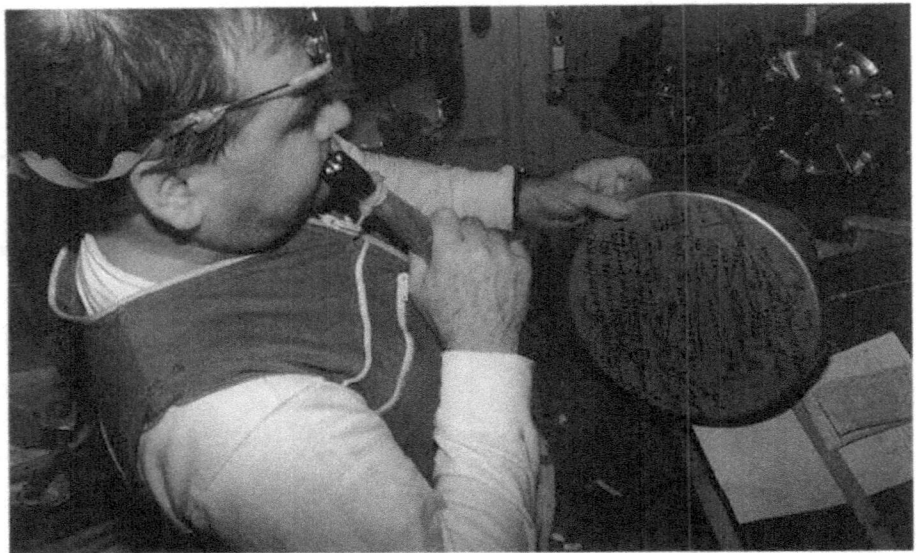

Strekalov (above) draws schematics on the lid of a can due to a shortage of paper on the Mir space station. Thagard (below) exercises on the bicycle ergometer as support teams in the Russian Mission Control Center monitor his activity (medical test) during a communication pass.

Deprivation of another kind caused problems, too. This had to do with a human metabolism investigation and Thagard's diet. According to Thagard, the Mir-18 food supply consisted of a basic, repeating, six-day menu. Four of the entrees were canned fish, which Thagard loathed. All the basic foods were bar-coded so crewmembers could record with a scanner exactly what they ate. Also onboard was a supply of more flavorful supplementary foods. But, these were not bar-coded; and crewmembers had been asked to record everything they ate. However, such a dearth of paper existed onboard Mir that none was available to keep a meal log. Thagard gave this example: "Later on, when we moved the solar battery and had to reroute the electrical cables, ... Gennady took marker pens and wrote out the new schematics on an aluminum can lid." Meals and their recording became a situation of diminishing returns. "The upshot," Thagard said, was that "the food supply was not adequate for any of us." Dezhurov and Strekalov basically quit the food program. Strekalov told Thagard that half his food was

coming from the supplementary supply. "But, I," Thagard said, "religiously adhered to the requirement, and I was constantly hungry." Worse, he was losing weight.

Ground controllers finally realized that Thagard had lost 17½ pounds. During a Mir-to-ground medical conference, Thagard attributed his weight loss to his faithfulness in following the scientific protocol. Russian doctors told Thagard,

"With that much weight loss, you're not just losing fat, You've lost muscle mass." According to Thagard, "They told me that I was free to eat anything onboard, other than my crewmates—and that's the way they put it." Trying to control everything in a closed and confined environment had led to undesired consequences.

Some time in the first week of May, while doing some cleaning work, Strekalov accidentally scratched his arm. The arm became inflamed, and this caused concern about Strekalov's ability to do the planned spacewalks. Physicians on the ground viewed downlinked video of the injury and prescribed a medication, which physician Thagard administered. The injury healed and extravehicular activity plans proceeded.

On May 12, Dezhurov and Strekalov conducted their first spacewalk to prepare the station for Spektr's arrival. They exited the Kvant-2 airlock and moved to the Kvant astrophysics module, where they installed electrical cable attachments and adjusted solar array actuators. They then moved to the Kristall module and practiced folding three solar panels of the solar array to be moved to Kvant. Thagard supported the crew from inside Mir by relaying instructions from the ground and by consulting reference manuals when Mir was out of range of ground communications. The spacewalk lasted six hours and 15 minutes—past the allotted time—so the cosmonauts had to postpone another task: the

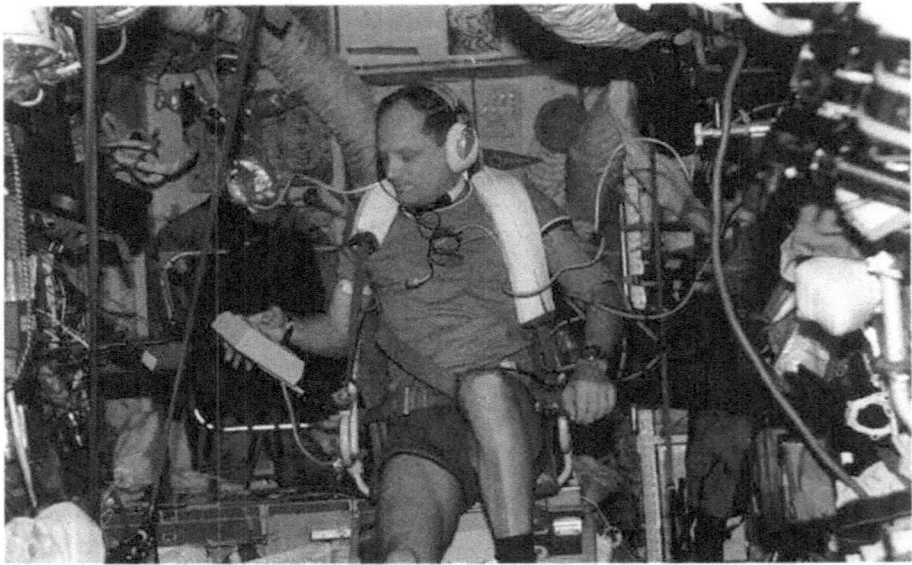

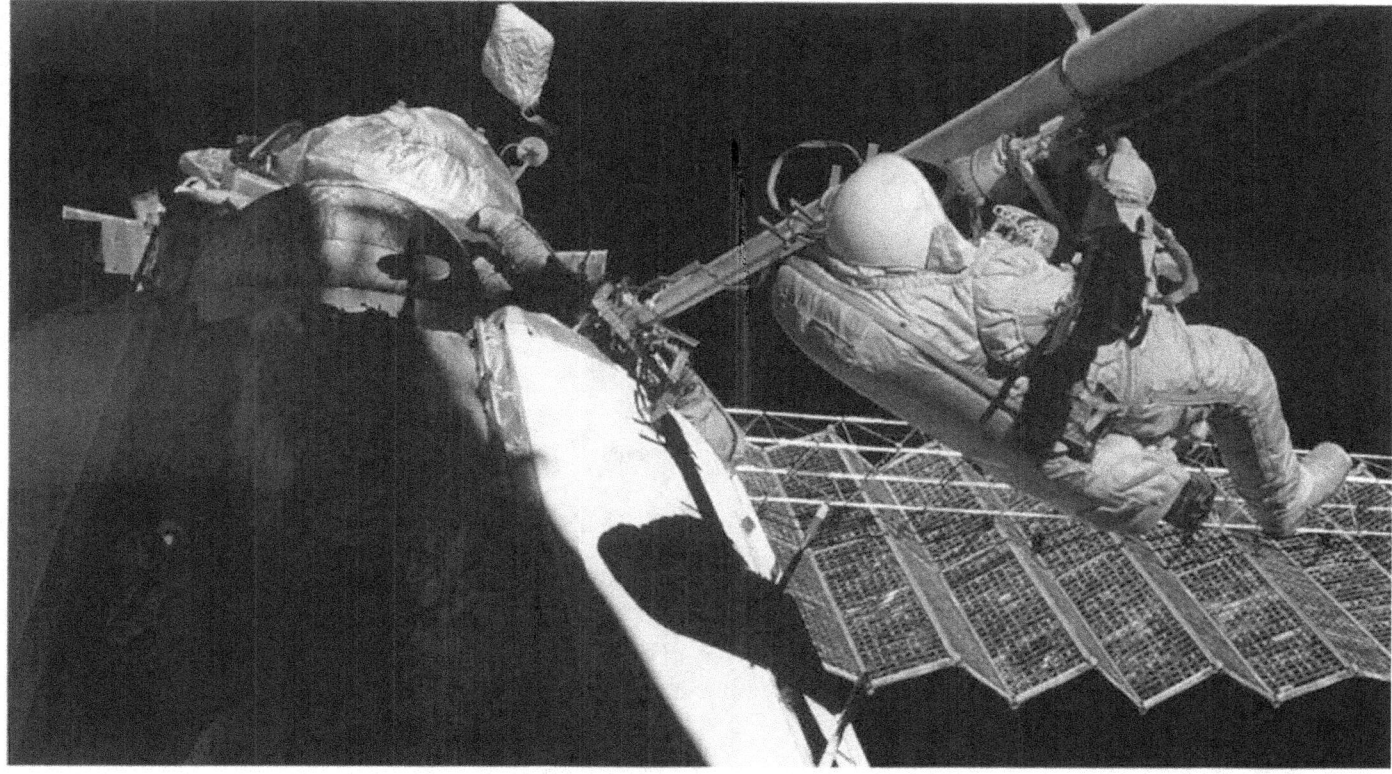

removal of an American experiment's space radiation detectors.

In their second spacewalk, on May 17, the two cosmonauts successfully folded the solar array panels while Thagard controlled servomotor switches from inside the Kristall module. The spacewalkers then disconnected the array from Kristall, attached it to the Strela boom, and moved it to Kvant. But, the work used up so much time and oxygen that they were forced to use their tool tethers to tie off the array to Kvant, and they had to postpone making the electrical connections. This marathon extravehicular activity lasted over 6½ hours.

The next day, while exercising, Thagard suffered an eye injury. He was doing deep knee-bends, using a device with elastic straps, when one end of a strap slipped off his foot and flew up and hit him hard in the right eye. "I was pretty sure for a while that I had done some serious damage to the eye," Thagard said. Even small amounts of light caused him pain, and using the eye was "like looking at the world through gauze."

Thagard patched his eye. When he told Strekalov what had happened, Strekalov joked, "Oh, yes. Those things are dangerous. That's why I don't use them."

"Thanks, Gennady, for the heads-up on that one," Thagard responded.

After a consultation with an ophthalmologist at Mission Control-Moscow, Thagard applied steroid drops, and the eye healed.

On May 20, the Spektr geophysical research module launched from Baikonur on top of a Proton rocket. Two days later, Dezhurov and Strekalov conducted a five-hour spacewalk. Working more efficiently than on their two previous excursions, the cosmonauts successfully connected the solar array to Kvant, and Thagard commanded its redeployment from inside the station. The cosmonauts then returned to Kristall, where they retracted 13 panels of another solar array to provide clearance for rotation of Kristall during its relocation to make room for Spektr.

Back inside Mir, the crew moved umbilicals, cables, and spacesuit control panels from Kvant-2

to the Base Block transfer compartment, which would be depressurized and used as the airlock for the next two extravehicular activities. On May 26, Commander Dezhurov used a remote manipulator system to relocate Kristall. On May 28, both cosmonauts performed an intravehicular activity inside the depressurized transfer compartment. They relocated a docking cone to serve as the docking receptacle for Kristall in its next move, and they moved the module again on May 30. The new Spektr module docked successfully on June 1. Thagard called the docking "an awesome sight."

On June 2, with the Mir crew and ground controllers in joint control of Spektr's Lyappa manipulator, the module was moved to its final

position. Later, only one of Spektr's two auxiliary solar array panels unfurled successfully. Planning began for a sixth spacewalk.

During this busy time, on June 6, Thagard surpassed the American single-mission duration record of 84 days, which had been set in 1974 by the Skylab-4 crew of Gerald Carr, Edward Gibson, and William Pogue.

The Mir crew completed their final reconfiguration of the station on June 10, using the Lyappa arm to move Kristall once more. Thagard began activating the American equipment inside Spektr, including two freezers for biomedical sample storage.

At about this time, all three crewmembers were able to talk to their families; but overall, communication with the ground was far from perfect. According to Thagard, "There were days ... when we had as little as about 42 minutes of communication time for the whole 24-hour period. That's for everything. Obviously, the stuff I was doing can't have priority over stuff that you need to do to keep the Mir station running. I think there were four times during the flight when I went 72 hours without talking to anybody in the Mission Control Center [-Moscow]."

On June 15, the scheduled sixth extravehicular activity of the Mir-18 mission was canceled, partly because Strekalov thought there was insufficient planning and lack of proper tools. The spacewalk was rescheduled for the next Mir crew, which would be trained in the use of the tools before they launched to the station.

The Mir-18 crew began getting ready for their departure from Mir aboard the U.S. Space Shuttle. They packed up experiments, biomedical samples, and other items to take onboard *Atlantis*. They also concentrated on microgravity countermeasure exercises and spent time preparing their cardiovascular systems for return to normal gravity by wearing a Chibis suit, the Russian lower body negative pressure device. Thagard spent several communication passes with ground controllers, inventorying the contents of the Spektr lab.

Flight Engineer Strekalov performs his tasks after being moved into position by the Strela arm during one of the five spacewalks conducted by the Mir-18 cosmonauts during their 115-day mission.

Thagard also conducted a number of experiments to help investigators characterize the microbial environment on Mir. He began collecting air, water, and surface samples and preparing them for return to Earth. These experiments had been scheduled late in the Mir-18 mission so that the samples would remain fresh until they were returned to Houston.

June 18 was a sad day onboard Mir when Commander Dezhurov learned that his mother had died. He received two days "off" for mourning; grief, undoubtedly, is just as heavy in space.

Preparations for *Atlantis* continued but, before its arrival, the space station crew successfully commanded a solar array into the desired position for a Shuttle docking. They also worked on disassembling Spektr's remote control unit, which had provided a backup capability for commanding Spektr during docking. After being returned to Earth on *Atlantis*, the unit would be used on the Priroda module, which would arrive

at Mir during Shannon Lucid's increment. Later, this remote control system would be involved in two dangerously close calls during Jerry Linenger's and Mike Foale's increments.

Thagard's historic flight onboard Mir was coming to an end. He and his two Russian crewmates were ready to return on STS-71 *Atlantis*, but not before four days of delays caused by bad weather at Kennedy Space Center in Florida. Thagard later told about learning of the postponement. "Velodya and Gennady and I had been looking out the window as we passed over the Cape. It was just solid clouds ... over the whole Southeast ... just solid cloud cover. And, Gennady says the weather was fine anyway.

"Velodya and I were wondering what window Gennady had been looking out, because he certainly didn't see the same weather that we had seen.

"But then, Gennady says, 'Well, that's fine.' He says, 'Just don't even let it launch. We'll just wait until September, and we'll come home in the Soyuz.'

"And, this was not really in character with Gennady, because throughout that mission Gennady was just the nicest guy in the world. There was never a time when I was doing something and he passed through the area that he wouldn't ask if he could help. I mean, that's just in Gennady's nature. But, he wanted to go

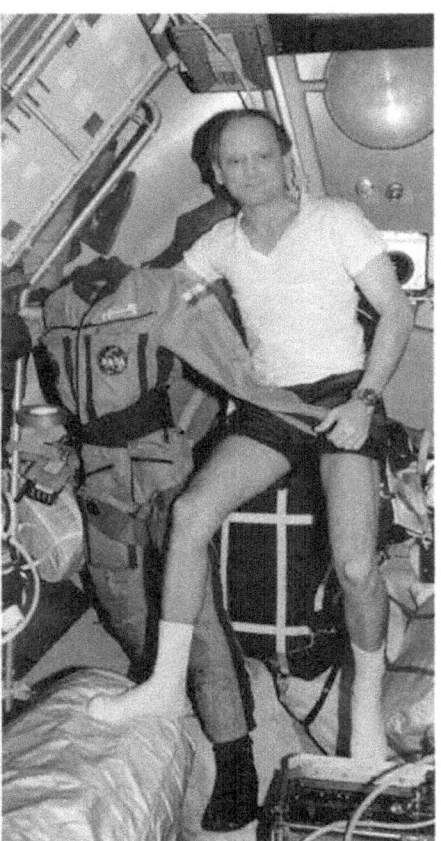

The Mir-18 crew watches the approach of *Atlantis* with its payload bay open (above) in preparation for docking. STS-71 ferried the Mir-19 crew to the space station and welcomed Thagard and his two Russian crewmates aboard for the ride back to Earth. While on the U.S. Space Shuttle, Thagard (below) displayed his Mir-18 flight suit that he wore during his record-breaking stay on the Mir.

home, clearly, and he was not happy with the Shuttle delays. So, he says, 'Well, we'll just wait, and we'll come home on the Soyuz.'

"And, I looked at him, and I said, 'That's fine with me, Gennady.' I said, 'I've never ridden on the Soyuz [back to Earth]. It's an experience I'd like to have. You, on the other hand, are going to miss your daughter's wedding,' because his daughter was supposed to be married in August. So, that kind of shut him up."

The Space Shuttle *Atlantis* docked successfully with Kristall's docking port on June 29, delivering the Mir-19 crew of Anatoly Solovyev and Nikolai Budarin.

On July 7, the U.S. Orbiter returned all three of the Mir-18 crew to Earth. After 115 days in orbit, Thagard was the first to unstrap from his seat and stand up.

Meanwhile on Earth
NASA-1

Early during Norm Thagard's stay onboard Mir—on March 30, 1995—NASA announced four Mir astronaut assignments. Shannon Lucid would be the second American to be a Mir crewmember. Jerry Linenger was, at first, scheduled to be the third. However, John Blaha was already training as Lucid's backup in Star City, so he was soon moved into the NASA-3 slot. Scott Parazynksi was selected as backup for Linenger. He and Linenger began training in May.

While Thagard was being first in so many ways in orbit, NASA's people on Earth were breaking new ground too. James Van Laak, who would become Frank Culbertson's Deputy, joined Tommy Holloway's Phase 1 staff in April 1995. Van Laak later described the Program (at that point), "There really wasn't any specialization in the staff … with one or two exceptions. Essentially, we all did whatever we were assigned [to do], and pulled together as a little core team to try and implement the Program. That was shortly after Norm Thagard was launched, and it was the very beginning of the operational phase. As such, it was a time of great challenge for the Program and a time when this 'flat' organization without specialization reached its limit very rapidly. It became clear that some specialization was required. We were in the beginnings of defining or deciding what that was going to be." With his U.S. Air Force experience and his time in Johnson Space Center's Missions Operations Directorate, Van Laak was basically "sliding into an operational or semi-operationally focused job."

Dave Ward and Mike Barratt were the Flight Surgeons during much of Thagard's training and flight. As such, they tended to Thagard's health and needs, managed much of NASA's communications to Mir, and found common ground with their Russian medical counterparts. Barratt later said that this was "a little awkward at first," in particular with the medical standards. The American and Russian medical teams had different philosophies on astronaut selection and certification, and on the medical monitoring of astronauts. According to Barratt, the Russian philosophy was more toward "functional loading," putting potential Mir crewmembers under

different situations—such as low atmospheric pressure, high temperatures, sleep deprivation, and centrifuge-induced g-loads—and then seeing how they responded physiologically. "Our philosophy was more to look for health factors, not necessarily fitness…. And, we would differ quite a bit on what test result would declare a person healthy or certified for training or spaceflight…. We have learned a lot about each other since then." More than anything, the two medical teams learned how to make their two systems work together.

Thagard launched with the idea that he was flying and working on Mir as a cosmonaut, as a member of the Mir-18 crew. That limited the amount of voice communications with him. NASA's managers in Russia realized that the people who did talk directly with Thagard on Mir would have to know him and his medical state, be familiar with his science program, and speak passable Russian as well. The best candidates for that job naturally happened to be Thagard's Flight Surgeons—Ward and Barratt. As Barratt put it, "By default, we became CAPCOMs [Capsule Communicators]—the last thing on our minds when we went over there."

Helping to set the stage for the Phase 1 Program were pioneers (left to right) NASA-1 Astronaut Norm Thagard, Flight Surgeon Mike Barratt, Shuttle-Mir Program Scientist Peggy Whitson, and Flight Surgeon Dave Ward.

Because the communications system was so awkward and limited, Barratt said, "It became much more efficient for us to speak English to Norm, to try to get experiment results and procedures up and down the best we could." The Flight Surgeons prebriefed the Russian flight control team on everything they were going to say, and then would speak in English to Thagard. "Then," Barratt said, "we'd have to debrief everybody, one by one, of what we said. So, just getting some words back and forth in English was a very difficult thing."

According to Barratt, the Russians saw this speaking in English as "uncontrolled communication to their space platform, so it was not something that they took lightly. Nor did we." The concern eventually melted away.

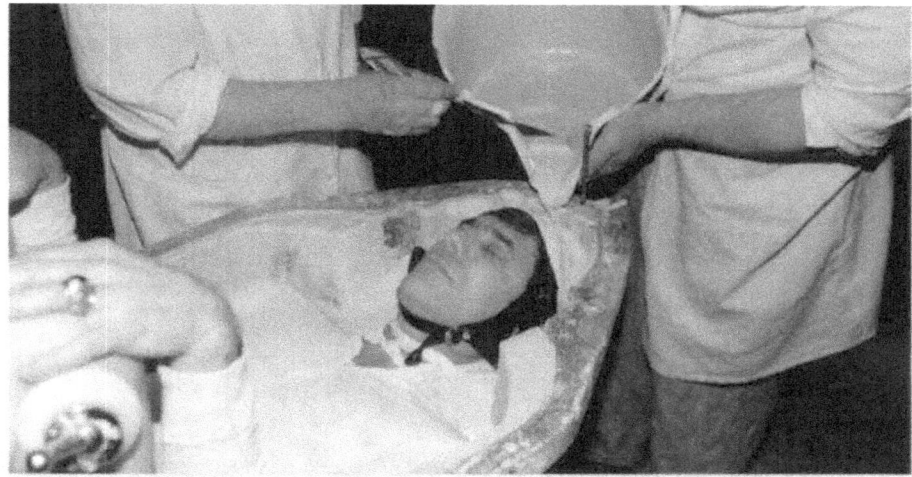

Barratt and Ward were also the American physicians responsible for covering the health of all NASA employees in Russia—both in Star City and at Mission Control-Moscow, the Russian flight control center. According to Barratt, "Whichever one of us was assigned to the flight control center, we would get there before the first communication with the crew and leave after the final one. So, there were very long days. We would typically be there at eight or nine [o'clock] in the morning and typically leave at eleven [o'clock] or midnight."

Quick Work on the Ground

A t times, managing the long-haul Shuttle-Mir Program required quick thinking and action. A good example occurred in the summer of 1995, when a solar array on the newly arrived Spektr module failed to deploy as designed. An aluminum tube needed to be cut. However, the array was so close to the space station that space-walkers could not get close enough to do the work. The STS-71 *Atlantis* mission to Mir was about to launch. Here was an opportunity—if a special tool could be developed in time.

In Moscow, Russian engineers went to work on a tool of their design; and at Johnson Space Center (JSC), a "Tiger Team" assembled to see if they could design, build, test, and certify a tool before launch. A JSC machinist suggested trying an automobile steering wheel cutter, and found one at a local fire department. The Tiger Team members lengthened the arms of the cutter so that it would work at a distance like a tree limb lopper. They adapted the tool so that a spacesuited cosmonaut could use it. They then segmented the arms so that the tool would fit in a compact box. The whole development process took about six days. Meanwhile, the Russians had developed a tool that worked somewhat like a scissors-jack. After review, the JSC tool was selected as more suitable, and it was packaged and put onboard *Atlantis*.

The Mir-19 crew of Anatoly Solovyev and Nikolai Budarin was trained on the tool's use while Solovyev and Budarin were waiting in quarantine for their launch. One week after the Space Shuttle departed Mir, the cosmonauts performed an EVA and tried out the tool on the solar array. It worked like a charm.

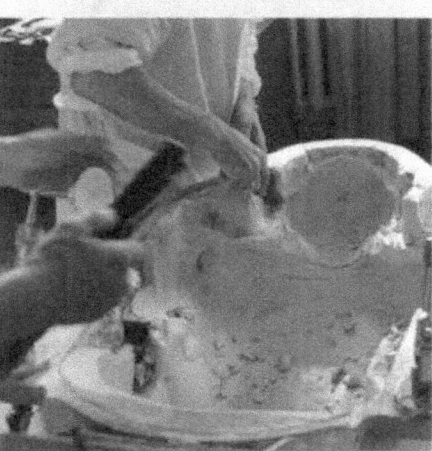

The Creation of a Soyuz Seat-Liner

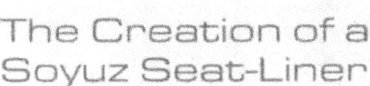

Every individual traveling in a Soyuz capsule must first endure the process of creating the flight-critical seat-liner. The process takes two to three hours and involves a number of crucial steps to ensure an exact fit. Pictured is NASA-3 Astronaut John Blaha, wearing a protective body suit while technicians pour the plaster around him that creates the mold of his body. With the aid of an overhead crane, the astronaut is lifted out of the partially set plaster mold. Technicians with artisan skills sculpt the rough edges to provide a smooth, form-fitting

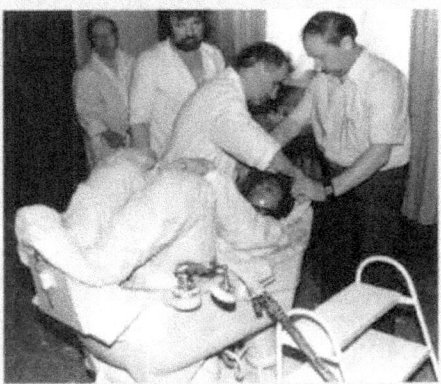

contour. Again, Blaha is lowered into the mold, this time wearing a Sokol launch and entry suit (covered with a protective body suit) so final adjustments can be made. From the mold, technicians create the synthetic seat-liner. Those Mir residents arriving by Space Shuttle transferred their unique seat-liner to the space station.

Space Shuttle *Atlantis*

100th U.S. human spaceflight mission

Launched:	June 27, 1995, 3:32 p.m. EDT Kennedy Space Center, Pad 39-A
Orbit:	170 nautical miles
Inclination:	51.6 degrees
Landed:	July 7, 1995, 10:54 a.m. EDT Kennedy Space Center
Mission:	9 days, 19 hours, 22 minutes

STS-71 CREW

Commander Robert L. "Hoot" Gibson
Fifth Shuttle flight

Pilot Charles J. Precourt
Second Shuttle flight

Mission Specialist Ellen S. Baker, M.D.
Third Shuttle flight

Mission Specialist Bonnie J. Dunbar, Ph.D.
Fourth Shuttle flight

Mission Specialist Gregory J. Harbaugh
Third Shuttle flight

Cosmonaut Anatoly Y. Solovyev
Russian Space Agency
First Shuttle flight; remaining on Mir

Cosmonaut Nikolai M. Budarin
Russian Space Agency
First Shuttle flight; remaining on Mir

Astronaut Norman E. Thagard, M.D.
Fifth Shuttle flight; returning from Mir

Cosmonaut Vladimir N. Dezhurov
Russian Space Agency
First Shuttle flight; returning from Mir

Cosmonaut Gennady M. Strekalov
Russian Space Agency
First Shuttle flight; returning from Mir

PAYLOADS

Spacelab/Mir Science
Metabolic Research
Cardiovascular and Pulmonary Research
Neurosensory Research
Hygiene, Sanitation, and Radiation Research
Behavior and Performance Research
Fundamental Biology Research
Microgravity Research
Protein Crystal Growth Experiment
IMAX Cargo Bay Camera
Shuttle Shortwave Amateur Radio Experiment

STS-71
First Docking
June 27 – July 7, 1995

With STS-71, NASA staged a space opera that outdazzled the popular movie *Star Wars*. This mission presented gleaming spacecraft, galactic vistas, precision flying, nerves of steel, and onboard diplomacy between two superpowers who had recently been at odds. It also included a cast of ten crewmembers in space (six astronauts and four cosmonauts—the most for one mission) and hundreds of team members on Earth, speaking two languages.

STS-71 featured the first Shuttle "changeout" of Mir crews and the return of NASA-1 Astronaut Norm Thagard from his U.S. record-setting, paradigm-breaking, nearly four months onboard the Russian space station. And, to share the experience with audiences on Earth, STS-71 went "wide screen" and captured the adventure on large-format IMAX film.

It was serious business, this first docking of a Space Shuttle to the Mir, but the drama and thrill of this historic event could not be denied. Indeed, Pilot Charlie Precourt said the sight of

Inside the Spacelab science module of *Atlantis* (above), the crews of STS-71, Mir-18, and Mir-19 pose for this historic in-flight photo. Clockwise from left are Solovyev, Harbaugh, Gibson, Precourt, Budarin, Baker, Dunbar, Thagard, Dezhurov, and Strekalov. At right, STS-71 crewmembers viewed this scene of the Kristall module while the Mir and the Shuttle were docked. Opposite page, bottom right, STS-71 Commander Hoot Gibson and Mir-18 Commander Vladimir Dezhurov exchange greetings in the docking tunnel hatch.

arriving at Mir reminded him of "science-fiction movies, where Luke Skywalker is flying in on his ship and lands on the big station … because that's really what you're seeing—this outpost—against black space [and] this huge planet's horizon in the background. And, if you just take your imagination one step further, you are arriving from another galaxy to this space station outpost in some other solar system."

Contact and Capture: Emotions of the Moment

"I remember sitting there the day that Hoot [Gibson] maneuvered the docking system in and called down, 'Contact and capture.' I had a lump in my throat for the rest of the day. I really did. It was ... so historical.

"The last time we actually docked [during] Apollo-Soyuz, it was a different time, it was a different place, and we weren't necessarily friendly. This time it just had an overwhelming impact on me. I was amazed at my emotional response.... The rest of the day—the training team— we just walked around. We were all kind of in a daze because we had trained these people to do this, and they had gone up and done it perfectly....

"All of the hard work and everything paid off."

– Lisa Reed, Lead Docking Trainer for STS-71 at Johnson Space Center

According to Precourt, "You can put yourself in that little world fairly easily. To see ... a little speck grow into something that has shape and realize there are people living in that thing is pretty phenomenal."

For the docking, Commander Hoot Gibson positioned *Atlantis* directly below Mir, so that the Earth's gravity naturally braked the Orbiter's approach "up" to Mir. He and his crew executed a nearly perfect docking, off by less than one inch and one-half degree. Together, *Atlantis* and Mir totaled almost 500,000 pounds, the largest structure ever assembled in space.

After docking on June 29, the two commanders met in the docking tunnel for handshakes and greetings. Shortly after, both crews gathered in the Mir Base Block for a ceremony. On Flight Day 4, in Spacelab with the Russian and U.S. flags as a backdrop, the STS-71 and Mir crews exchanged ceremonial gifts. Together, they rejoined a halved pewter medallion bearing a relief image of a docked Shuttle and Mir space station.

After transferring responsibilities from the Mir-18 to the Mir-19 crew, joint operations included scientific investigations and the transfer of gear and supplies between the two spacecraft. The *Atlantis* crew retrieved from Mir samples of

urine, saliva, blood, and water as well as a broken Salyut-5 computer. They transferred to Mir custom spacewalking tools to repair a jammed solar array on the Spektr module; nitrogen and oxygen to raise Mir's air pressure; and more than 1,000 pounds of water for waste system flushing and electrolysis.

Under Payload Commander Ellen Baker's direction, STS-71 conducted a full-scale scientific campaign, using nearly three tons of science equipment in Spacelab. Mir-18's research in seven medical and scientific disciplines concluded during STS-71. Eleven experiments remained on Mir to be conducted by the Mir-19 crew. These experiments would investigate the human cardiovascular and pulmonary systems; neurosensory effects; hygiene, sanitation, and radiation; human behavior and performance; fundamental biology; and microgravity.

When the time came for undocking on July 4, 1995, the Mir-19 crew temporarily left Mir in their Soyuz spacecraft to videotape the separation, which looked like a "cosmic

ballet," according to Commander Gibson. For reentry, Mir-18 crewmembers Thagard, Dezhurov, and Strekalov lay in the mid-deck of the *Atlantis* in custom-made Russian seats designed to ease their transition back into gravity. The Mir-18 crew had exercised intensively to prepare for the stresses of reentry and gravity after more than three months in space. Their changes in pulse, blood pressure, voice, and posture were monitored during the reentry portion of STS-71.

Rendezvous & Docking

Moving a spacecraft around in orbit is not as easy as one might think. A physical principle called orbital mechanics makes getting around in space very different from getting around on Earth. Astronauts cannot simply point the spacecraft in the direction it needs to go and then fire the rockets.

Furthermore, the nearly 100-ton Shuttle has a lot of momentum. A Shuttle may appear weightless in microgravity, but its mass makes it act somewhat like a barge on a river—hard to change direction and hard to stop. Also, the masses of Mir and of a Shuttle are great enough that, according to calculations NASA-5 Astronaut Mike Foale once did, even if the two spacecraft were "parked" in orbit 30 feet apart, their own gravity would pull them together in the span of a few hours.

Basically, an Earth orbit can be compared to "falling" around the Earth. The spacecraft's flight path can be described as continuously arching downward at precisely the curvature of the Earth.

Orbits of different heights have different set speeds. Lower orbits are faster; higher orbits are slower. This can be compared to the different speeds of a leaf going down a drain. The leaf circles the drain slowly at first; then it spirals faster and faster as it gets nearer the drain. Mir's orbit was higher—and slower—than a typical Shuttle orbit.

Once a spacecraft has achieved an orbit, that orbit will not change unless the upper atmosphere slows it down or some energy is applied with rockets or jets. Therefore, to match the Shuttle's orbit to Mir's, astronauts must use the principles of orbital mechanics.

STS-71's June 29, 1995, rendezvous and docking with Mir actually began with the precisely timed launch of *Atlantis* on a course for the station. *Atlantis* had a "launch window" of exactly 10 minutes and 19 seconds. Missing this window would have made it impossible to link up with Mir. Over the first two days of the mission, Commander Hoot Gibson and Pilot Charlie Precourt made periodic small engine fir-

ings, gradually bringing *Atlantis* to a point eight nautical miles behind Mir. On the third day, they fired a terminal phase initiation burn, and the Shuttle's rendezvous radar system began tracking Mir to provide range and closing rate information to the Shuttle crew. As *Atlantis* neared Mir, the trajectory control sensor—a laser ranging device mounted in the payload bay—supplemented the Shuttle's onboard navigation information.

Onboard *Atlantis*, the crew started watching for Mir. According to Precourt, Mir "starts out as a little star, so it's kind of exciting to see who's going to be the first on the crew to spot [it]. We'll be maneuvering, and we know where it should be, so I'll yell out to everybody, 'Okay! Look out this window. See if you can see it.' Then we see it, and we call the ground, 'Hey, we've got them in sight. We see them'—this twinkling little star out there. Then you can follow them as you go around from the sunlit side to the dark side, and you can see, with each successive [revolution], them starting to get larger and larger."

Atlantis proceeded to a point directly below Mir, along the Earth radius vector (R-bar), which is an imaginary line drawn between the Mir center of gravity and the center of the Earth. Approaching Mir from directly below allowed natural gravitational forces to slow the approach more than would occur during a typical Shuttle approach from directly in front of a satellite. The R-bar approach also reduced the need for jet firings close to the Mir's solar panels and the Soyuz capsule's navigation surfaces.

About one-half mile below Mir, STS-71 Commander Gibson took over manually. He flew the Shuttle using the aft flight deck controls, located below the windows looking back into the payload bay. Using a center-line camera fixed within the docking system in the payload bay, Gibson centered the Shuttle's docking mechanism with the Mir's docking device, which was positioned on the end of the Kristall science module. He continually refined this alignment as the

Shuttle approached to within 300 feet of the station.

Gibson held *Atlantis* at a "station-keeping" distance, 250 feet from Mir, while flight directors in Houston and outside Moscow reviewed the status of the two space vehicles. When he got a "go" for final approach, Gibson maneuvered *Atlantis* at a rate of 0.1 feet per second until the Shuttle was 30 feet away. During the station-keeping period and the final approach, the Shuttle crew radioed the Mir crew, keeping them informed of events from that point on. The final approach ended 216 nautical miles above Russia's Lake Baikal region, with a nearly perfect docking—off by less than one inch and 0.5 degree.

On June 29, 1995, two spacecraft from two nations took center stage in world history.

The U.S. Space Shuttle *Atlantis* docked for the first time with the Russian Space Station Mir, bringing the leaders of space exploration together for a new phase of space cooperation.

The commander of the STS-71 mission, Robert L. "Hoot" Gibson, positioned the American Orbiter directly below the station. He centered the Shuttle's docking system, located in the payload bay, with the Mir's docking device, located on the end of the Kristall science module. When he got the "go" for final approach, he maneuvered the American spacecraft towards Mir for the docking that created a 500,000-pound complex in space.

On July 4, 1995, Mir-19 Commander Anatoly Solovyev and Flight Engineer Nikolai Budarin temporarily unparked the Soyuz spacecraft from the cluster of Mir elements and performed a brief fly-around to capture the undocking (pictured at right) of this technological feat. The Russian cosmonauts took pictures while the STS-71 crew and the three Mir-18 crewmembers (aboard *Atlantis*) undocked to return to Earth.

Solovyev and Budarin had been taxied to Mir by STS-71 during the ascent portion of its mission.

35

Space Shuttle *Atlantis*

Launched: November 12, 1995, 7:30 a.m. EST
Kennedy Space Center, Pad 39-A

Orbit: 213 nautical miles

Inclination: 51.6 degrees

Landed: November 20, 1995, 12:01 p.m. EST
Kennedy Space Center

Mission: 8 days, 4 hours, 31 minutes

STS-74 CREW

Commander Kenneth D. Cameron
Third Shuttle flight

Pilot James D. Halsell, Jr.
Second Shuttle flight

Mission Specialist Chris A. Hadfield
Canadian Space Agency
First Shuttle flight

Mission Specialist Jerry L. Ross
Fifth Shuttle flight

Mission Specialist William S. McArthur, Jr.
Second Shuttle flight

PAYLOADS

Orbiter Docking System (Docking Module)

Solar Arrays

IMAX Cargo Bay Camera

Shuttle Glow Experiment

Photogrammetric Appendage Structural
Dynamics Experiment

Shuttle Shortwave Amateur Radio Experiment

STS-74
A New Docking Module
November 12 – 20, 1995

The STS-74 crew, pictured in the docking module delivered to the Mir space station, are (bottom center) Mission Specialist McArthur; (clockwise) Pilot Halsell, Mission Specialists Hadfield, Ross; Commander Cameron.

STS-74 was the first Space Shuttle mission to actually help build a space station. Its payload included two solar arrays and a module essential for future dockings of the Russian space station and the U.S. Shuttle.

For the first Shuttle-Mir docking (STS-71), the Russian cosmonauts, with the aid of the Lyappa manipulator arm, relocated the Kristall module to allow ample clearance for *Atlantis*. After the Orbiter departed, the Mir crew had to return the Kristall to its original location to provide Russian Soyuz and Progress vehicles access to the station.

To avoid future movements of the Kristall, STS-74 ferried a Russian-built docking module and Orbiter docking system to Mir for installation. The new mechanism would provide the means to effect Shuttle dockings without interference.

STS-74's journey was delayed one day as the crew waited for weather to clear at a trans-Atlantic emergency landing site. Soon after reaching orbit, the crew began setting up for business.

On Flight Day 1, the astronauts powered up the docking module, which was stowed on its side inside the Shuttle's payload bay. On Flight Day 2, they examined the robotic arm and installed a guidance camera in the Shuttle's docking system. On Flight Day 3, astronauts Chris Hadfield and Bill McArthur used the robot arm to grapple the docking module, swing it out of the payload bay, and position it on the end of the docking system at a right angle out of the payload bay. Commander Ken Cameron and Pilot Jim Halsell then fired Shuttle thrusters to "bump" the docking assemblies together. After checking for leaks, the crew entered the

Workers at Kennedy Space Center prepare to close the doors on the payload bay of *Atlantis* that contains the Russian-built docking module.

docking module and moved the guidance camera to help in docking with Mir.

Flight Day 4 marked the second Shuttle-Mir docking. Commander Cameron did not have the good view that STS-71 Commander Hoot Gibson had enjoyed during the inaugural docking of the two spacecraft. Stacked together, the combined docking assembly and module measured almost 20 feet. According to Cameron, it was "like looking at the top of a building from the ground floor. You can see that it's up there but you really can't … accurately judge position or orientation." For the successful docking, Cameron and his crew used several aids, including the camera inside the docking module, another camera on the outside of the module, a "wrist" camera on the robot arm, and a laser system that worked in conjunction with a series of reflectors mounted on Mir's Kristall module.

During the three days of docked operations with Mir, *Atlantis* took onboard U.S., Russian, and European Space Agency (ESA) equipment and samples. The crew delivered water, supplies, and equipment, including two new solar arrays—one Russian built and one built in the U.S.

Mir's residents at the time were Commander Yuri Gidzenko, Flight Engineer Sergei Avdeyev, and Cosmonaut Researcher Thomas Reiter of Germany.

This mission marked the first time astronauts from the U.S., Russia, Canada, and the ESA were in space on the same complex at one time—an example of future international cooperation.

Atlantis lingered near Mir and made two fly-arounds at a distance of about 400 feet while the crew filmed the station with the large-format IMAX camera.

Flying Through the Furlough

Much has been made of the Russian Government's difficulties in meeting the financial demands of its space program. However, the Russians constructed the docking module in a remarkably short time. It was the U.S. Government that made efforts difficult for Shuttle mission STS-74.

Congress' failure to put together a budget deal caused "nonessential" Government workers to be furloughed from November 14 through November 19, 1995—right in the middle of the STS-74 flight. While those NASA employees critical to mission success kept working, other employees were sent home and NASA was unable to release public information.

The STS-74 crew kept flying of course, and they kept their sense of humor. On a private Web site titled "The Utterly Unofficial STS-74 Mission Guide," they and their friends on Earth kept information flowing to the public. They posted: "NOTE: A number of NASA [Web] servers were shut down due to the budget idiocy in Washington, D.C.… Meanwhile, please be patient with NASA as they get back into the swing of things. It certainly wasn't their fault." To serve Canadian followers of Mission Specialist Chris Hadfield, an "equally unofficial" Canadian Web site was maintained.

On Earth, Russian Team Leader Oleg Lebedev commented to NASA's Spektr Module Manager Charles Stegemoeller, "I've never talked to an unemployed American before."

The Russians could appreciate the Americans' difficulties.

Training and Operations

An old spaceflight saying goes, "You fly as you train."
Prepare well, and you'll perform well. And you'll get it
right the first time.

Training

With the Shuttle-Mir Program, astronaut training took on many new dimensions. NASA's Mir astronauts, who had previously prepared for Shuttle spaceflights of up to two weeks, now spent many months in Russia preparing to spend up to six months in orbit—on a Russian space station, with Russian crewmates, speaking the Russian language.

The Mir astronauts trained to: ensure crew safety (including a possible emergency descent in a Soyuz capsule); support Mir's systems and equipment; perform workstation organization; communicate with Mission Controls in both Houston and Moscow; conduct scientific research; and perform station maintenance and physical exercises using Mir's onboard facilities.

Meanwhile in Houston, the NASA Shuttle astronaut crews selected for the Mir docking missions now trained for new tasks. These included flying with Russians as crewmates, rendezvousing and docking with the Russian space station, and working onboard Mir. Five Russian cosmonauts also trained at NASA's Johnson Space Center to fly onboard Space Shuttles as members of American Shuttle crews.

This training and preparation for the Shuttle-Mir missions required frequent travel, lengthy relocations, and innovative institutional support. It also required that the two space

agencies adapt to each other's techniques and philosophies.

For the most part, the Russian and American space agencies used similar training methods, but the mixes of methods differed. In general, NASA required its crews to have more hands-on training and relied more on "sims"—the realistic simulations of spaceflight situations. The Russian Space Agency also used simulations, but it did not *veri-simulate* as many situations and activities.

Training for the Shuttle-Mir Program brought new dimensions to participants serving as crewmembers and to those who trained them. Travel between the two principal training centers, the Gagarin Cosmonaut Training Center (GCTC) in Star City, Russia, and the NASA Johnson Space Center (JSC) in Houston, Texas, became a necessary element to provide essential preparation for all of the types of spacecraft involved in the Phase 1 Program. Photos include: (above) the Star City complex featuring the Prophylactory (building on left), partially occupied by American participants; (bottom left) Russian microgravity aircraft IL-76MDK "flying laboratory;" (bottom right) NASA-2 Astronaut Shannon Lucid and trainer floating in Russian aircraft; (opposite page, top left to right) Astronaut Bonnie Dunbar in exercise room in Star City gym; underwater training at the GCTC; Mir-18 crewmembers Norm Thagard (left) and Vladimir Dezhurov during medical operations training at JSC; STS-91 Mission Specialists Valery Ryumin (left) and Franklin Chang-Diaz performing a fitcheck in their Launch and Entry Suits in the mid-deck of the Shuttle mock-up at JSC; NASA-4 Astronaut Jerry Linenger (left) with a Russian Space Agency trainer at the GCTC; NASA JSC, Building 9A, Training Facilities with the Space Shuttle Full Fuselage trainer and the remote manipulator system mock-up.

NASA's Mir astronauts in Russia spent a lot of time in classroom training before they took final oral examinations. (Until recently, American astronaut candidates were not given formal examinations.) As NASA-3 Astronaut John Blaha said, "In Russia, they do it the old-fashioned way. A person takes a piece of chalk and he goes to a chalkboard, and you're sitting as one student or two students, no more. That piece of chalk goes to the chalkboard, and the man starts teaching you a particular system in a Soyuz or on Mir. And, you take notes and you ask questions. When the course is complete, the Russians have another team administer an oral exam to the student."

William C. "Charlie" Brown, cochair for the training working group, said this training system works for the Russians partly because they have a lower turnover in staff. Once NASA establishes a program, it systematizes the training and relies more on written materials, part-task trainers, and full simulations. The Russians, however, have historically resisted writing things down, partly for security reasons. According to Brown, "Their advantage has been [that] they keep people around for 20 or 30 years, doing the same kind of thing. So, they don't have that problem of turnover. But at the same time, they don't have the portability of, say, a workbook—handing it to somebody and telling them to go home and study it."

Regarding classroom-style training, Shuttle-Mir Program Manager Frank Culbertson said, "The impression of a lot of our people, particularly initially, was that it was not very efficient—that we could come in here, clean house, and do all this in a couple of days and everybody would be thoroughly trained.

"The Russians, for their own reasons … based on practical experience, didn't agree with that, [and they] still don't agree with that. They still believe

that you need a certain amount of this type of training to be really ready to do what we're asking you to do." Culbertson said he received a post-flight note from NASA-6 Astronaut David Wolf that read that in retrospect the Russian training methods made Wolf learn the systems in a different way than he would have otherwise. Wolf's note said, "It made me think about [the systems] rather than just read about them, and it made me exercise the language and the jargon." Understanding the Russian space jargon was crucial, said Culbertson, "because you're not dealing just with the Russian language versus the English language…. It's like when you first come to NASA. I mean, how many of us understood the first lecture we sat through at NASA? It was like a totally new language."

Financial costs also factored into the differences in training techniques. The American space program was much better funded than the Russian space program at the time of Shuttle-Mir, and it could better afford the expensive simulators.

But, a big reason for the differences in training was the basic differences in the missions that the two space programs were required to carry out. The Americans typically trained for two-week-long Space Shuttle missions on which everything had to go right the first time. Also, NASA gave its Shuttle crews more operational responsibility than the Russian Space Agency gave its Mir crews. Therefore, almost every activity on a Shuttle mission was rehearsed, as exactly as possible, on the ground. The Russians, on the other hand, trained their Mir crews for the long haul. Mission Control-Moscow may have controlled many of Mir's orbital adjustments, but cosmonauts had to be well-trained in the systems and in maintaining the systems because—unlike on the Space Shuttle—malfunctioning equipment could not be brought down to be fixed on the ground.

Also, sometimes activities such as spacewalks became necessary after a Mir crew was in orbit. For those, the cosmonauts had to do their training onboard Mir.

Another training method the Russians employed was that of putting cosmonauts through weeks-long sessions in an isolation chamber to experience psychological, social, and culture deprivation. No NASA Mir astronaut took part in this isolation training, but NASA later incorporated this training method for its International Space Station crew candidates.

In sum, having to learn two systems—the Space Shuttle's and the Mir's—practically doubled the Mir astronauts' required knowledge base. Having to train in different countries, time zones, languages, and cultures made the training aspect of the Shuttle-Mir Program one of the most difficult jobs NASA has ever attempted. As NASA-5 Astronaut Mike Foale said, "Flights are hard; but, believe it or not, the training is harder."

Making this "hard" but effective training possible were crew training working group co-chairs Aleksandr P. Aleksandrov, Charlie Brown, Yuri N. Glaskov, Yuri Kargopolov, and Donald Puddy. Tommy E. Capps was Johnson Space Center's Training Manager.

During the Shuttle-Mir Program, nine NASA astronauts trained in Russia at the Gagarin Cosmonaut Training Center (GCTC). Five Russian cosmonauts (Krikalev, Titov, Kondakova, Sharipov, and Ryumin) underwent training at JSC for their Shuttle flights as part of the American crews. (Titov flew twice.) Nine Shuttle crews took one week of training in Russia for joint activities with Russian crews on Mir. The Russian primary and backup crews of Mir-20 through Mir-25 underwent one week of training at JSC.

The Mir simulators used for training at Star City are shown above. Pictured top (left), NASA-5 Astronaut Mike Foale trains in Russia, and (right) NASA-6 Astronaut David Wolf and American Astronaut Wendy Lawrence light flares to attract rescuers during a simulated emergency Soyuz capsule landing during Arctic survival training.

Training in Orbit

An astronaut's training for a Mir increment was so extensive, and the time to train was so limited, that NASA engineers and scientists developed a training system to be used onboard the space station.

The Crew On-orbit Support System (COSS) consisted of a laptop personal computer with a CD-ROM drive. Compact disks held training lessons, experiment instructions, and background information as well as psychological support materials such as greetings from family members. In many cases, ground team members videotaped the Mir astronauts' last training sessions so that later on orbit they could see themselves asking questions and operating the equipment. Shuttle-Mir Mission Scientist John Uri called the system "a perfect refresher."

An operations lead in Mission Control-Moscow might say to the Mir astronaut, "Okay, on Tuesday you're going to do the experiment; but on Monday, we want you to watch the video (on the CD-ROM), so by the time you get to the experiment, you know what you're doing."

Gagarin Cosmonaut Training Center

NASA's Mir astronauts trained for their Shuttle-Mir missions mainly at the Gagarin Cosmonaut Training Center at Star City, about 30 kilometers north of Moscow.

Like NASA's astronaut training center at Johnson Space Center in Houston, Texas, the Gagarin Cosmonaut Training Center offers high-tech training facilities. These include: integrated simulators for the Soyuz spacecraft and the Mir space station; a 5,000-cubic-meter water tank for spacewalking training; an IL-76MDK aircraft "flying laboratory" for simulating microgravity; and large (TsF-18) and small (TsF-7) centrifuges for simulating g-loads during launch.

The Center also conducts survival training for many possible landing situations, including mountains, woodlands, marshes, deserts, the Arctic, and open water. Russian cosmonauts

The Gagarin Cosmonaut Training Center in Star City, Russia, features a statue of its namesake, Yuri Gagarin, the world's first space traveler.

have experienced several rough Soyuz landings. For example, cosmonauts Belyayev and Leonov landed on permafrost in 1965; and Lazarev, Zudov, and Rozhdestvensky splashed down in Lake Tevngiz in 1976. All seven Shuttle-Mir astronauts returned to Earth onboard Space Shuttles, but an emergency evacuation of Mir in a Soyuz was always a possibility.

The Soviet decision to construct a cosmonaut training center was made on January 11, 1960. In 1968, it was named for Yuri Gagarin, the first human in space. On May 15, 1995, after the collapse of the Soviet Union, the Russian government established the Russian State Scientific Research Center of Cosmonaut Training, also named after Yuri Gagarin. It was placed under the authorities of the Russian Ministry of Defense (Air Force) and the new Russian Space Agency. Beyond training Soviet and Russian cosmonauts, the Gagarin Cosmonaut Training Center had trained, by April 1, 1996, 25 international crews, including 24 astronauts from 17 countries.

Survival Training

NASA's Mir astronauts underwent survival training in Russia to learn how to deal with emergency Soyuz capsule landings, both on land and in water.

Land survival training took place during winter in the forest near Star City. Crews practiced getting out of a capsule, still hot from reentry, and then keeping warm in subfreezing temperatures. Water survival training took place in the Black Sea. For a cold-water landing, the crews had to get out of their reentry suits before they put on four layers of Arctic winter survival clothing, followed by an orange drysuit. For a warm-water landing, crewmembers stayed in their reentry suits, put on life jackets, and gathered all their survival gear.

According to NASA-5 Astronaut Mike Foale, "This is all done with the hatch closed, in extremely close quarters, so a significant heat load builds up, added to by the continuous bobbing about in the waves. It takes one to two hours to get ready, then quickly open the hatch, and then jump out with the proper equipment.

"Jumping out is the key to the exercise.

The capsule does not float level, and there is great danger of the first person rocking the capsule so that water comes in through the top and sinks the others. We were told to simply fall, and not [to] push off in any way with our legs."

Finally, after the launching of signal flares, the crew was picked up. Foale said, "I had five liters or more of water in my suit by this time." Reassuringly, "the training suits leak worse than real ones."

Bonnie Dunbar said the land survival training was perhaps "the most comfortable thing" that she had done in her training for Mir. That was because "I grew up in the Northwest. My dad was a World War II Marine, and he used to teach us survival. You know, the snow's on the ground, you're out with your horse. You're checking the fence and it gets foggy, and you lose your way. How do you survive? I learned survival from the time I was a kid. So, being out there in the woods in the snow and taking care of the fire actually was kind of fun." Russian trainers were always nearby for safety.

Future Mir resident David Wolf appropriately dressed for his Arctic survival training in Siberia.

Americans and 'Warm Fuzzies'

Training Cosmonauts for the Space Shuttle

Lisa Reed, Training Lead at Johnson Space Center, was partly responsible for training of the Russian cosmonauts who were scheduled to fly on Space Shuttles. Later, Reed talked about working with Sergei Krikalev and Vladimir Titov, and dealing with the peculiarities of the American language.

"In order to get those guys trained by February [1994], so they could join the rest of the [STS-60] crew in the training flow … we instructors spent eight hours a day with them. For example, Monday on their schedule would be electrical power day. I would start in the morning with a briefing for two hours. I'd have a [Russian language] translator sitting with me, and Sergei and Vlodya [Titov] sitting across from me.

"It was difficult, because [we had] to pause with the translators. If it took an hour to teach it to an American, it took three hours [with the Russians] because we had to translate everything.

"I actually felt sorry for those guys, because they were getting a lot. It must have been like drinking from a fire hose for them

because … every day, it was a different system. But, they rose to the task and did very well."

The trainers also ran into problems with slang. According to Reed, "We Americans like our slang. We don't realize we use it as much as we do. So, [the Russians] would stop us. I remember one day, I was teaching Sergei about the hydraulic system.

"I was explaining to him that … [the Shuttle crew] will start one [hydraulic system] prior to the deorbit burn … basically just to make everybody feel good that you have one running. The term that all the instructors here use [for this reassurance] is 'warm fuzzy.'

"I remember [Krikalev] stopped, and he looked at me, and he goes, 'What is this warm fuzzy?'

"And, I couldn't explain it to him because I tried to separate the words. 'Warm,' to him, was like 'near hot.' Then, [there was] 'fuzzy.' We got off into talking about teddy bears and fur, and it just degraded from there.

"I don't think that he ever quite understood what a 'warm fuzzy' was."

Cosmonauts Vladimir Titov and Sergei Krikalev traveled to the NASA Johnson Space Center in Houston, Texas, for extensive training in preparation for their Space Shuttle missions. They learned basic Shuttle operations and procedures in a limited amount of time, while at the same time coping with the language and culture of the United States. Titov is pictured left at the NASA Sonny Carter Training Facility Neutral Buoyancy Laboratory, Houston, Texas.

Life in Star City, Russia, and "Space City," Texas

NASA's astronauts train at the Lyndon B. Johnson Space Center (JSC) in Houston, Texas. Established as the Manned Spacecraft Center in 1961 and renamed in 1973, JSC was built on former ranch and farm land just north of Clear Lake and near one of the largest petrochemical complexes in the world. The area has since become a suburban center, with shopping malls, office buildings, and housing developments. The climate is hot and humid for much of the year.

Russia's cosmonauts live and train at a place called Star City, home to the Gagarin Cosmonaut Training Center. Star City lies in a forest 30 miles outside of Moscow. Star City has a northern continental climate with cold winters and brief, warm summers. Many Shuttle-Mir team members lived or stayed in Star City during the Phase 1 Program, and several of their family members joined them there.

In 1960, when Star City was established, housing consisted of two 5-story buildings. Now, thousands of people live and work at Star City. Many NASA Shuttle-Mir team members lived in the Prophylactory, a three-floor dormitory the Soviets built for Americans during the Apollo-Soyuz Test Project and that has since been used for cosmonauts who have come back from a flight. NASA leases the second floor for housing and for the office of the Director of Operations-Russia. NASA also built several duplexes for housing.

"Star City really is a little tiny city," said astronaut William Readdy, who served as a Director of Operations-Russia. "It's self-contained, and all the services are there. And, there are several generations [of Russians] there … the original cosmonauts and their trainers, and then … kids and grandkids and grandparents," all living in the self-contained town.

Mike Barratt, a NASA Flight Surgeon during Shuttle-Mir, said. "It's a very historical place…. Star City has always been kind of … the 'forbidden city' or the 'hidden city.' It wasn't on any maps…. It was a secret cosmonaut training base. Of course, everyone knew where it was, but it was considered a closed and secure city." Barratt said that early in the Program in 1994, Star City was a "little overwhelming" to the Americans who had just arrived there. "There

considerable time living and training at the Johnson Space Center in Houston, Texas. They also had experiences adapting to a different culture. But for them, early in the program the changes were in the other direction—from a situation of fewer material goods to more. In 1998, Titov looked back on the "difficult time" of the early '90s in Russia. He said, "Moscow was not very comfortable, and when we arrived [in] Houston, here was another life, another style, and other stores. And, everything was for us like a little bit new."

Travis Brice, a key Shuttle-Mir Program Manager, recalled that, in late 1992, the Krikalevs and Titovs wanted to go to Tampa, Florida, to visit French Astronaut Jean-Loup Chrétien. Brice suggested driving.

"We can do that?" the Russians asked.

"Sure," Brice said, "You can do it.... You just get in your car and you go down to I-10 and turn right."

"We can do that?"

"Yes."

"We won't get arrested?"

"No. Just go. Obey the speed limits ... and you'll probably be okay."

The Russians asked, "How about gasoline?"

"Plenty of gas stations along the way," Brice told them. "Just watch your tank and don't get low. Keep your car full of gas and just go."

The Gagarin Cosmonaut Training Center, Star City, Russia, (above), and the NASA Johnson Space Center, Houston, Texas, (below) served as the two primary training sites for participants in the Shuttle-Mir Program.

were only four or five of us English speakers and an awful lot of ... Russian speakers. I think we were all looking at each other with a certain amount of curiosity, maybe suspicion. But, that gave way relatively quickly.... It was a very short time later that it became a very comfortable place—a second home, really.... Star City is a very friendly place and a very beautiful place—peaceful, quiet, forests, trees, fresh air."

Early in the program, and largely due to the problems Russia faced after the fall of the Soviet Union, many aspects of life and work at Star City did not match those of Houston.

Communications and customs were big problems. Some foods were available only in the little store inside the U.S. Embassy in Moscow. But, according to NASA-5 Astronaut Mike Foale, "Things got steadily better. In fact, Russia has changed incredibly in terms of services.... Moscow is unrecognizable compared to the way it was" in 1996.

Often, work took up almost all of the Shuttle-Mir team members' time. After his Mir flight, NASA-3 Astronaut John Blaha recalled looking at albums of photos of Russia that his wife, Brenda, had put together. He remembered he told her, "Brenda, these are really great books, but that's not what I saw of Russia." Blaha said that what he remembered was "a desk in a little room in my apartment, where I was studying my lesson material."

Blaha went on to describe a typical workday at Star City. "I would get up in the morning, sometimes at four-thirty ... to study for my classes that day. Somewhere around [eight o'clock] ... [my wife] would walk into my little study room. She would say, 'Your breakfast is ready.' ... I'd eat breakfast with her.... Then, I would go off to class.... There was a one-hour lunch break, [then] class from two to four [o'clock and] from four to six [o'clock]. It was like going to college. The instructors used a blackboard and a piece of chalk. At six o'clock, I would arrive home. I'd walk in, I'd relax a little bit, [then] I'd start studying until ten-thirty, 11 o'clock at night. Somewhere in between, Brenda would say, 'John, dinner's ready.' And, we'd go and sit down; and we'd eat dinner together and talk."

Shuttle-Mir Cosmonauts Vladimir Titov and Sergei Krikalev—with their families—spent

Bilingual Blues

anguage is the lyrical heartbeat of a culture. It was also at the heart of many challenges during Shuttle-Mir. To learn to speak and read Russian, Americans had to master the strange-looking Cyrillic alphabet, train their ears and mouths to encompass awkward new sounds, and grapple with a vocabulary rich with meanings. Protocols and agreements had to be translated from English to Russian, and then back into English to make sure that translators had not skewed the information. NASA Flight Surgeons put together a bilingual onboard medical manual so that in an emergency the cosmonauts and astronauts would be, in effect, "reading from the same page."

Astronaut Bonnie Dunbar said that language was the hardest part of training in Russia "because you had to learn all of these things ... and take oral exams and sit in lectures that were only in Russian." Dunbar said the experience must be similar to a first-grader going to graduate school. Although "you knew the answer, you didn't know how to say it in Russian. For about six

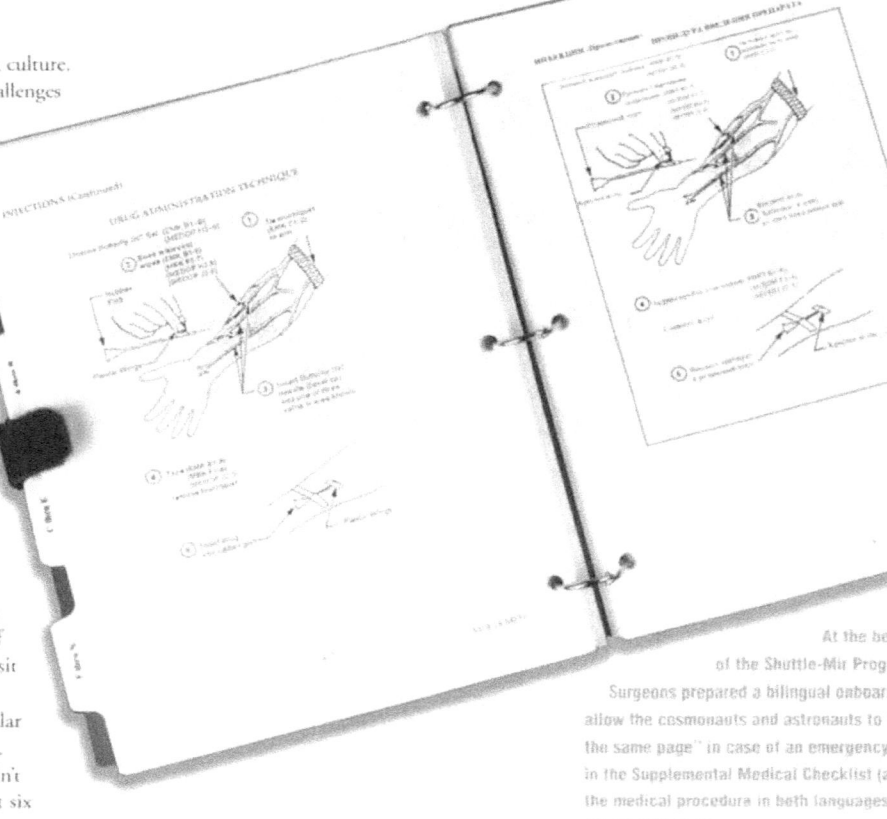

At the beginning of the Shuttle-Mir Program, Flight Surgeons prepared a bilingual onboard manual to allow the cosmonauts and astronauts to "read from the same page" in case of an emergency. Facing pages in the Supplemental Medical Checklist (above) display the medical procedure in both languages and include detailed illustrations.

months, I felt like a small child." Then, "all of a sudden," about six months into the program, the language came together for Dunbar when she reached a level of fluency.

The Mir astronauts had varying levels of Russian language preparation. Some began their training at the Defense Language Institute in Monterey, California; others studied mainly at Johnson Space Center in Houston, Texas, before moving to Russia. NASA-3 Astronaut John Blaha thought that inadequate language preparation was perhaps NASA's "biggest mistake" of the Shuttle-Mir Program, although he became comfortable with Russian while onboard Mir.

Although some people seemed to have a better knack for learning languages, NASA-5 Astronaut Mike Foale attributed his relative fluency to the hours he spent working on it. In his youth, Foale's approach to foreign language had been "fairly disdainful." But, when he learned from

NASA that he was going to Russia, he had "the cold realization that I was going to have a miserable time if I didn't learn the Russian language." He said to himself, "You have got to stop doing all those things you like doing in your free time." And from then on, he did even his pleasure reading in Russian.

NASA-2 Astronaut Shannon Lucid's crewmates, Yuri Onufriyenko and Yury Usachev, spoke no English. As Lucid's Russian continued to improve, "We made a lot of jokes about it," according to Lucid. "Yury, as a joke, said, 'Well, we're developing a new language, a cosmic language.' And, there was a fair amount of truth to that … because both Yuri and Yury very rapidly got to understand how I was talking.

"I mean, if a Russian teacher had been listening to me, she would have stuck her hands over her ears. But, I got the point across. Onufriyenko and Usachev never said, 'Oh, Shannon, just be quiet. We can't stand to listen to your Russian anymore!'"

The news media magnified Russian "complaints" that NASA-7 Astronaut Andy Thomas did not speak Russian well enough, but Thomas said that his technical Russian was fine. He wanted to work on his conversational abilities.

Onboard Mir, the continuous and nearly exclusive use of Russian could add to an American astronaut's sense of isolation. The communications passes were conducted nearly entirely in Russian, so this important verbal connection to the Earth could seem foreign

and strange. Further, important communications could be imperiled because one language's sounds might be hard for another's ears to decipher. Once, NASA-3 Astronaut John Blaha was training to help with his crewmates' spacewalk. During the spacewalk, Blaha was going to have to throw some switches. As Flight Surgeon Tom Marshburn told the story, the Russian word for "on" is включить (vkloo-*cheet*), and the word for "off" is отключить (ot-kloo-*cheet*), two very similar sounds.

Although Blaha was enunciating the words very precisely, the Russian ground controllers wanted to double-check. They turned to Marshburn and asked him how to say the words in English. "It's *on* and *off*," Marshburn told them, and the Russians burst out laughing. Those two English words sounded nearly identical to the Russians' ears.

The Russian alphabet's 33 letters (right) were developed 1,000 years ago from Greek and Hebrew letters by St. Cyril and his followers.

NASA-6 Astronaut David Wolf (above) reads his Russian-English dictionary while aboard the Mir. NASA-2 Astronaut Shannon Lucid (opposite page) reads the English version of a St. Petersburg paper during her residency on the Russian space station.

Cyrillic Alphabet

Russian	English
Аа	Ah
Бб	beh
Вв	veh
Гг	gheh
Дд	deh
Ее	yeh
Ёё	yo
Жж	zheh
Зз	zeh
Ии	ee
Йй	eey
Кк	kah
Лл	el
Мм	em
Нн	en
Оо	oh
Пп	peh
Рр	ehr
Сс	ess
Тт	teh
Уу	oo
Фф	ef
Хх	khah
Цц	tseh
Чч	cheh
Шш	shah
Щщ	shyah
Ъ	yehr
Ы	yehry
ь	yehr'
Ээ	eh
Юю	yu
Яя	yah

Operations

A Tale of Two Systems

"Operations," in essence, means managing a space-flight while it is taking place. Originally, the Space Shuttle and the Mir were not compatible vehicles because they were developed independently for different purposes. During Shuttle-Mir, the differences required many adaptations to allow the joint operation of the docked vehicles and to support an astronaut on a long-duration mission.

Mission Control Center-Houston, located at the NASA Johnson Space Center, directed the efforts of the 11 Space Shuttle missions connected with the Shuttle-Mir Program.

The joint missions required coordination between two control centers thousands of miles apart from each other, in different time zones, and with different native languages.

Communications links, processes, and procedures were developed to exchange information between control teams, coordinate decisions, and accommodate changes. An American, who served as a Russian Interface Officer (RIO) in Houston, and a Russian equivalent, the "PRP," in Moscow, acted as communicators between the NASA Flight Director at Mission Control Center-Houston and the Russian Flight Director at Mission Control Center-Moscow.

During docked operations, neither Mission Control Center was fully in charge of joint activities. The Mission Control Center-Moscow had authority for Mir. The Mission Control Center-Houston was responsible for the Space Shuttle. Similarly, the Shuttle commander was responsible for the Shuttle and its crew, and the Mir commander was responsible for the Mir and its crew.

"Joint Flight Rules," developed before each Shuttle-Mir mission, detailed both planned operations and practical responses to unusual situations. These rules minimized the need for quick decisions and ensured that each course of action had been reviewed and agreed to by both NASA and the Russian Space Agency.

The differences between a typical Space Shuttle mission and a Mir expedition are noteworthy. For a typical Shuttle mission, Mission Control Center-Houston runs operations and directs practically every activity onboard the spacecraft. The short duration of a Shuttle mission requires that the crew's time be closely scripted and monitored to get everything done. Further, with the aid of tracking satellites, Mission Control Center-Houston enjoys nearly constant contact with its Space Shuttles, so up-to-the-minute instructions can be radioed to the crew.

With much less voice communication than Houston has, Mission Control Center-Moscow typically uplinked a written flight plan to the Mir crew each day. This plan outlined a single day to five days in the future, and it gave a high degree of crew autonomy in the actual scheduling of activities. Furthermore, the long duration of a Mir expedition required more routine and normalcy for both its crews and its controllers. A "crew day" on Mir started at 8 a.m. Moscow time and extended to 11 p.m., allowing nine hours off for personal time and sleep. Usually, Mir crews worked five days per week and took two days off. During their days off, crewmembers were required only to do light housekeeping and perform their physical exercises, which took about two hours each day. Mission Control Center-Moscow personnel likewise had more regular hours. According to Flight Director Phil Engelauf, this gave people the feeling "that you get up in the morning and you have some time

to yourself. You have a workday. Then you're off in the evening, and you have your weekends to yourself." Of course, hardware failures and other contingencies frequently changed the routines and lengths of days worked, both for the crews onboard Mir and for the operations people on the ground.

During the course of Shuttle-Mir, NASA adapted its onboard astronaut operations—especially those of its science program—to be more flexible, like the Russians'. For example, according to the Shuttle-Mir Program Deputy Manager James Van Laak, "[We] might tell the crewmember, 'Today we'd like you to accomplish the following three things.' Then, if the astronaut began the first procedure and the hardware did not work well, he or she could postpone it and start a second project until the next communications pass came up. After discussing the situation with Mission Control, the astronaut might then go back to the second experiment because it might take Mission Control some time to figure out what to do next. Then, at the following communications pass, Mission Control would uplink a solution to the problem, and the astronaut might go back to the first experiment."

Flight Director Robert Castle said that, with Mir operations, "Things are a little slower-paced." He gave as an example an episode when an Elektron oxygen generator malfunctioned. "[The Americans] said, 'Okay, you're going to bring in a bunch of people to go work on this, right?' [The Russians] said, 'Well, we're really not going to bring anybody in. We'll start working on [the malfunction] Monday.' And, this was a Friday afternoon. Then our own environmental people said, 'With the current pressure of oxygen in the Mir, and the volume of the Mir, and the number of people onboard [the Mir], it will take 10 days to breathe the oxygen down.' ... Well, I can see why you might want to just take the weekend off, let people come in fresh, and work on it Monday, because you've still got a lot of time to work on it—whereas in the Shuttle Program, ten days is an entire mission."

Except in rapidly evolving situations, controllers could assume that Mir would "be there in the morning." So, they could take the time needed to do things in the best way overall.

Operations Leads and Russian Interface Officers

Early on, NASA managers realized that a Flight Surgeon alone could not handle all the communications with the astronaut onboard Mir. So, they created the position of Operations Lead—or "Ops Lead"—to act both as a CAPCOM [capsule communicator] with the Mir astronaut and as the leader of a NASA operations team in Russia. The goal, according to Program Manager Frank Culbertson, was for the Mir astronaut, the Flight Surgeon, and the Ops Lead to become "a good, solid team and stick together through training, know the mission very well, and be able to execute it together."

Jeffery Cardenas, the Ops Lead for Norm Thagard's Mir increment, later said that he typically arrived at Mission Control Center-Moscow about an hour before the first communications pass, or "com pass," with Mir. He talked to the Russian shift flight directors for that day, asked them what had happened onboard Mir overnight,

and checked updates on any of the Mir systems, especially anything that would affect events for that day. He then read over the prepared script of what he was planning to say because he would have, at most, 10 minutes of communications' time. When the com pass with Mir occurred, Cardenas would go over the script with Thagard onboard Mir, "trying to give him a thumbnail sketch of what's coming up on the day." For the rest of the day, Cardenas kept current on Thagard's activities and prepared for com passes three or four days in the future.

Ops Leads during Shuttle-Mir included Cardenas, Christine Chiodo, Scott Gahring, William Gerstenmaier, Isaac "Cassi" Moore, Patricia "Patti" Moore, Anthony Sang, and Keith Zimmerman.

Other unique Mission Control positions created for the Shuttle-Mir Program were NASA's Russian Interface Officer (RIO) at Mission

Control Center-Houston and its Russian Space Agency equivalent—the PRP—at Mission Control Center-Moscow. According to Flight Director Phil Engelauf, the RIO served as the focal point for communications, or facilitator for communications, between the two control centers. He or she managed fax traffic back and forth to the control centers as well as the voice conversations. Flight directors and specialists still communicated directly; but, as RIO Sally Davis said, "As far as marching down the road of executing the mission, you need somebody that's going to keep each control center in sync, because you can go off [in] 40 directions with everybody [involved doing their own tasks]. You have that problem anyway with one flight control team. So, with two [control centers operating], you needed. I guess you'd call it, a 'sync pulse'… [to keep] the control teams in line."

DOR: Director of Operations-Russia

In early 1994, the Shuttle-Mir Program created its Director of Operations-Russia position and named astronaut Ken Cameron as the first. Cameron accompanied Astronauts Norm Thagard and Bonnie Dunbar to Russia to manage NASA's operations at Star City and at the Russian control center at Kaliningrad.

The Director of Operations-Russia's responsibilities comprised a huge task, including supervising NASA astronaut training at Star City; developing Mir rendezvous training materials for Shuttle crewmembers; coordinating training for scientific experimenters; and establishing and maintaining operations, including procedures to support joint flight operations between NASA and the Russian Space Agency. In practice, he or she personally handled many problems that arose in Star City. Astronauts who served in that position for about six months each included: Cameron, William Readdy, Ronald Sega, Michael Baker, Charles Precourt, Wendy Lawrence, Michael Lopez-Alegria, Brent Jett, and James Halsell.

Readdy later talked about his early experiences. He said, "The objective was to prepare the ground and make sure that the two initial crewmembers that were over there, Norm Thagard and Bonnie Dunbar, got the support they needed—basically from soup to nuts." That meant everything "from

everyday living accommodations all the way through all the technical and professional training, and all the travel and support and everything. Star City is not inside Moscow. Depending on the roads and the weather, it can be a couple of hours outside of Moscow, and so the logistics are daunting.

"But, the idea was to support them. Of course, that means also supporting the trainers and experimenters and all the other NASA folks

that kind of go along with that."

As NASA's organization built up in Russia, the Director of Operations-Russia's duties evolved to become more managerial and less "Jack-of-all-trades."

Renaissance Hotel (Pentat, Moscow, Russia, housed the original Communications Center for American operations.

Mission Control Center-Houston

Many of the personnel from the Mission Control Center-Houston who assisted with the successful STS-79 mission are pictured above. Hundreds of others in numerous "ground support" teams supported the Shuttle-Mir Program.

At the moment the Shuttle's two solid rocket boosters ignite, vehicle responsibility transfers from the Kennedy Space Center launch team to Johnson Space Center's Mission Control Center, where a team of experienced flight controllers works in shifts to monitor and direct Shuttle operations.

Flight Director (FLIGHT)
Has overall responsibility for the conduct of the mission.

Spacecraft Communicator (CAPCOM)
By tradition an astronaut; responsible for all voice contact with the flight crew.

Flight Activities Officer (FAO)
Responsible for procedures and crew timelines; provides expertise on flight documentation and checklists; prepares messages and maintains all teleprinter and/or Text and Graphics System traffic to the vehicle.

Integrated Communications Officer (INCO)
Responsible for all Orbiter data, voice, and video communications systems; monitors the telemetry link between the vehicle and the ground; oversees the uplink command and control processes.

Flight Dynamics Officer (FDO)
Responsible for monitoring vehicle performance during the powered flight phase and assessing abort modes; calculating orbital maneuvers and resulting trajectories; monitoring vehicle flight profile and energy levels during reentry.

Trajectory Officer (TRAJECTORY)
Aids the FAO during dynamic flight phases; responsible for maintaining the trajectory processors in Mission Control and for trajectory inputs made to the Mission Operations Computer.

Guidance, Navigation, & Control Systems Engineer (GNC)
Responsible for all inertial navigational systems hardware such as star trackers, radar altimeters, and the inertial measurement units; monitors radio navigation and digital autopilot hardware systems.

Guidance & Procedures Officer (GPO)
Responsible for the onboard navigation software and for maintaining the Orbiter's navigation state, known as the state vector; monitors crew vehicle control during ascent, entry, or rendezvous.

Rendezvous Guidance & Procedures Officer (RENDEZVOUS)
Monitors onboard navigation of the Orbiter during rendezvous and proximity operations.

Environmental Engineer & Consumables Manager (EECOM)
Responsible for all life support systems, cabin pressure, thermal control, and supply and wastewater management; manages consumables such as oxygen and hydrogen.

Electrical Generation & Illumination Officer (EGIL)
Responsible for power management, fuel cell operation, vehicle lighting, and the master caution and warning system.

Payloads Officer (PAYLOADS)
Coordinates all payload activities; serves as principal interface with remote payload operations facilities.

Data Processing Systems Engineer (DPS)
Responsible for all onboard mass memory and data processing hardware; monitors primary and backup flight software systems; manages operating routines and multi-computer configurations.

Propulsion Engineer (PROP)
Manages the reaction control and orbital maneuvering thrusters during all phases of flight; monitors fuel usage and storage tank status; calculates optimal sequences for thruster firings.

Booster Systems Engineer (BOOSTER)
Monitors main engine and solid rocket booster performance during ascent phase.

Ground Controller (GC)
Coordinates operation of ground stations and other elements of worldwide space tracking and data network; responsible for Mission Control computer support and displays.

Maintenance, Mechanical, Arm, & Crew Systems (MMACS)
Monitors auxiliary power units and hydraulic systems; manages payload bay and vent door operations; handles in-flight maintenance planning; oversees Orbiter structure, tiles, blankets, etc.

Extravehicular Activities (EVA)
Monitors and coordinates spacewalks, including extravehicular activity suit and hardware performance.

Russian Interface Officer (RIO)
Acts as the communicator between the NASA Flight Director in the Mission Control Center in Houston and the Russian Flight Director at the Russian Mission Control Center in Kaliningrad; coordinates joint activities between the Shuttle and the Mir; implements joint decisions made by the two flight directors.

Payload Deployment & Retrieval Systems (PDRS)
Monitors and coordinates operation of the remote manipulator system.

Flight Surgeon (SURGEON)
Monitors health of flight crew; provides procedures and guidance on all health-related matters.

Public Affairs Officer (PAO)
Provides the media and public with explanations of mission events during all phases of flight.

Mission Control Center-Moscow

During Shuttle-Mir, the Russian Space Agency had three control rooms in a single complex in Kaliningrad, near Moscow. Mission Control Center-Moscow could process data from as many as ten spacecraft, although each control room was dedicated to a single program: one to Mir; one to Soyuz; and one to the now discontinued Russian space shuttle, *Buran*.

Flight control personnel were organized into teams, similar to NASA's system at the Mission Control Center-Houston. The Flight Director provided policy guidance and communicated with the mission management team. The Flight Shift Director was responsible for real-time decisions, within a set of flight rules. The Mission Deputy Shift Manager was responsible for the control room's consoles, computers, and peripherals. The Mission Deputy Shift Manager for Ground Control was responsible for communications. The Mission Deputy Shift Manager for Crew Training was similar to NASA's "CAPCOM," or capsule communicator. Generally, this person had served as the Mir crew's lead trainer.

Mission Control Center at Russia includes consoles (shown below) and display screens with Soyuz spacecraft and orbit trajectory (above). Known as the "TsUP," the MCC-Moscow maintained 24-hour operations throughout the Shuttle-Mir Program.

Long-Duration Psychology: The Real Final Frontier?

Spaceflight offers astronauts immense psychological rewards. However, long-duration spaceflight also poses great psychological risks. Dangers, deprivation, isolation, and confinement helped make the Mir residencies—in the words of some U.S. astronauts—the "hardest thing" they had ever done.

Moreover, their spaceflights came on the heels of difficult periods in Russia, where NASA astronauts trained immersed in a foreign culture and language. Add to this the fact that many NASA managers and support people were likewise experiencing lifestyle disruptions and heavy workloads, and the psychological aspect became as important as any physical aspect.

Up until Shuttle-Mir, most NASA astronauts were able to consider the physical risks before their flights. They could put [the risks] into perspective and launch with happy hearts. When dangerous situations occurred suddenly, they were usually over quickly. Even in the frightening case of Apollo 13, the crisis lasted only a few days.

With long-duration spaceflight, danger always exists; and living long in danger's presence increases one's awareness of it. Yet, for the most part, the Mir astronauts were able to adjust to this awareness—and even to add to their confidence in Mir's overall safety. For example, about the collision incident that occurred during NASA-5, Mike Foale described his feelings.

"It was frightening for one or two seconds," he said. "The first thought was—are we going to die instantly because of air rushing out so that we couldn't control it? It was obvious within two or three seconds that the air wasn't rushing out. Then we thought we had time, and I heard the pressure dropping. Immediately from that point, I thought 'Oh, this is a surprisingly robust station.'"

Meanwhile on the ground, Frank Culbertson and other managers faced the danger from their perspectives. They not only had to assess the risks, but they had to assure critics in Congress and elsewhere that their assessments were correct.

Sensory deprivation is another key factor in long-duration spaceflight. It manifests itself in many ways. The first sensory element to go is the pull of gravity and its physical comfort of rootedness. On shorter Shuttle flights, the lack of gravity often remains a pleasant novelty. But, combined with the other factors, microgravity could add to a long-duration astronaut's discomfort. However, most of the Mir astronauts reported enjoying microgravity. John Blaha said he didn't miss the pull of gravity at all. Andy Thomas wrote that microgravity was "the one thing that makes spaceflight both interesting and, at the same time, very frustrating … It can be a joy to experience, but [it] also can really make your work day difficult."

An astronaut's sense of time could also be affected. Sunrise and sunset alternate every 45 minutes. The sleep-and-wake cycle could come to feel arbitrary. And, when one is working long hours without refreshing breaks, the passage of time could seem to expand or contract. Worse, the amount of time left in a mission could become difficult to gauge. Yet, the Mir astronauts did not relate any real problems. Jerry Linenger wrote from Mir that "life in space is never monotonous."

But, the kind of deprivation that most affected Mir astronauts was social deprivation—being away from one's culture and family for such an extended period. Norm Thagard talked about missing his American cultural and linguistic environment. Blaha said what he really missed most was his wife.

Deprivation of meaningful work—or, conversely, of refreshing rest—also affected the U.S. Mir astronauts. Norm Thagard had to wait for many of his science experiments to arrive, and the Russians did not allow him much interaction with Mir's control systems. In the words of NASA psychologist Al Holland, "The situation of work underload is one of the worst situations you can ask a high-achieving, bright, interested astronaut to subject himself to."

Other astronauts went many days without much rest, working mainly on menial tasks, such as cleanup, when they weren't working on their American science projects. Those experiments were very important to the American astronauts. Foale said, "I loved the greenhouse experiment. It didn't matter that the shrubs were tiny … I enjoyed being a bee pollinating plants. I enjoyed

looking at [the plants] every morning for about 10 to 15 minutes. It was a moment of quiet time, almost. It was a moment where it was nice and bright and almost sunny in a module [Kristall] that had no power for about two months." Foale was dubbed "Farmer Foale" by the ground-based science team for his persistence in keeping the plants alive under trying conditions.

Mir astronauts did engage in exercise and recreational activities. Shannon Lucid read books. Blaha watched videos. Thomas sketched.

With long-duration spaceflight, an astronaut does not have the freedom to go where he or she wants to, and when he or she wants to, and no one can "drop in" for a visit. A spur-of-the-moment walk becomes a thing of the past — and, hopefully, of the future. Linenger wrote in a letter to his son, "A simple walk would be fine. Or a paddle in the canoe. Indoors won't do. Need fresh air. Need to feel a breeze ... the sound of wind through the trees overhead."

Before Shuttle-Mir, the Russians had years of experience with long-duration spaceflight. They had developed rigorous methods of selecting Mir crews and were able to give their selected crews psychological training. General Yuri Glaskov, Deputy Commander at the Gagarin Cosmonaut Training Center, described one of the training methods. He said, "We put our crewmembers into ... an isolation chamber. I had to myself be in this chamber for 14 days.

It is called ... 'alone in public.' Everybody is watching you, but you can't see anybody. There are certain psychological nuances there because you fight yourself."

Glaskov also experienced a 35-day ground test of the Mir orbital station. He said, "At that time, there were two of us; but the hatches were closed, and we were absolutely alone for 35 days. This experience created different problems. Here, we had to tolerate each other, forgive each other, and supplement each other's faults or experiences.... One person doesn't like certain traits of another [person], and so you have to learn to adapt to each other."

NASA's Shuttle-Mir astronauts were basically volunteers. While that eliminated the value of a selection process, it did give NASA psychological scientists the opportunity to observe — and to support — a range of personalities during the seven-mission program. NASA psychologist Holland said, "It's really probably good that we weren't allowed to do selection in our usual manner beforehand, but we had to work with the people who were assigned to us to fly — because in that way we learned a lot more."

NASA's "flight docs" and managers worked to make the Mir astronauts' missions as normal as possible, with things like weekly talks with family and friends; "surprise packages" coming up on Progress resupply vehicles; ham radio conversations with friends, family, and even strangers; and the crew on-orbit support system,

a laptop computer and compact disks that included items such as special greetings that were timed to coincide with an astronaut's birthday.

So, what was learned about psychology during the Shuttle-Mir experience? In the opinion of both the Russian crewmembers and American astronauts who served during Shuttle-Mir, greater attention needs to be given to matters of the psychological compatibility of crewmembers. For this, a longer training period should be carried out for each crew. And, joint training sessions for survival under extreme conditions would also help.

Holland pointed to the entire supporting organization. "One of the things that was astounding to me was that, traditionally, we had this focus on the individual.... We were thinking that's where you need to put your effort. In the Mir series, what was so striking was the influence of the organizational policies and the organizational context on the individual's psychological health.... There were just so many organizational lessons that were learned ... in terms of policies and procedures," he said.

"Basically, NASA had to learn how to deploy people and their families, and [to] make sure that people got back and forth without a lot of problems. NASA's not like the military. It never had before deployed people for long periods of time in foreign countries, so there was no infrastructure at all to do that. We just gave them a ticket and sent them over there."

STS-76
Starting a Continuous U.S. Presence
March 22 – 31, 1996

Space Shuttle *Atlantis*

Launched: March 22, 1996, 3:13 a.m. EST
Kennedy Space Center, Pad 39-B

Orbit: 160 nautical miles

Inclination: 51.6 degrees

Landed: March 31, 1996, 8:28 a.m. EST
Edwards Air Force Base

Mission: 9 days, 5 hours, 16 minutes

STS-76 CREW

Commander Kevin P. Chilton
Third Shuttle flight

Pilot Richard A. Searfoss
Second Shuttle flight

Mission Specialist Linda M. Godwin, Ph.D.
Third Shuttle flight

Mission Specialist Michael R. Clifford
Third Shuttle flight

Mission Specialist Ronald M. Sega, Ph.D.
Second Shuttle flight

Mission Specialist Shannon W. Lucid, Ph.D.
Third Shuttle flight; remaining on Mir

PAYLOADS

Space Habitation Module

Shuttle Shortwave Amateur Radio Experiment

Mir Environmental Effects Payload

Trapped Ions in Space Experiment

Mir Wireless Network Experiment

KidSat Educational Activity

When STS-76 took Shannon Lucid up to Mir, the U.S. began a continuous presence in orbit that would last for more than two years. The mission also marked the first U.S. spacewalk around the two mated spacecraft, and the first time the SPACE-HAB pressurized module was used to carry supplies and equipment to and from Mir.

After a day's delay because of a high wind forecast, *Atlantis* launched in the cold predawn morning, with light from the engines reflecting off the clouds. During ascent, a sensor detected a leak in one of *Atlantis'* hydraulic systems, but NASA determined it was not a danger to the Orbiter. Over the next two days, the Russian flight controllers were satisfied that Mir was not at risk, and they gave their go-ahead to continue the joint mission. On Flight Day 2, rendezvous and docking came with future Mir resident David Wolf serving as CAPCOM in Houston and cosmonaut Sergei Krikalev sitting next to him. Onboard *Atlantis*, Rick Searfoss took the Commander's seat, and Kevin Chilton controlled

For the first time, a Shuttle crew completed its mission with fewer astronauts than when the mission began. Before Shannon Lucid transferred to the Mir, the STS-76 crewmembers posed for a photo in the mid-deck of *Atlantis* (top, left to right) Mission Specialists Clifford, Lucid, Sega; (bottom) Mission Specialist Godwin, Commander Chilton, Pilot Searfoss.

the Orbiter from the aft flight deck windows. Former NASA Director of Russian Operations Ron Sega handled computations on the laptop computer. As the two spacecraft closed, the Sun set and Mir turned a rosy color. Then, Mir's rusty-hued docking module inched down toward the Shuttle's docking system. After "capture," a slight glitch with the fasteners caused the two spacecraft to sway slightly, effectively demonstrating the robustness of the docking system.

Lucid linked up with her Mir crew, and everyone began transferring equipment and

Mir Environmental Effects Payload

Mission Specialists Godwin and Clifford attached the Mir Environmental Effects Payload (MEEP) to the Shuttle-Mir docking module during an STS-76 extravehicular activity. (During an STS-86 spacewalk, MEEP would be retrieved by NASA Mission Specialist Scott Parazynski and Cosmonaut Vladimir Titov.)

MEEP studied the frequency and effects of both human-made and natural space debris striking Mir, capturing some debris for later study. This payload also exposed International Space Station materials to the effects of space and orbital debris.

MEEP consisted of four separate experiments. The Polished Plate Micrometeoroid and Debris Experiment studied how often space debris hit the station, the sizes and sources of the debris, and the damage the debris might do on hitting a space station. The Orbital Debris Collector Experiment captured orbital debris for return to Earth to determine the possible origins and components of that debris. The Passive Optical Sample Assembly I and II Experiments tested various materials intended for use on the International Space Station, including paint samples, glass coatings, multilayer insulation, and a variety of metallic samples.

The four MEEP experiments were contained in four passive experiment carriers. Each carrier consisted of a sidewall attachment to the Orbiter's payload bay, a handrail clamp for attachment to Mir's docking module, and an experiment container to house individual experiments.

supplies. The U.S. astronauts discovered that Mir was, in different ways, both large and small. Searfoss observed that, at first, it was possible to get disoriented in Mir's transfer node, and that moving around in Mir is kind of like being a spelunker. Sega later noted the tightness of some passageways: "After hearing of the experiences of the two previous [STS] crews that had been on Mir … we had a pretty good idea of what the Mir would be like. Some of the tools were placed in more tenuous situations … so you were cautioned, correctly, by the cosmonauts to be careful … not to dislodge pliers and screwdrivers and … things that were tacked along the sides." During five days of docked operations, approximately 1,500 pounds of water and two tons of scientific equipment, logistical material, and resupply items were transferred to Mir while experiment samples and miscellaneous equipment were brought over to *Atlantis*.

On Flight Day 6, NASA Mission Specialists Linda M. Godwin and Michael "Rich" Clifford conducted the first U.S. extravehicular activity around the two mated spacecraft. This was also

the first time since Skylab—22 years before—that U.S. astronauts had spacewalked outside a space station. According to Godwin, the Russians were understandably concerned about the two Americans getting too close to solar arrays and other fragile equipment, but it was a thrilling experience for the spacewalkers and for those watching them. During their six-hour extravehicular activity, Godwin and Clifford attached four Mir Environmental Effects Payload experiments to Mir's docking module.

Flight controllers shortened STS-76 by one day due to weather concerns at Kennedy Space Center. After leaving Lucid on Mir and undocking, *Atlantis* performed a fly-around about which Searfoss later commented: "The whole station is chameleon-like, with different orbital lighting conditions and Sun angles…. I was enthralled by the constantly changing and indescribably beautiful hues of white, tan, gold, and blue."

Commander Chilton landed *Atlantis* at Edwards Air Force Base in California and, for the first time, a Space Shuttle had returned with fewer astronauts than it had launched with.

Shannon Lucid

NASA-2

March 22 – September 26, 1996

Enduring Qualities

She returned from space a hero and a teacher of patience, but Shannon Lucid went up to Mir more as a student. Her aims were not so much to endure space hardships as to do her science, to learn all she could, and to get along with her crewmates. These goals paralleled several of NASA's own goals: to conduct scientific research; to learn how to manage a space station; and to get along with international partners.

Much has been made of Lucid's record-setting 188 days in orbit, with her last six weeks added because of an unexpected delay. But, Lucid's mission was also an early test and a tentative triumph for NASA's Shuttle-Mir managers. It provided good reason for cautious optimism, as Lucid's six-month posting to the Russian space station began a continuous, 2½-year American orbital presence.

Lucid's own sense of presence and stick-to-it permanence are something to consider when following the story of her NASA-2 increment. What kept Lucid going—and doing so well—had several varied aspects.

First was Lucid's innate patience, faith, and good humor. She likely learned these qualities from her parents at a very young age. She also worked hard herself at developing them as she grew. Lucid was born in Shanghai, China, to Baptist missionary parents. When she was six weeks old, her family became Japanese internees. After a year in a internment camp, they were released in a prisoner exchange and returned to the United States. After the war was over, they returned to China. But, they had to leave again when the Communists took control.

Already introduced to living in different cultures, the well-traveled youngster wanted to see and learn more. "I was interested in exploring," Lucid has said. But, when she was young she worried that "by the time I grew up, the world would be explored, so what would be left for me to do?" Then, she read about American rocket pioneer Robert Goddard and she "started reading a little bit of science fiction, and it just sort of clicked." Well, she thought, "You can go explore the universe. That wouldn't get used up before you grew up." She became a scientist, and then an astronaut. By the time of her Mir experience, Lucid had flown on four Shuttle missions, including the one that released the Galileo spacecraft on its journey to Jupiter.

A second interesting aspect of Lucid's mission was gender—including its biological, psychological, and societal implications. Lucid had been a member of NASA's

1978 astronaut class—the first to include females. These women distinguished themselves in space, and they helped make NASA one of the more enlightened U.S. workplaces. Did being a woman give Lucid a special ability to endure? Are women better candidates for long-duration spaceflight? Did Lucid's male Mir crewmates treat her differently than they might have treated an American male, or than they might have treated any other male? During Shuttle-Mir, a few Russian space officials made comments concerning women's "traditional" roles, and, by and large, Russian society lags behind American society in women's rights. Although Lucid has said that she always felt "like I was just being treated as an individual, as myself," both in Russia during training and in orbit with her crewmates, her Mir experience may point to the wisdom of selecting crews of mixed genders to staff the International Space Station.

A third consideration is one of Lucid's favorite Earthly pastimes: reading. Not only did reading give Lucid a healthy method of psychological escape, but her reading materials— Charles Dickens, for example—may have given her insights into psychology and the human condition, which in turn helped her endure and even enjoy her experience.

A fourth necessity, when considering Lucid's Mir experience, is to keep in mind a comment made by NASA-3 Astronaut John Blaha, who said

that the seven NASA Mir astronauts were actually much alike; however, their situations were different. Indeed, Lucid and those who followed her had already had their ways paved for them by NASA-1 Astronaut Norm Thagard. Shuttle-Mir operations had much improved. To keep Lucid from repeating Thagard's cultural deprivation, NASA worked to keep her in contact with family, friends, and events on Earth. This included sending up books and tapes, and arranging weekly radio links and biweekly video contacts with her family.

> To keep Lucid from repeating Thagard's cultural deprivation, NASA worked to keep her in contact with family [and] friends ...

Furthermore, Lucid's crew turned out to be particularly harmonious. The Russian media had affectionately dubbed her crewmates "the two Yuris." About her crewmates, Mir-2 Commander Yuri Onufriyenko and Flight Engineer Yury Usachev, Lucid said, "They're both very, very nice people and I've enjoyed working with them very, very much. They have different personalities.... I think the personalities mesh quite well together. Yuri the commander tends to be a little more

quiet, and Yury the [flight] engineer always has something to say. And so, that works out real well."

Still further, Lucid's flight did not have any especially dangerous and frightening events. Certainly, there were problems and inconveniences, and plenty of Mir housekeeping to keep up with. But, Lucid's day-to-day experiences were not as stressful as some that occurred on other NASA Mir missions.

Regardless, Lucid's mission was extremely difficult to undertake and complete. She spent by far the most time of any American onboard Mir—with six weeks of her mission being thrust upon her unexpectedly. Already high in orbit, she rose to the occasion.

Her residency began on Saturday evening, March 23, 1996, when *Atlantis* and the STS-76 crew delivered Lucid to Mir to begin the increment known as NASA-2. She said later, "It was just pretty neat to look out the window and see Mir, and know that [it] was going to be your home.... It was great to see Yuri and Yury. They'd been up there a month before I got there. They acted very happy to see me. I believe that they really were. So, as soon as the hatch opened, I moved over and became part of the Mir-21 crew."

While *Atlantis* was docked to Mir, the crews transferred supplies and equipment, and Astronauts Rich Clifford and Linda Godwin conducted a six-hour spacewalk to install MEEP (Mir Environmental Effects Payload) exposure panels on the Mir docking module. Soon after arriving, Lucid had to put to rest a minor controversy in the media that had been started by a comment of a Russian space official. General Yuri Glaskov, Deputy Commander of the Gagarin Cosmonaut Training Center at Star City, had earlier hinted that the two male cosmonauts would welcome Lucid "because we know that women love to clean." He also said, "The simple presence of a lady onboard the Mir station helps ... because [our crewmembers] simply pay more attention to the way they behave, they act, they speak, and so on.'"

Lucid kindly responded to this in a news conference from orbit. "That kind of thinking doesn't bother me. We all work together to keep the place pretty tidy." Commander Onufriyenko said that all three crewmates would share in Mir

Mir-21's particularly harmonious crew, shown in the Mir Base Block, had trained together for months before their mission.

housekeeping and that Lucid would improve the "cultural level" onboard the station.

Mission Control-Moscow coordinated the Mir crew's activities, laying out the crew's work schedule in the form of a "cyclogram"—comparable to a Shuttle Flight Plan. The cyclogram was prepared four days ahead of schedule, but it was modified by both Mission Control-Moscow and the crew via radiograms and separate messages. A group of scheduling experts from NASA served as consultants to the Russian control team and was on duty in Mission Control-Moscow throughout the crew's workday. The workday typically began with wake-up around 8 a.m. Moscow time and ended at about 11 p.m. This translated in Houston to midnight to 3 p.m. Central Daylight Time and often caused difficulty for NASA people in both Moscow and Houston to comfortably schedule their own workdays.

A week into Lucid's residency, the Mir-21 crew had settled into an on-orbit routine of experiment work, including material and life sciences research and Earth observations. The crew focused early on the Optizon Liquid Phase Sintering Experiment, the first American experiment designed to be conducted in the Russian furnace onboard Mir. The Optizon furnace processed metals at high temperatures for future study on the ground, with hopes to improve industrial technologies such as cutting tools. Lucid also worked to monitor long-term protein crystal growth experiments, and preserved samples for a quail egg experiment that studied embryonic development at various stages in microgravity.

Lucid and her crewmates ended the week of April 12 by observing Russia's Cosmonautics Day. They celebrated the 35th anniversary of the first human in space—Yuri Gagarin in 1961—and the

15th anniversary of the first U.S. Space Shuttle launch—*Columbia* in 1981. The cosmonauts talked with family and friends on a two-way video link, and the crew took part in press conferences with both the Russian and American media. Questions during the Russian press conference ranged from where they slept to what language they spoke on the station. The latter question elicited a joking response from Usachev, who said "mainly Russian, but we try to learn more English words so Shannon won't forget her English." Lucid also said that Russian space food was "pretty good"—with beef stew being her favorite—but she really missed M&M™ candies.

Questions about food and democracy came during another radio hookup with high school students at the Ulyanovsk school, about 200 miles outside of Moscow. The students asked Lucid what beverage she liked on Mir. She answered, "Cherry," echoing the fondness that other American Mir astronauts had for the Russian fruit drinks. The students also asked Usachev and Onufriyenko whom they planned to vote for in the June presidential election. The election pitted Boris Yeltsin against his Communist Party opponent Gennady Zyuganov, and there was concern that Zyuganov did not support Russian-American cooperation. The two cosmonauts diplomatically told the students that they planned to vote absentee, but they had not yet

decided on their candidate. Interestingly, the students asking this question about Russia's new democracy were calling from the boyhood school of the first Soviet dictator, Vladimir Lenin. Indeed, Lenin's original family name was Ulyanovsk.

The week of April 19 was spent conducting science experiments, documenting Earth observation sites, performing small-scale maintenance procedures, and searching—without success— for a small, troublesome leak in one of the thermal cooling loops in Mir's core module (the Base Block). Similar leaks would grow into major problems for future Mir crews.

The crew began preparations for the arrival of the Priroda science module, which would complete Mir's configuration. The launch was supposed to have taken place six weeks earlier on March 10 so that Priroda would be ready for Lucid when she arrived. However, the module had been delayed twice—once because of its late delivery from the Khrunichev factory, and once because a commercial launch took priority. Ironically, not only was Russia practicing democracy but it was embracing capitalism.

The crew continued work with the Optizon experiment, and they kept repositioning the space acceleration measurement system to measure the slightest movements of the Mir station. These measurements would help experimenters on Earth explain any changes noticed in their

data after the mission. The crew also took periodic radiation measurements by repositioning a dosimeter around the station. They took blood samples, and they were able to repair a body mass measuring device used to record any changes in crewmembers' body masses due to microgravity.

On the radio, Lucid's Mir-21 crew talked with Russian cosmonauts Gennady Manakov and Pavel Vinogradov, and French cosmonaut Claudie Andre-Deshays, who were the designated Mir-22 crew scheduled to launch to the station in July. They also discussed training with the designated Mir-24 crewmembers Valeri Korzun and Alexander Kaleri. Unknown to anyone at the time, Korzun and Kaleri would later take Manakov's and Vinogradov's places on Mir-22 because of a last-moment heart problem for Manakov.

On April 23, 1996—one month into Lucid's mission—the Priroda module launched aboard a Proton rocket and took a rapid course to Mir to save battery power. One of its two battery systems dropped "off line" because of overheating, and it was later learned that the malfunction had caused a fire. On April 26, a flawless automated docking occurred. However, because of potential dangers due to the malfunction, Lucid was instructed to stay in the Soyuz escape capsule during the docking.

The next day, the cosmonauts used the Lyappa robot arm to move Priroda to its permanent berthing port, opposite the Kristall module. With Priroda's addition, Mir was now complete. It was bigger, newer, and more capable; and it carried five major U.S. science facilities, weighing about a ton, with additional Earth observation equipment mounted on the outside. Furthermore, in the words of Shuttle-Mir scientist Tom Sullivan, "One of the more mundane things about a space station is that there is never enough closet space. Priroda does bring a great deal of additional stowage volume for science hardware that will be brought up on future Shuttle flights."

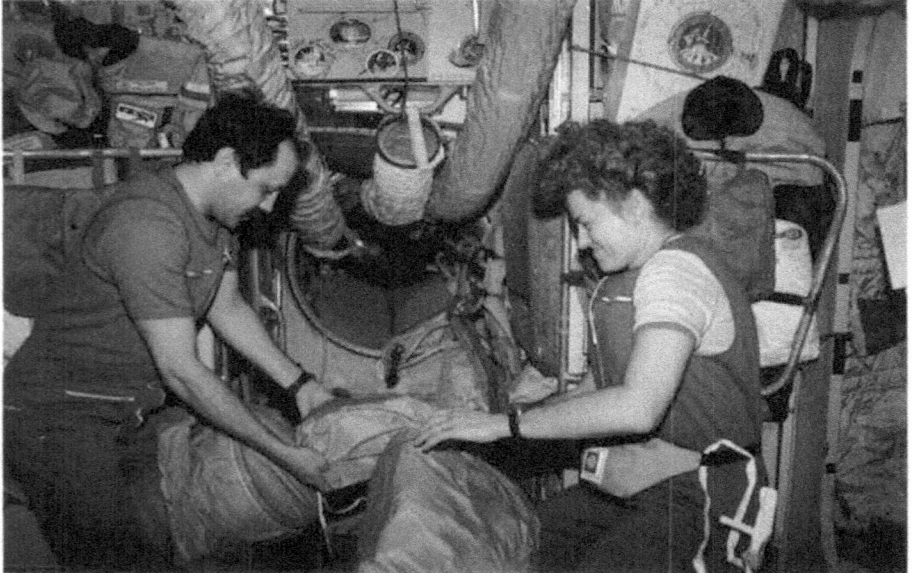

Usachev (left) and Lucid work together in the Mir Base Block module during the NASA-2 increment.

Shannon W. Lucid

Shannon Lucid was born in Shanghai, China, the daughter of missionaries. She attended the University of Oklahoma and earned a Bachelor of Science in chemistry and a Master of Science and a Ph.D. in biochemistry. She is a licensed pilot.

Lucid was one of the first women admitted into the U.S. astronaut program and became a NASA astronaut in 1979. Before Shuttle-Mir, she flew on four Space Shuttle missions, including STS-34, which released the Galileo spacecraft on its long journey to explore Jupiter and its moons. She also flew on STS-58 with John Blaha and David Wolf, who also participated in the Shuttle-Mir Program.

President Bill Clinton awarded Lucid the Congressional Space Medal of Honor after her Shuttle-Mir mission. She is the only woman to have received this award.

Russian President Boris Yeltsin awarded her the Order of Friendship Medal, the highest Russian award that can be presented to a noncitizen.

Lucid is working with the International Space Station Program while she prepares to be assigned to another flight. Among her preparations are completing winter survival training at Cold Lake, Canada.

Second U.S. astronaut to live onboard Mir • Set American space endurance record of 188 days • Set world space endurance record for women
Launched with STS-76 on March 22, 1996 • Returned with STS-79 on September 26, 1996

During the week of the Priroda docking, Lucid conducted an inventory of all U.S. hardware onboard Mir. The crew completed the Optizon Liquid Phase Sintering Experiment and proceeded with the protein crystal growth experiments. They also kept working with the space acceleration measurement system, recording the slightest movements of the space station.

In early May, the Mir-21 crew worked to prepare the new Priroda module. This included cleaning up contamination due to the fire and wrapping in plastic the 168 big batteries that had powered Priroda while on its way to Mir. This activity, expected to take six days, was accomplished by the crew in less than two days. They also connected Priroda to the Mir's power system and began troubleshooting a Priroda power system problem detected during the rendezvous.

In a letter to Earth, Lucid described the crew's work in Priroda: "After a lot of work, the batteries on the floor were unbolted and I thought the job was complete. Then, Yuri opened a panel that revealed more rows of batteries to be unbolted. Another opened panel revealed yet more batteries; there were batteries without end!!!

And, each battery had to be unbolted, plastic caps had to be put on the four 'feet' and on the connectors, and then each battery had to be bagged and tightly tied. Talk about a lot of work!!!! To even reach the batteries, some of the equipment had to be unbolted and the supporting metal framework taken apart.

"So there the three of us were, floating in Priroda surrounded by floating batteries, bagged batteries, equipment, and scrap metal. At times I thought that there was enough scrap metal floating there to build station *Alpha*!!! Periodically, free-floating metal pieces would impact each other, creating clear metallic tones like cathedral bells in the module and we joked with each other about the 'cosmic music' that we were hearing. We devised an assembly line to clean up the mess and got so efficient that we finished the task in one-sixth of the time that the ground expected and earned ourselves a holiday."

As part of her Earth observations work, Lucid took photographs of the massive wildfires burning out of control in Mongolia. Lucid reported that she had not seen such large fires from space during any of her previous four Space Shuttle flights.

On May 7, 1996, the unmanned Progress docked automatically with Mir on the first attempt. The crew got busy unloading food, supplies, and equipment, and they "made some good inroads" into their fresh supply of candy. The Progress arrival helped the crew in preparing the Priroda module for the startup of science activities. These activities had been slowed, partly owing to all the packing materials that came out of Priroda because, until Progress arrived, there was no place to temporarily store trash.

The crewmembers also continued troubleshooting the Priroda power system. They replaced three nickel-cadmium batteries, which seemed to fix the problem. Lucid prepared and successfully tested the Mir interface payload system. She also conducted blood analysis using the portable clinical blood analyzer. And, she reviewed experiment procedures and her pre-mission training using a new audio-video compact disk system called the crew on-orbit support system.

On May 9, Lucid made a telephone call to her parents in Oklahoma in honor of her mother, who was celebrating her 81st birthday. The next day, Lucid "saw" and spoke to her family in

Houston through a two-way video conference in advance of Mother's Day.

By mid-May, the Mir-21 crew was gearing up for spacewalks by Onufriyenko and Usachev. In the new Priroda, Lucid set up the Biotechnology System, designed to support long-duration cell culture experiments in microgravity. The Mir crew also prepared two Canadian experiments: the microgravity isolation mount and the Queen's University Experiment in Liquid Diffusion (QUELD). Although bad nickel-cadmium batteries had been replaced the previous week, another power controller on Priroda had failed. As a result, flight controllers at Mission Control-Moscow would now monitor and control battery charging from the ground. The crew continued to search for the source of the coolant leak.

On May 20 and 24, with Lucid assisting from inside Mir, Onufriyenko and Usachev conducted two spacewalks lasting 5½ hours each. They removed the Mir cooperative solar array from its stowed position on the exterior of the docking module at the base of Kristall. They used the Strela boom to reach and move the array to the Kvant module. They also deployed

> Although the whole world seemed interested in her, … Lucid's … world had shrunk.

an aluminum and nylon pop-up model of a Pepsi Cola® can, which they then filmed against the backdrop of Earth. The soft drink company paid for the procedure and planned to use the film in a television commercial. However, the commercial never aired—reportedly because

Pepsi® later changed the design of the can.

NASA science activities continued even while the crew prepared for and conducted the spacewalks. Lucid worked inside Priroda to complete verification of the microgravity isolation mount facility and to check out the microgravity glovebox. Using magnetic levitation, the microgravity isolation mount could isolate experiments from vibrations and other disturbances, and the glovebox would be used for many microgravity experiments. Lucid also performed several life sciences experiments, including studying changes to the human immune system in space. Lucid continued her Earth observation activities and photographed the eruption of the Montserrat volcano in the Caribbean.

Late at night on May 30, Onufriyenko and Usachev conducted another extravehicular activity of four hours, 20 minutes, to install the modular

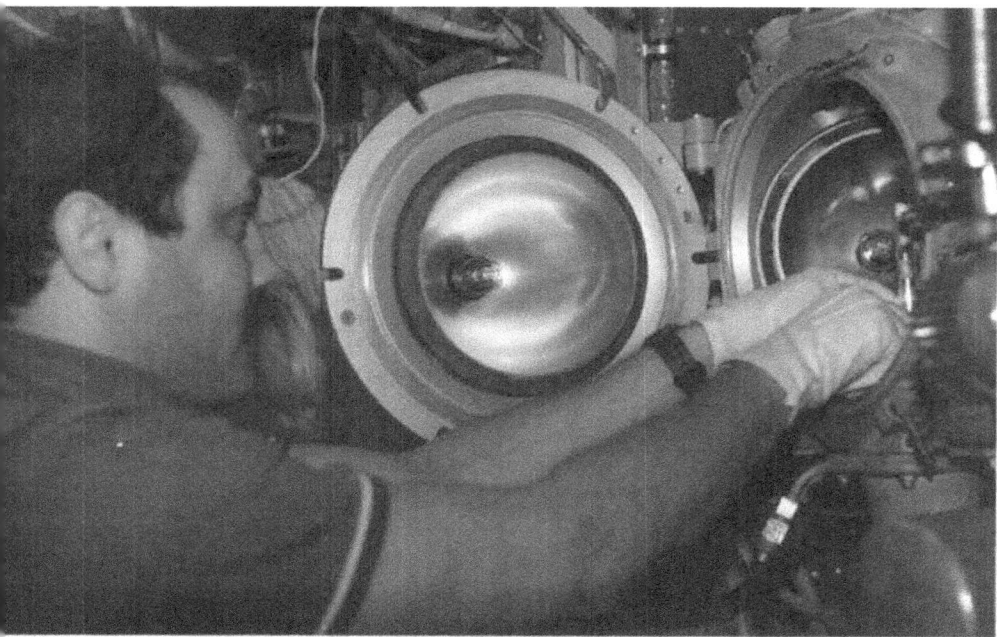

Mir-21 Commander Onufriyenko (photo left) and Flight Engineer Usachev (in suit with blue stripes) traverse an existing truss on the Kvant module during one of their six spacewalks. (Above) Usachev inserts metal canisters into the QUELD furnace. (Below) Lucid chooses reading material from the Mir library located in the Spektr.

optoelectrical multispectral scanner outside Priroda. The scanner, which flew on Shuttle missions STS-7 and STS-41B, would be used to study the Earth's atmosphere and environment. From inside Mir, Lucid sent commands to power up the system once the spacewalkers finished installing the hardware. The two spacewalkers also installed a new handrail on the Kvant-2 module to facilitate moving around outside the station during future extravehicular activities.

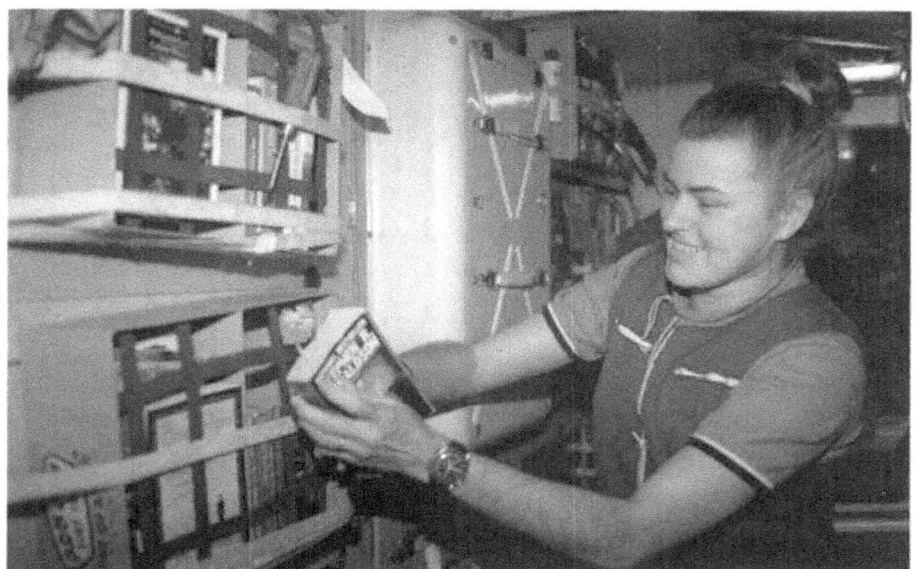

Books Onboard Mir

For recreation, Mir crewmembers exercised, conversed, played computer games, watched movies, observed the Earth, wrote e-mails, and read books. Shannon Lucid said the Russians had quite a library on Mir. As stowage was always a problem on the space station, some of the books had to be stuffed behind a panel.

Regarding her own Mir reading experience, Lucid said, "Here I was, reading *David Copperfield* and *Bleak House*. I thought, Wow, here was this guy," Charles Dickens, who "lived in a totally different era.... And, it had never ever crossed his mind that his book would be read ... by an American on a Russian space station. I mean, that would have just absolutely blown his mind—that the words that he penned way back there in England, I was reading on Mir."

Lucid thought often "about the power that authors have," and about Dickens' ideas and how "his story was transcending the centuries—transcending culture" and improving her life in space.

When Lucid left Mir, she left most of her books behind. Unfortunately for some of her successors, her books, stored in the Spektr module, became inaccessible because of the June 1997 Progress collision.

NASA-7 Astronaut Andy Thomas took many paperbacks to Mir, including science fiction and some classics such as Edgar Allen Poe and Mark Twain. Thomas said, "I'd always wanted to read *Huckleberry Finn*, since it's a landmark book ... and very controversial." And so, Twain's story of Huck and Jim floating on their raft on the mighty Mississippi was read by a man floating in outer space.

Another book Lucid took to Mir was a small Bible, which she always carried when she traveled. She said that she noticed a Gideon's Bible in one of the cosmonaut's cabinets. "It was in Russian," she said, "a little New Testament."

The Space Shuttle *Atlantis*, undergoing preparations for STS-79, returned to the Vehicle Assembly Building twice, when hurricanes approached Kennedy Space Center, Florida.

As Lucid passed through the halfway point in her mission, she told reporters during a space-to-ground news conference, "I couldn't ask for anything more out of a flight than what I've gotten out of this flight so far. If the second half is as good ... all I can say is you just can't beat it."

About her relationship with her two crew-mates, she said, "I think maybe we laugh a little more together now than we did at the very beginning because we're more comfortable with each other and we understand each other."

Although the whole world seemed interested in her, and although she was zipping around the entire planet every 90 minutes, Lucid's personal and professional world had shrunk. Later, she would remark, "There I was on Mir, and on a daily basis I talked with the American support group in [the] Russian control center ... and most days it was twice a day. I talked with Bill Gerstenmaier, who was in charge of the science experiments ... and Gaylen Johnson, the Flight Surgeon who also worked with Bill. So, a lot of times it was just the three of us, and the world seemed to shrink down to that.'

Onboard Mir, a data card failed within the Mir interface to the payload systems computer. A new card would be delivered on a Progress vehicle in July, but that card would not work either and Lucid would have to record all the data onboard.

On June 6, Onufriyenko and Usachev conducted a spacewalk of three hours, 34 minutes. They replaced cassettes in the Swiss/Russian Komza experiment and installed the Particle Impact Experiment, the Mir Sample Return Experiment, and the SKK-11 cassette, which exposed construction materials to space conditions.

During the week of June 14, Lucid completed the Humoral Immunity Experiment, which measured the effects of spaceflight on the human immune system. Previous investigations had suggested that perhaps the human immune system is suppressed during long-duration space missions. For this experiment, Lucid injected herself with an immune system stimulant. She then collected blood and saliva samples that would be compared to samples taken before and after her stay on Mir to measure changes in her body's response to the stimulant. Results later indicated that there was no immune system suppression. Also, Lucid performed an experiment designed to measure the forces generated as a crewmember pushes off the surfaces of the spacecraft to move about.

Onufriyenko and Usachev performed the sixth in their series of spacewalks, installing a truss structure called Rapana to the Kvant-1 module. Rapana took the place of a similar structure named Strela as a mounting point for future experiments. Strela could now be used better as a spacewalker's moveable "ladder." Onufriyenko and Usachev also manually deployed the saddle-shaped traverse synthetic aperture radar antenna on Priroda. The large antenna had failed to open fully after receiving commands from inside Mir.

Big news came up to Mir on June 21, fore-shadowing the announcement that Lucid would have a longer-than-planned spaceflight. Russia's news agency Interfax reported that Onufriyenko and Usachev would be on Mir until August 30. Yuri Koptev, the Director General of the Russian Space Agency, was quoted as saying that there wasn't enough money to build the Soyuz booster rockets necessary for ferrying cosmonauts to and from the Mir. At this moment, 90 days into her stay onboard Mir, however, Lucid's future remained according to plan. She was slated to return to Earth as scheduled on a U.S. Space Shuttle in early August.

The Mir crew continued with their scientific duties, including running the Canadian Queen's University Experiment in Liquid Diffusion and sampling the air in Mir with the solid-sorbent air sampler and the grab sample container. The solid-sorbent air sampler was designed to sample air quality over 24 hours; the grab sample container took quick, snapshot-type readings of air quality.

> On July 12, NASA announced that Lucid's mission to Mir would be extended to mid-September....
> Lucid ... took the information in her stride.

As July passed, the other shoe dropped for the Mir-21 crew. Lucid would also remain on Mir. And, she would stay for an unknown period—at least until STS-79 could be cleared for launch. NASA engineers at Kennedy Space Center had observed unusual soot patterns in the joints of the Shuttle solid rocket boosters used on STS-78. NASA Shuttle-Mir Program Manager Frank Culbertson told his Russian counterpart Valery Ryumin about the situation in an informal July 3 memo noting that "the worst case … is a potentially serious problem for our joint schedule. You should be aware of the situation." Compounding matters, a very active Atlantic hurricane season was in progress. Hurricane Bertha threatened the Florida spaceport, and the Orbiter *Atlantis* was moved for protection back into the Vehicle Assembly Building. NASA engineers decided to replace the possibly flawed solid rocket boosters.

On July 12, NASA announced that Lucid's mission to Mir would be extended to mid-September. Culbertson had already contacted Lucid with the news, who took the information in her stride. Three days later, while considering the receding time horizon in front of her, Lucid sailed through Norm Thagard's 115-day record for the longest American time in orbit. In press conferences that week, she told reporters, "The two things I had planned on being home for were my son's birthday and my daughter's birthday … but I told them we'd make it up to them when I do get home." Her son Michael would turn 21 on August 22, and her daughter Kawai would turn 28 on September 19. Lucid also said she would continue to miss things "like going to the bookstore … potato chips and junk food. And, … feeling wind and the Sun." On the other hand, she said, "You know, that's life. And, we'll just go on and I'll continue to have an enjoyable time."

Lucid also recorded an address to be played

at part of the opening ceremonies at the XXVI Olympic Games in Atlanta, Georgia. The Olympics were about running marathons and breaking records, and Lucid was now being seen as a star athlete in space.

She later joked that one of her first thoughts upon hearing of the extension was "Oh, no, not another month-and-a-half of treadmill running!" But, Lucid persevered. She persevered with her science, too. Only one of her 28 scheduled experiments failed to yield results because of equipment breakdown. In the Candle Flame in Microgravity Glovebox Experiment, she had burned 51 candles to study the complicated physiochemical process of combustion. The original plan was based on a total of 60 candles, but Lucid would burn a total of 79 candles of varying size, wick diameter, and length. Along with the candle flame sessions, the crew collected other microgravity data with the enhanced dynamic load sensors and space acceleration measurement system.

During the week of July 19, the crew began assembling the Russian/Bulgarian Svet facility in preparation for the Fundamental Biology Greenhouse Plant Experiment. By studying the chemical, biochemical, and structural changes in plant tissues, researchers hoped to understand how processes such as photosynthesis, respiration, transpiration, stomatal conductance, and water use are

Astronaut Shannon Lucid is shown using the Russian lower body negative pressure suit, also called the Chibis suit, that acts as a microgravity countermeasure by preparing the cardiovascular system for return to Earth's gravity.

The Mir-21 crew assembled the Russian/Bulgarian Svet facility (shown left) in the Kristall module for a greenhouse plant experiment.

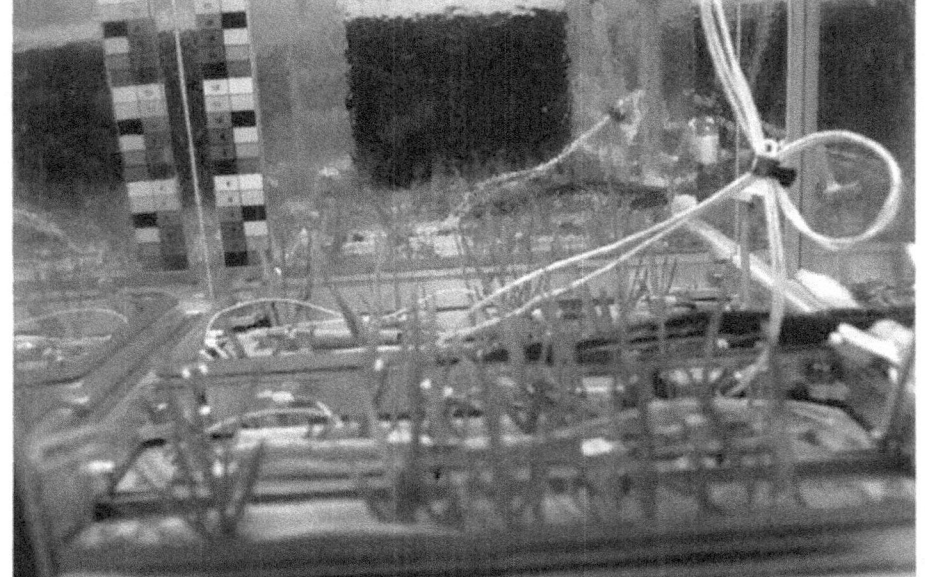

affected by microgravity. Plants could eventually be a major contributor to spaceflight life support systems because they produce oxygen and food while eliminating carbon dioxide and excess humidity from the environment. Although it had not been planned for Lucid to perform this experiment, she was able to do so, demonstrating how well the on-orbit training of astronauts can work.

During the next few weeks, until the arrival of the Mir-22 crew and French cosmonaut Claudie Andre-Deshays, the Mir-21 crew continued with a full scientific program, including Earth observations and studies of radiation, tissue growth, and neuromuscular activity. They finished fabricating the Svet greenhouse and planted the first wheat seeds, after using an extension power cord to plug the greenhouse into a steadier power supply. They also continued troubleshooting the biotechnology system to ensure its readiness for the next Mir mission.

Progress-232 launched from Baikonur Cosmodrome on July 31 and docked with Mir on August 2. The resupply vehicle brought more than two tons of supplies for the crew, including fresh food, oxygen, and experiment hardware for the upcoming French-Mir mission. Charles Stegemoeller, Shuttle-Mir Research Implementation Manager, talked about the process of getting extra supplies to Mir for Lucid's extended stay. He said, "The Russians are used to taking care of the crew at the last minute. So, they understood what we were trying to achieve by getting additional items up to Shannon. It was just a question of what could we send, and what would she want to have. So, we had a conversation with the crew, and we gathered items together from the family and the Flight Surgeons, and we passed those on to Moscow." Books were already taken care of in the family packages that were ready to go on the Progress, but Stegemoeller said that Lucid had commented that "they didn't have enough sweets on orbit or salty items, so we packed up a good grocery bag full of stuff ... items that they don't

normally get because there's no snack vending machine around the corner."

The crew busied themselves unloading and stowing the food and equipment. In early August, they continued troubleshooting the Elektron oxygen-generating system. For three days, the crew turned off the gyrodyne system, which provided control of the Mir, to allow the

> When Lucid broke the record for longest stay in space by a woman, Yuri Glaskov said, "I don't think you've taken the record from us. We have offered this record to you."

cosmonauts time to refurbish the system. Mir maintained its attitude using thruster firings.

On August 16, Soyuz TM-24 launched from Baikonur with the Mir-22 crew of Valeri Korzun, Alexander Kaleri, and French cosmonaut Claudie

Andre-Deshays. On August 19—Lucid's 150th day in orbit—the Soyuz spacecraft docked with Mir. Mir would be home to six cosmonauts and researchers from three different countries—Russia, France, and the United States—until Onufriyenko, Usachev, and Andre-Deshays returned to Earth. NASA's Shuttle-Mir science team coordinated with the French so that there was no impact to Lucid's science program.

Meanwhile, Lucid prepared for the end of her own stay and the arrival of *Atlantis* by conducting a thorough inventory of experiment supplies and equipment in the Spektr and Priroda modules of Mir for her handover to John Blaha. Lucid had so far packed seven bags of completed experiment samples, data, and equipment from her six months in space to be transported aboard *Atlantis* back to scientists on Earth. And, she reported that the dwarf wheat crop was about two inches tall, three weeks after planting.

At the end of August, Lucid was nearing an all-time record for the length of time a woman had spent in space on a single flight. She told reporters, "My family would be surprised at the patience I've developed in space. I hope I can bring some of that back with me."

On September 2, the Soyuz TM-23, with Lucid's Mir-21 crewmates and Andre-Deshays, undocked. Lucid remained onboard the space station with the Mir-22 crew of Korzun and Kaleri. On September 7, Lucid broke Elena Kondakova's 169-day record for longest stay in space by a woman. During a NASA news conference at about this time, Yuri Glaskov, Deputy Commander of Russia's Gagarin Cosmonaut Training Center, was asked what he thought about Lucid's taking the Russian record. Glaskov chuckled and said, "I don't think you've taken the record from us. We have offered this record to you." He also said, "As far as

Shannon Lucid floats through a transfer tunnel as she prepares to leave the Mir space station.

Pink Socks and Jell-O

May 19, 1996

Here it is, another Sunday on Mir!!! And how, you might ask, do I know that it's Sunday? Easy!!! I have on my pink socks and Yuri, Yury, and I have just finished sharing a bag of Jell-O!!!

When light follows darkness every 45 minutes, it is important that I have simple ways of marking the passage of time. The pink socks were found on STS-76 and Kevin [Chilton], the commander, said that they were obviously put on as a surprise for me, so I took them with me over to Mir and decided to wear them on Sundays.

And the Jell-O? It is the greatest improvement in spaceflight since my first flight over ten years ago. When I found out that there was a refrigerator onboard Mir, I asked the food folks at JSC if they could put Jell-O in a drink bag. Once aboard Mir, we could just add hot water, put the bag in the refrigerator and, later, have a great treat. Well, the food folks did just that and sent a variety of flavors with me to try out. We tried the Jell-O first as a special treat for Easter. It was so great that we decided the Mir-21/NASA-2 crew tradition would be to share a bag of Jell-O every Sunday night. (Every once in a while, Yury will come up to me and say, "Isn't today Sunday?" and I will say "No, it's not. No Jell-O tonight!!!")

There have been a lot of changes here on Mir since I arrived. And no, the changes were not because I am here!!!

The first big change was the arrival of Priroda, the final segment that is to be added to Mir. This segment is called Priroda because that's the Russian word for nature and there are sensors on the outside of the segment to study the Earth. The U.S. science equipment is located inside this segment.

As a graduate student years ago, I fantasized about having my own laboratory. I must admit, though, that in none of my fantasies was I gazing out the window of a space station watching "my laboratory" approach like a gigantic silver bullet moving in slow motion toward the station's heart!! Reality is indeed stranger than fiction!!!!

There had been a power problem on Priroda after its launch, so there was some concern about SO_2 leaking from the batteries into the atmosphere. When it arrived, we had to wait and check out the air quality before opening the hatch. Yuri checked the air and pronounced it good. After listening to the hissing air as the atmospheric pressure was equalized between Priroda and Mir, the hatch was opened. And yes, it was a dramatic moment! There it was, all bright, shiny, and new.

The other big change, although it is not permanent, was the arrival of Progress, the resupply vehicle. Usually about every six weeks one is sent to Mir with food, equipment, clothes and everything that, on Earth, you would have to go to the store and buy in order to live. Because it had

deployed solar batteries, it was easier to spot while approaching the station than Priroda had been. I saw it first. There were big thunderstorms out in the Atlantic, with a brilliant display of lightning like visual tom-toms. The cities were strung out like Christmas lights along the coast and there was the Progress like a bright morning star skimming along the top!!! Suddenly, its brightness increased dramatically and Yuri said, "The engine just fired." Soon, it was close enough so that we could see the deployed solar arrays. To me, it looked like some alien insect headed straight toward us. All of a sudden I really did feel like I was in a "cosmic outpost" anxiously awaiting supplies—and really hoping that my family did remember to send me some books and candy!!!

Soon after it docked, the three of us began opening the hatch. When Yuri opened a small valve to equalize the pressure, we could smell the air that was in Progress. Yuri said, "Smell the fresh food." I will admit it was a fruit smell, but I thought it smelled more like the first time you open your refrigerator after a 2-week vacation only to discover you had forgotten to clean out the vegetable compartment.

The first things we took out were our personal packages and, yes, I quickly peeked in to see if my family had remembered the books and candy I'd requested. Of course they had. Then we started to unpack. We found the fresh food and stopped right there for lunch. We had fresh tomatoes and onions; I never have had such a good lunch. For the next week we had fresh tomatoes three times a day. It was a sad meal when we ate the last ones!!!

A hand-embroidered patch arrived aboard a Progress as a gift to Shannon Lucid from a Russian team member who worked in Mission Control Center-Moscow.

After our impromptu lunch, we took the rest of the afternoon off, looking at our mail that was in the packages and enjoying the apples and oranges that were also onboard. Yuri commented that for the first time all six of the docking ports were now occupied—a Guinness Book record!

Like I said, I had a wonderful bag of new books on Progress. My daughters had hand-selected each one, so I knew I'd enjoy them. I picked out one and rapidly read it. I came to the last page and the hero, who was being chased by an angry mob, escaped by stepping through a mirror. The end. Continued in Volume Two. And, was there Volume Two in my book bag? No. Could I dash out to the bookstore? No. Talk about a feeling of total isolation and frustration!!! You would never believe that grown children could totally frustrate you with their good intentions while you were in low Earth orbit, but let me tell you, they certainly can. Suddenly, August and home seem a long way away!!!!

Shannon

Dr. Shannon Lucid is concerned, I would like to extend my sincerest thanks to the management of the Program for making such a selection. Because everybody's fond of [her].... Everybody loves her."

In a radio phone linkup, NASA Administrator Dan Goldin congratulated Lucid and asked what lessons could be taken from her experience. Lucid responded that on Shuttle missions, less crew involvement with experiments was better because of the press of time. But on long-duration flights, it was important to increase the crewmembers' involvement with experiments and to improve and expand communications with the experiment's principal investigators on Earth. That would help keep a scientist happy.

On September 16, *Atlantis* with the STS-79 crew launched from Kennedy Space Center, bringing NASA-3 Astronaut John Blaha to Mir. Lucid returned to Earth on September 26, 1996, after completing 188 days in Earth orbit and a record-breaking duration for the U.S.

Meanwhile on Earth
NASA-2

Bill Gerstenmaier was the Operations Lead during Shannon Lucid's mission. Later, he talked about how he adapted to supporting a long-duration astronaut onboard Mir. He said that one difference between the American and Russian programs was that short-duration flights require a "very concise, very 'one-answer'" way of operating, "with no creativity—whereas long duration allows you a lot of creativity."

Gerstenmaier followed the Russians' example, in that, when he spoke to Lucid during the communications passes, he began slowly. "When I would talk to her on [a communications pass], my first question to her would be, 'How are you? What's going on? Is there anything you need to tell me?' Even though I may have had a huge list of 50 items I've got to tell her, my first thing was

always nice and calm. I didn't use the official NASA radio language, 'Over,' and 'Roger,' and 'Out,' and all the short abbreviation stuff.... I wanted it to come across as, 'We've got forever.... We're going to just do this nice and easy, and then we'll work it out.'"

As soon as Lucid had her say, Gerstenmaier would "start through my list of 50 things. And, I would get through as many of them as I could in the com pass.... I would tell her steps of the experiment she was going to be doing that were critical—that had to be done a certain way." As for procedures that allowed some variability, "I let her know that she had free-form to go do those any way she wanted.... I tried to give her enough information that she could go run the experiment autonomously without me being around."

Women on Mir
Women in Space

Shannon Lucid was the third woman to live onboard Mir.

The first woman, Helen Sharman, arrived on Mir via a Soyuz rocket with future Shuttle-Mir STS-60 Cosmonaut Sergei Krikalev. A British food scientist and the first Briton in space, Sharman won a contest and got to visit Mir for six days in May 1991. Since the British contest sponsors were not able to raise all of the required money, the Soviet Government had to subsidize her flight.

Cosmonaut Elena Kondakova spent 169 days onboard Mir in 1994-95. Kondakova set the space endurance record later broken by Lucid. In 1997, Kondakova visited Mir again, as a Mission Specialist on the NASA STS-84 mission. Frenchwoman Claudie Andre-Deshays became the fourth woman to live onboard Mir when she visited for two weeks during Lucid's residency.

During Shuttle-Mir, nine American female

Eileen Collins

astronauts visited Mir. Ellen Baker, Bonnie Dunbar, Linda Godwin, Marsha Ivins, Wendy Lawrence, Shannon Lucid, Janet Kavandi, and Janice Voss served as Mission Specialists. Eileen Collins served as the Pilot on the "near Mir" STS-63 mission, and later docked with Mir as Pilot on STS-84. Dunbar and Lawrence visited Mir twice.

The first-ever woman in space was Soviet Cosmonaut Valentina Tereshkova, who flew in 1963. In 1982, Svetlana Savitskaya was the second woman in space and the first female to conduct a spacewalk. In 1983, Sally Ride became the first American female astronaut to fly in space. Kathryn Sullivan made the first spacewalk by an American woman in 1984.

When Cosmonaut Kondakova flew on STS-84 in 1997, she became the 29th woman to fly onboard an American spaceship. As of April 2000, 108 men and 32 women were in NASA's astronaut corps, and NASA was preparing to hire new astronaut candidates. In Russia's cosmonaut corps, there were 43 men and two women, Kondakova and Nadezda Vasilyevna Kuzhelnaya.

During Lucid's mission onboard Mir, future NASA Mir astronauts continued to prepare for their spaceflights at the Gagarin Cosmonaut Training Center in Star City. Astronaut Wendy Lawrence took over from Charlie Precourt as Director of Operations-Russia after Precourt returned to the United States to train as Commander of STS-84, which would fly to Mir in 1997. In addition to overseeing astronaut training activities at Star City, Lawrence was taking Russian language classes several times a week.

During May 1996, NASA established an office at RSC-Energia, the Russian rocket corporation, and opened a communications center on the sixth floor of the Renaissance (Penta) Hotel in downtown Moscow. The communications center was equipped with computer, fax, copier, and telephone, and greatly improved communications for NASA and its Shuttle-Mir contractors. The STS-81 *Atlantis* crew arrived for four days of training with the Mir-22 crew and U.S. Astronauts John Blaha and Jerry Linenger. The STS-81 crew received classes on Mir's construction, components, and life support and communication systems; and they had several sessions with the Mir-22 crew and their backups, going over docking and transfer procedures. Blaha trained in Star City on Mir equipment for his upcoming NASA-3 mission. NASA-4 and NASA-5 Astronauts Linenger and Mike Foale participated in their first training in the Russian Orlan extravehicular activity spacesuit. Mir backup Astronaut Jim Voss participated in language and physical training. Blaha worked with several science experiments he would conduct onboard Mir.

Linenger's training included emergency evacuation procedures used onboard Mir, and he undertook two sessions in the altitude chamber. Foale spent a session in the Hydrolab, the Russian swimming pool that was used to simulate the weightless environment of space. He then learned about the construction and components of Mir, including its control panels and life support system.

In late May, Blaha conducted his final training sessions for the U.S. experiments he would perform as part of the Mir-22 crew. He also allowed scientists to take some of his physical measurements, to be compared to measurements taken during and after his Mir mission. Linenger's training focused on both Mir systems and U.S. science experiments, while Foale spent the week doing water survival training in the Black Sea.

Backup Voss began his first classes on the Soyuz transport module.

In early June, Blaha had a chance to talk to Lucid about lessons she had learned so far in her mission. She told him she was pleased with the progress of her flight and suggested that he learn more about the workings of Mission Control-Moscow. Linenger completed water survival training, and then he participated in science training sessions while Foale focused on Mir's control panels and life support system.

As Lucid's Mir mission continued, Blaha and Linenger took medical examinations for their Russian certifications for spaceflight. Linenger and Foale completed a four-hour extravehicular activity training session in the Hydrolab.

The Russian space station's solar cell arrays, clearly visible in this photo, converted sunlight to electricity, providing power for Mir.

Meanwhile, funding was approved for modifications to the Space Shuttle *Discovery* to enable it to perform the last Shuttle docking (STS-91) with Mir. And, NASA's Shuttle-Mir Program Manager, Frank Culbertson, had to inform his Russian counterparts that STS-79 would have to be postponed until September.

As Lucid's journey was ending, Russian and U.S. space officials signed a formal agreement making the U.S. astronauts an integral part of the Mir crew. Work had already begun to modify the training program to allow for expanded duties.

Space Shuttle *Atlantis*

Launched: September 16, 1996, 4:54 a.m. EDT
Kennedy Space Center, Pad 39-A

Orbit: 196-245 nautical miles

Inclination: 51.6 degrees

Landed: September 26, 1996, 8:13 a.m. EDT
Kennedy Space Center

Mission: 10 days, 3 hours, 19 minutes

STS-79 CREW

Commander William F. Readdy
Third Shuttle flight

Pilot Terrence W. Wilcutt
Second Shuttle flight

Mission Specialist Thomas D. Akers
Fourth Shuttle flight

Mission Specialist Jay Apt, Ph.D.
Fourth Shuttle flight

Mission Specialist Carl E. Walz
Third Shuttle flight

Mission Specialist John E. Blaha
Fifth Shuttle flight; remaining on Mir

Mission Specialist Shannon W. Lucid, Ph.D.
Fourth Shuttle flight; returning from Mir

PAYLOADS

Space Habitation (Double) Module

IMAX large-format camera

Shuttle Shortwave Amateur Radio Experiment

Biotechnology Systems Experiment Facility

Material in Devices as Superconductors Facility

Commercial Generic Bioprocessing Apparatus

Extreme Temperature Translation Furnace

Commercial Protein Crystal Growth Experiment

Mechanics of Granular Materials Experiment

STS-79
First American Handover
September 16 – 26, 1996

STS-79 and Mir-22 crews join together for an in-flight photo during the first American "handover." Pictured are (front, left to right) Mir-22 Flight Engineer Kaleri; STS-79 crewmember Apt; NASA-3 Astronaut Blaha; Commander Readdy; Lucid; (back row) Akers; Walz; Mir-22 Commander Korzun; STS-79 Pilot Wilcutt.

STS-79 faced several delays, causing Shannon Lucid to remain in space for a total of 188 days and to set records for the longest spaceflight by an American and by a woman of any nationality. Lucid's replacement, John Blaha, ferried up to Mir onboard STS-79 to continue the U.S. presence on the Russian space station.

Originally slated for launch on July 31, 1996, *Atlantis* had to be rolled back twice into the Vehicle Assembly Building due to threats from Hurricanes Bertha and Fran. Also, the Orbiter's solid rocket boosters had to be replaced because of abnormal sooting on the solid rocket joints. (A 1986 joint failure had caused the Space Shuttle *Challenger* to explode after liftoff.) An analysis showed that the probable cause of the sooting was a new adhesive used for the first time on STS-78. For this mission, managers decided to replace the motors with a new set of motors using the old adhesive material.

Another predawn Shuttle launch to Mir had been scheduled for STS-79. The countdown proceeded smoothly to an on-time liftoff. Onboard Mir, Lucid was able to see *Atlantis* during the final

portion of its ascent. Approximately 13 minutes into flight, an auxiliary power unit shut down prematurely, but the Mission Management Team at Johnson Space Center in Houston analyzed the situation and concluded that the mission could proceed as planned.

STS-79 was the first Shuttle to carry a SPACEHAB double module, which Astronaut Tom Akers called "a lot bigger than my JSC office." SPACEHAB was berthed near the back of the Shuttle payload bay to balance the Orbiter's mass. The forward section carried crew experiments. The aft section carried mainly food, clothing, supplies, and spare equipment for Mir. Soft storage bags were arrayed on walls and the floor. Colored labels indicated their uses: pink would go to Mir; blue would hold materials from Mir;

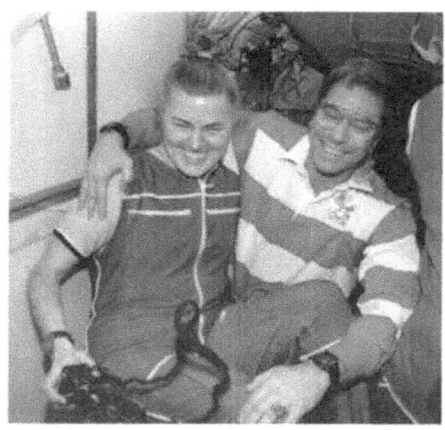

the 120-ton *Atlantis*, the 121-ton Mir, the seven-ton Soyuz, and the seven-ton Progress comprised a mass of 255 tons—the heaviest combined spacecraft ever to orbit the Earth.

During five days of docked operations, the two crews transferred more than 4,000 pounds of supplies to Mir, including water generated by Orbiter fuel cells. Three experiments also were transferred: the Biotechnology System for study of cartilage development; Material in Devices as Superconductors to measure electrical properties of high-temperature superconductor materials; and the Commercial Generic Bioprocessing Apparatus, containing several smaller experiments, including self-contained aquatic systems. About 2,000 pounds of experiment samples and equipment were transferred from Mir to *Atlantis* for a total logistical transfer of more than 6,000 pounds; the most extensive to date.

and white would stay on *Atlantis*. Before docking, the *Atlantis* crew filled eight of the 15 planned 100-pound water bags for Mir. In microgravity, the bags reminded the astronauts of sea cows.

STS-79 performed the first docking with Mir in its completed configuration. The newest and final Mir module, Priroda, had arrived at the station during Lucid's stay. *Atlantis'* payload bay floodlights illuminated Mir as *Atlantis* made its final approach, and sunrise occurred as Commander Bill Readdy brought the Orbiter to within 15 feet of Mir. Docking took place over the Carpathian Mountains, west of Kiev, at the beginning of the 12-minute communications window over Russian ground stations. Together,

Three experiments remained on *Atlantis*: Extreme Temperature Translation Furnace, a new furnace design allowing space-based processing up to 871°F (1,600°C) and above; Commercial Protein Crystal Growth, a complement of 128 individual samples involving 12 different proteins; and Mechanics of Granular Materials, designed to study "loose" materials that could lead to a better understanding of how Earth's surface responds during earthquakes and landslides.

After *Atlantis* landed back at Kennedy Space Center, Lucid walked from the Orbiter to the

crew transport vehicle, and later the same day she received congratulations from U.S. President Bill Clinton.

Back at Johnson Space Center, Commander Readdy discussed his job of picking up Lucid—late. He said, "Depending on whether you want to look at the glass as half full or half empty, it meant that Shannon had to spend an awful lot longer on Mir than she planned ... [But] I think Shannon's girlhood dream to run her own laboratory was such that actually, you know, it fit right in with her plan; and it also allowed her to set the world record for time in space."

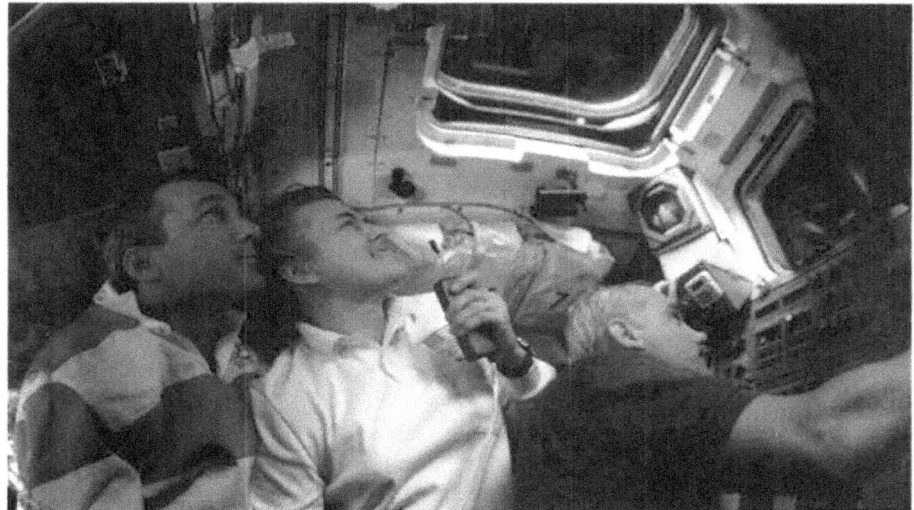

Shannon Lucid greets NASA-3 Astronaut John Blaha in the docking module (top left). STS-79 Mission Specialist Tom Akers (top right) reviews stowage items (bags with pink tags) to be transferred to the Mir. Water transfer bags surround Pilot Terry Wilcutt (bottom right) before being moved to the Russian space station. Lucid (bottom left) watches with Commander Bill Readdy (right) and Wilcutt (left) as *Atlantis* undocks from Mir.

John Blaha

NASA-3

September 16, 1996 – January 22, 1997

Pulling It Together

John Blaha accomplished an important yet seldom remarked first. He was the first astronaut to directly follow a previous U.S. Mir resident. He thus forged the first link in a six-flight, 2½-year chain of Shuttle-Mir missions.

Before, during, and after his mission, Blaha worked to make sure all future missions would go as smoothly as possible. This included improving the "handover" from one increment to the next and working on communications—between Mir and the visiting Space Shuttles, between the NASA astronaut and the ground, and between Phase 1 and Phase 2 of the International Space Station Program.

It also meant speaking directly about situations and conditions on Mir. Before launching to Mir, Blaha said, "Every Shuttle flight I've flown, I've never wanted to come home on entry day. I really enjoyed being in orbit. I've always said I would stay there forever. I think that on this mission, I will define what 'ever' is."

For Blaha, the definition of "ever" was given a new meaning just weeks before he flew to Mir when he heard the announcement that heart problems had grounded Gennadi Manakov, the Russian Mir Commander with whom Blaha had trained. Both Manakov and Blaha's other crewmate, Flight Engineer Pavel Vinogradov, were replaced by their backup crew of Valeri Korzun and Alexander Kaleri. Over the years of operating their space stations, the Russians had learned the value of keeping long-duration crews together. They believed that replacing an entire crew was better than replacing a single crewmember. At this point, NASA was not in a position to do likewise and wanted Blaha to fly the Mir-22 mission with Korzun and Kaleri. Blaha traveled to Kazakhstan and spent three days with the two Russian cosmonauts before their launch to the Mir aboard the Soyuz.

This crew change presented Blaha with a new challenge. Besides spending four months in orbit, immersed in another culture and a new language, he would be working and living with two men who—although professional and personable—were, in effect, strangers. Blaha had only taken a two-day winter survival training with Korzun.

Blaha's road to Mir began with a personal realization about the future of space-flight. In October 1991, while he was in Berlin, Germany, for a conference of the Association of Space Explorers, the possibility of joining the U.S. Space Shuttle and

NASA-3 Astronaut John Blaha, shown (page 70) at the Kennedy Space Center's Shuttle Landing Facility, Florida, participated in training procedures in a number of locations prior to his departure to the Mir space station. He spent many months at the Gagarin Cosmonaut Training Center in Star City (photo above) training with Pavel Vinogradov (left, standing) and Gennadi Manakov (seated). However, due to a medical condition that was detected in Manakov, Valeri Korzun and Alexander Kaleri replaced the cosmonauts to become the Mir-22 crew. French Cosmonaut Researcher Claudie Andre-Deshays, also pictured, journeyed to the Mir on the Soyuz with Korzun and Kaleri and returned with Mir-21 crewmembers. Blaha arrived at the Russian space station as part of the STS-79 crew and, during rendezvous, communicated from *Atlantis* with the Mir crew (bottom right).

the Mir space station was discussed. Later, Blaha said, "I remember returning from that conference and thinking to myself, 'We ought to be doing that right now.'" He considered his age—he was then 49—and his prospects for living aboard an American space station. And, he realized, "If I'm ever going to fly on a space station, I'm going to have to fly on the one that's really up there." When Norm Thagard left for training in Star City, Blaha recommended that Shuttle pilots as well as mission specialists ought to experience Mir, and he volunteered for the Shuttle-Mir Program when the opportunity arose.

This decision certainly fit in as the next step in his professional life, which had been an archetypal pilot's climb through propeller trainers to jet fighters to rocket-propelled spacecraft. Born in 1942 in San Antonio, Texas, Blaha graduated from a Norfolk, Virginia, high school and received an appointment to the U.S. Air Force Academy, where he earned his bachelor's degree in engineering science. He then earned his master's degree in astronautical engineering from Purdue University, a famous "finishing school" for future astronauts. For the Air Force, Blaha flew F-4, F-102, F-106, and A-37 aircraft, completing 361 combat missions in Vietnam. In 1971, he became a test pilot and coaxed an NF-104 research aircraft to 104,400 feet—nearly 20 miles high. He then served as an instructor pilot at the USAF Aerospace Research Pilot School and as a test pilot with the British Royal Air Force in Boscombe Down, UK. NASA selected Blaha for astronaut training in 1980. He piloted two Space Shuttle missions, and commanded two more Shuttle missions, both of whose crews included Shannon Lucid.

When asked later whether being a pilot—rather than a mission specialist—helped him during his stay on Mir, Blaha said, "No, I don't think it made any difference…. All crewmembers are the same, and everybody needs to pitch in and help—kind of like everybody pitches in and helps on a camping trip."

Blaha launched to Mir on September 16, 1996, aboard the Space Shuttle *Atlantis* on mission STS-79. He officially became a member of the Mir-22 crew, joining Commander Valeri Korzun and Flight Engineer Alexander Kaleri. They had arrived at Mir in August and had been working with NASA-2 Astronaut Shannon Lucid, who, due to her unplanned six-week extension, had begun some of the experiments originally scheduled for Blaha.

"After docking, we spent five days transferring about 4,000 pounds of supplies and science equipment to the Mir," said Blaha, "and about 2,000 pounds of supplies and equipment to the Shuttle." He added he was amazed at the incredible skill of the crews as they worked 18-hour days to accomplish all the work.

"Each evening, the STS-79 crew and the Mir crew met for dinner either on Mir or *Atlantis*. These were unforgettable times," he added. "I will always remember how they all helped me move into my home."

While the Space Shuttle *Atlantis* was still docked with Mir, Blaha noticed that the vessel for the Biotechnology Systems Experiment was not rotating. He alerted Mission Control-Houston. Since the Biotechnology Systems Experiment supported living cells, the rotation had to be restored quickly. The time was nearing for closing the hatches, and so was the possibility of having to return the Biotechnology Systems Experiment facility to *Atlantis*.

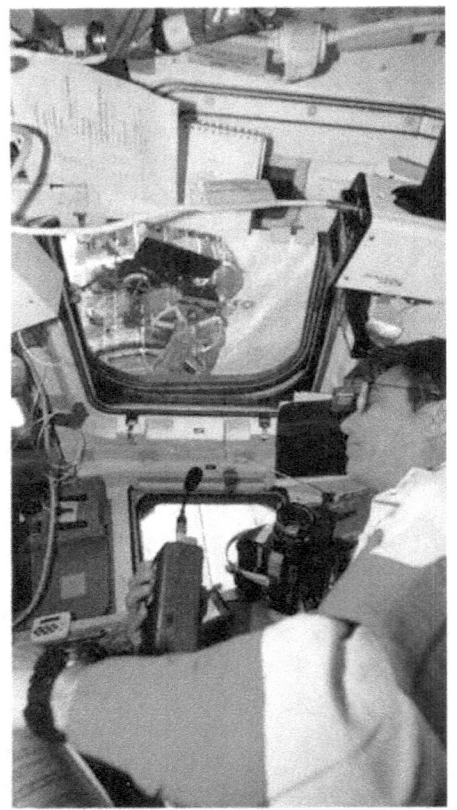

John E. Blaha

John Blaha was born in San Antonio, Texas. He earned a Bachelor of Science in engineering science at the U.S. Air Force Academy and a Master of Science in astronautical engineering from Purdue University. He has logged more than 7,000 hours of flying time in 34 different aircraft, completing 361 combat missions in Vietnam and piloting the NF-104 research aircraft to 104,400 feet. He earned two Air Force Distinguished Flying Crosses, the British Royal Air Force Cross, and the Vietnam Cross of Gallantry.

Blaha was selected as an astronaut in 1980. Before his Mir residency, he piloted two Shuttle missions and commanded two other Shuttle missions, including STS-58 (with Shannon Lucid and David Wolf). This seven-person life science research mission, lasting a record 14 days, was recognized by NASA management as the most successful Spacelab flight that NASA has flown.

About his four months on Mir, Blaha said, "The only thing I missed was my wife.... and I told people that all the time. If I had the choice, I would go to the Mir to work ... if you could beam me up on Monday morning and beam me back on Friday evening. I would go there, and it would be like going to any job that anyone goes to in America. I would do it as a profession forever."

John Blaha retired from NASA in September 1997 to return to his hometown of San Antonio, Texas, where he now works as a management executive and volunteers with the Challenger Center for Space Education.

Third U.S. astronaut to live onboard Mir • Launched with STS-79 on September 16, 1996 • Returned with STS-81 on January 22, 1997

Moving quickly, Blaha tried numerous procedures to correct the Biotechnology Systems Experiment, while Astronaut Jay Apt took digital photos of the equipment and downloaded them to Johnson Space Center. There, the ground team worked feverishly. They compared the photos to a sister system, troubleshot the problem, and sent up instructions and ideas. Blaha tried them; they did not work.

It was time to close the hatches. The American Shuttle crew and Lucid bid farewell to the Mir-22 crew. But, Blaha continued working on the Biotechnology Systems Experiment problems and did not join the formal goodbyes so that he could "work with ground crews" while good communications remained intact through *Atlantis* to Mir. At Johnson Space Center, Allen Moore, Krug Life Science's lead engineer for the Biotechnology Systems Experiment facility, pored over the photos taken by Apt and discovered a control and data cable had become dislodged. While the Shuttle undocked and began its fly-around of Mir, the Biotechnology Systems Experiment team drew up another "fix." This one would ask Blaha to power down the equip-

ment so the cable could be remated without damaging the system.

In the nick of time, the ground team had arrived at a solution. But, all was not well—yet. These procedures were more complicated. Before they could be passed on to Blaha, they would have to be approved by Mission Control-Moscow. That could take days.

At that moment, communication between Moscow and Mir was suddenly established—again by way of *Atlantis*. Bill Gerstenmaier, leading the U.S. consultants group in Russia, broke in to announce that the Russian flight director had approved the procedures. Then, Gerstenmaier in Russia and Mission Scientist John Uri in the United States both talked with Blaha directly through the communications relay. After completing the procedure, Blaha called down to report that the cable had been secured and the Biotechnology Systems Experiment vessel was now operating properly.

"It ain't Apollo 13," quipped Uri, "but from a scientist's perspective, we pulled it off and saved that experiment."

Atlantis pulled away. A few days later, Blaha

wrote in an e-mail. "I will always remember the incredible sight as the *Atlantis* undocked and flew around the Mir. The views of *Atlantis* silhouetted against the darkness of space, the horizon of the Earth, or zooming over the top of Russia and China will never leave my memory. Wow, what an incredible spaceship America built."

From orbit, Blaha also described his impressions of Mir. "Actually," he said, "I was surprised. There was a lot of empty space. It may be five times the size of the volume of a Space Shuttle. The environment is actually very good. The air is very healthy. It's not dry. It's not humid. Nothing smells. Two of the modules are very new inside. The other four modules look a bit used—as you could imagine a house looking after people have lived in it in orbit for ten or 11 years, without having the advantage of bringing the vehicle home and letting it be cleaned up on the ground."

As his mission began, Blaha got right to work on the revived Biotechnology Systems Experiment, which studied the long-term growth of living mammalian cartilage cells suspended in microgravity. At scheduled times, Blaha sampled the cellular environment for postflight analysis and recorded

the progress of the experiment on video. He also used the samples to decide when to replace the medium that kept the cells growing.

Blaha also "fixed" several wheat plants from the Svet greenhouse experiment designed to study the effects of microgravity on plant growth, and he noted that the heads of the plants were maturing. He also took several physical measurements of himself to help researchers study the changes in his muscle mass during his stay on Mir.

His scientific regimen was now well under way. Into October, Blaha concluded work with samples of the binary colloidal alloy tests, which grew crystals of two materials together over time. He collected samples of the microbial environment around Mir, including air, water, spacecraft surfaces, and samples from the Mir crewmembers' skin. He appreciated Mir's several excellent windows, which had covers that opened and closed. He studied the geography and weather on the Earth below. Mir's high orbital inclination meant that the space station flew over nearly all of the inhabited regions of Earth—and all of the Earth's hustle and turmoil.

Blaha's scientific investigations included monthly photography of samples for the diffusion-controlled crystallization apparatus for microgravity, which slowly grew protein crystals to be compared to samples grown on Earth. All three Mir crewmembers exercised on the U.S. exercise

bicycle, while hooked to equipment that measured their breath through a metabolic analyzer. The crew was also quizzed by Russian psychologists who were interested in any changes that might occur in interpersonal relationships during long-duration space missions.

In an October 25, 1996, press conference, Blaha praised his hardworking shipmates, saying, "Valeri and Sasha [Kaleri]—they're incredible cosmonauts. We work about a 16-hour day— Sasha and Valeri, for certain. I'm a little older, so after about 14 hours I need to settle down a little bit and look at the stars or the Earth. I watch movies." He would later say that the crew worked so hard on their separate tasks that even his work-time interactions with his crewmates were limited. "Every now and then I would do something with one of the cosmonauts, but not often.... Maybe there were 15, 20 times in that 4½ months. The reason was—we all were too busy. We couldn't be together. All three of us had to be working on things to accomplish all of the work."

Blaha said that one lifestyle difference he made on the space station was bringing up movies, which allowed him "to settle down in

Mir-22 Commander Valeri Korzun (right) and Flight Engineer Alexander Kaleri in their Russian liquid cooling and ventilation garments prepare for an extravehicular activity.

Atlantis pulled away. A few days later, Blaha wrote in an e-mail, "I will always remember the incredible sight as the *Atlantis* undocked and flew around the Mir. The views of *Atlantis* silhouetted against the darkness of space, the horizon of the Earth, or zooming over the top of Russia and China will never leave my memory. Wow, what an incredible spaceship America built."

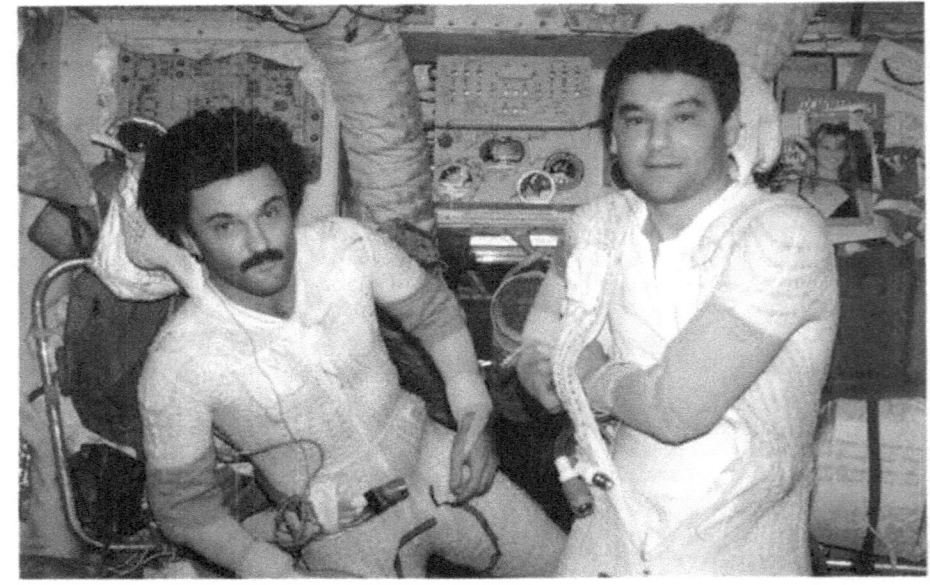

the evening. As a result, I [had] a fantastic night's sleep." Sleep had been one thing Blaha worried about before his Mir mission. He was eating well and enjoying his exercise on Mir, and "the movies have been helpful. They're like medicine to me." He added that the movies helped him relax and prepared him for a full seven hours of sleep.

One important relationship that would not change for Blaha was the one he had with his wife. During the October 25 news conference, U.S. reporters asked him what he missed most about being away from Earth for such an extended period. He didn't miss the pull of gravity at all, he said. However, "what I wish I had is my wife, Brenda. I miss her. We have a very good relationship. I miss talking with her and seeing her…. If she were here with me, I'd stay here for four or five years."

On November 1, as Blaha was completing six weeks aboard Mir, he had an interactive videoconference with crewmembers of the STS-81 Space Shuttle crew. They would launch to Mir in January with NASA-4 Astronaut Jerry Linenger. The crew used the two-way video as an opportunity for a brief Halloween party, and ground team members donned makeshift costumes to bring reminders of life at home to Blaha in orbit.

The crew's physical comfort was challenged by a breakdown in the system that recycled the crew's waste into Mir's cooling system. Early in November, the Russian news service *Itar-Tass* reported that waste reserve containers had nearly filled. The next Progress resupply ship would not launch for nearly three weeks. Also announced was that Russia's financial problems had slowed production of Soyuz booster rockets. This forced the postponement of the launch of the Mir-23 crew from December 15, 1996, to February 1997. Blaha was scheduled to depart Mir in January. He would now be able to serve out his mission with the same shipmates, Korzun and Kaleri.

When Blaha had "free time," he enjoyed talking to ham radio operators around the world, telling them about his Mir experiences and receiving news from them. Blaha had been an avid ham radio user during his previous missions with NASA. Due to his interest, Blaha had worked with NASA to set up ham radio conferences between the Mir crew and various organizations around the world.

Blaha's scientific investigations continued with the passive accelerometer system, which used a small metal ball inside a tube to measure minute

The crew of STS-79 captured this view of the Earth while *Atlantis* was docked to the Mir.

Ham on Mir
Amateur Radio

When a handheld amateur radio was first sent up to the Mir space station in 1988, ham radio became a popular and an important activity for Mir's space-bound crews. It provided a somewhat unofficial and often informal way to communicate with families, friends, students, and other ham operators. Amateur radio also functioned as a backup communications link between Mir and ground controllers, and it was highly appreciated by several of NASA's Mir astronauts. Not all of the Mir astronauts held amateur radio licenses, but they could use Mir's radio because the station qualified as an amateur radio "club" according to Russian amateur radio regulations.

The worldwide Mir Amateur Radio Experiment was formed to handle communications with radio amateurs around the world. The Mir International Amateur Radio Experiment coordinated activities with schools and students.

A similar Shuttle Amateur Radio Experiment continues to operate onboard Space Shuttle missions, and an amateur radio system is expected to be an important part of crew life onboard the International Space Station.

Life on Mir

Astronauts and cosmonauts living onboard Mir had to face isolation, confinement, and physical risk. But, they also got to experience long stays in microgravity, incredible views of Earth, and the adventure of being pioneers.

What was it like up there when things were going "normally"?

At different times, and by different people, Mir was described variously. Perceptions depended on expectations and on comparisons to other spacecraft, such as a Space Shuttle or a Soyuz capsule.

NASA-3 Astronaut John Blaha said, "I was surprised. There was a lot of empty space. It may be five times … the volume [of a Space Shuttle]. The environment is actually very good. The air is very healthy. It's not dry. It's not humid. Nothing smells. Two of the modules are very new inside."

But Mir was also very cluttered, mainly because of the limited ways of getting rid of unwanted equipment and supplies. Ten Progress resupply vehicles were loaded with trash during the American residencies, and nine visiting Space Shuttles brought back what they could. Even so, when STS-91 Shuttle Commander Charlie Precourt talked about the last Shuttle visit to Mir, he said that there were "still food supplies there that belonged to Shannon [Lucid] and John Blaha, and there were boxes [of food] with their names on them that nobody's ever going to eat."

NASA-5 Astronaut Mike Foale said that it was "very easy to lose each other … It's not that the Mir is such a big space—it's because it's such a cluttered space. You're basically winding your way through tunnels…. Equipment just isolates you from other parts of the station. So, … there may be food boxes stored, or spacesuits stored, or just trash…. And, you can't even be seen from the node area on Mir."

Foale said that there were times "when I would suddenly pop out of my warren—you know, out of my hole—and Vasily [Tsibliev] would say, 'Mike, have you seen Sasha [Lazutkin]? I haven't seen him all morning.'

"I'd say, 'No, I haven't seen him all morning.'

"Well, we knew he had to be on the station. But, we didn't know where he was. So, you could easily spend a day without talking to crewmembers—and that we considered not a good thing. So, we

made an effort to try and tag up, especially for lunch and often just for a 10-minute tea break."

NASA-2 Astronaut Shannon Lucid reported that her crew "ate all our meals together and spent a lot of time talking to each other over mealtimes." Other Shuttle-Mir astronauts, such as Blaha and NASA-4 Astronaut Jerry Linenger, reported working such long hours that days would go by without much interaction with their Russian crewmates.

This work was sometimes grueling physical labor. NASA-6 Astronaut David Wolf related how on his first day on Mir, he offered to clean up liquid that kept condensing on some heat exchangers. According to Wolf, "I didn't realize what I was getting into, because it took anywhere from four to eight hours a day, the rest of the mission, every single day except a few."

The Mir residents kept themselves clean with sponge baths and shampoo that they could towel out of their hair. The NASA Mir astronauts slept, generally, in Spektr or Priroda—the newer modules. Lucid said, "I put all my personal things in Spektr. I had a sleeping bag, [provided by the Russians]. Then I had one of the white collapsible bags that had all my personal things in there—my books and stuff.

"Then at the end of the day … I would go into Spektr and just unroll the sleeping bag and tie it off. Generally, we had about an hour in the evening to ourselves. That was just real nice. Then I would get in the sleeping bag, go to sleep, and wake up the next morning when the alarm clock went off." Several of the Mir astronauts reported excellent sleep.

Mir residents generally dressed in cotton T-shirts, shorts, and jumpsuits. Supplies of these were limited. Lucid said that during the first two-thirds of her flight, there were enough of the shorts and shirts "so that we could change twice a week—like on Wednesdays and Sundays." She wore her blue jumpsuit "every day for 188 days."

While the rewards of living and working on Mir were great, it was—all in all—tough duty. When Wolf was about to leave NASA-7 Astronaut Andy Thomas to his turn onboard the Russian space station, Wolf said in a news conference that his own time onboard Mir "has been an amazing experience. It's been one of the hardest of my life, and I think Andy can expect the same."

(Above) Mir-22 crewmembers (left to right) Kaleri, Blaha, and Korzun conducted many experiments during their mission including harvesting the first crop of wheat plants (below) grown in the greenhouse in the Kristall. (Photo at left) Blaha is pictured in his Russian Sokol suit, a garment tailored for each Mir resident and mandatory for travel in the Soyuz capsule.

residual gravity at space station altitudes. Blaha also ran another malfunction procedure on the Biotechnology Systems Experiment. This time, the experiment had developed an air bubble in the liquid growth medium and difficulty with the computer-controlled pump. Blaha replaced the growth medium and reset the computer, but problems continued. The experiment had shown increased metabolic activity in the cells, indicating a possible higher growth rate in microgravity.

Good news came finally when, after three postponements, a Russian Progress resupply spacecraft launched from Baikonur on November 20 and docked with Mir two days later. The Progress brought supplies, including Christmas gifts from Blaha's family, New Year's gifts for Korzun and Kaleri, and fresh fruit, clean clothing, and new equipment for all three men. Thanksgiving came, and Blaha watched the

beautiful Earth through the Mir windows rather than his usual viewing fare of football. He and his crewmates also worked on that holiday.

The first week of December aboard Mir began with a six-hour spacewalk performed by Korzun and Kaleri. Its main purpose was to complete connections of the cooperative solar array to provide more electrical power to the station. After the spacewalk, however, the crew reported that the ham radio was not working and may even have been damaged during the spacewalk. Further attention would be needed.

Onboard Mir, Blaha harvested the first crop of healthy plants grown through a complete life cycle in the microgravity of space aboard Mir. The plants were grown in the Svet greenhouse, a small growth chamber originally built in Bulgaria during the late 1980s. Svet had a compact growing area of about one square foot and could accommodate plants up to 16 inches tall. The wheat was grown in a material similar to kitty litter but was loaded with plant nutrients. Fluorescent lamps provided light. Water was injected directly into the growth material and transferred to the wheat seeds by a system of wicks. Blaha, on a daily basis, recorded critical experiment data and transferred the data files to the ground. Several times he made manual

changes to water and lighting cycle times. Day length and water injection were normally controlled automatically and adjusted throughout the experiment by project scientists. The next week, Blaha planted a new crop of wheat seeds.

It was also during this next week that the cosmonauts completed a second extravehicular activity to finish work on the solar array. In a video downlinked to Mission Control-Moscow, Blaha described an additional spacewalking activity. "Another thing Valeri and Sasha did on this [extravehicular activity] was they repaired our transceiver system that we use to talk to amateur operators all around the world.... They had quite a bit of equipment they were trying to move, and I was very impressed with all their work. They lived in those suits for nine hours and did a fantastic job." During their 6½-hour spacewalk, Korzun and Kaleri also completed connecting the solar array and installing a new Kurs antenna that would be used to guide Progress vehicles docking with Mir.

Shortly after the two cosmonauts returned inside, Blaha used the repaired shortwave radio to receive ham radio conversations over Brazil and to initiate conversations over Madrid. He later characterized the overall communications situation onboard Mir as "excellent."

Blaha also related his impressions of his crewmates' spacewalks: "I will forever have images implanted in my brain of Valeri and Sasha—working 18-hour days, preparing for the spacewalks, asking many questions to specialists on Earth, and probing every possible scenario. I will forever remember the incredible views of these two cosmonauts floating in space, silhouetted

NASA-4 Astronaut Jerry Linenger (left) joins Blaha at the table in Mir's Base Block to sample Russian food before officially becoming the newest Mir-22 crewmember.

against the black of space, with planet Earth rotating by us below. I will forever remember the sounds of strain in their breathing when the workload was intense. And, finally, I will never forget the incredible feeling of accomplishment after the job was complete, and everyone was safely inside the Mir Space Station."

On December 20, the Mir-22 crew held a news conference; and, naturally, several of the questions were about how they would celebrate the upcoming holidays. Blaha for the most part gave his answers straight, while Commander Korzun injected some humor and perhaps let the cat out of the bag about Christmas dinner:

Question: What plans do you have for your holidays in space?

Korzun: Maybe we could go for another spacewalk and get another new Christmas tree for Christmas this year!

Question: What will you miss about Christmas while you are there?

Blaha: As to spending Christmas here and

not with the family, I don't know how that's going to work yet…. We've been busy. I haven't had time to really think how I'm going to feel on Christmas Day.

Korzun: At a store, we have presents. We will get the presents from the store and give them to each other. John Blaha hasn't said what we really miss—which is a Christmas pie.

Question: What have you planned for your Christmas dinner?

Korzun: We're going to have an outstanding menu … both Russian and American products. We will have traditional cakes and other dishes, lamb, pork, and a wonderful dessert, as well as Italian food—macaroni and cheese.

Blaha: In six days, we're going to have quite a feast! I'm happy. This is the first time I've heard about that.

Besides celebrating the holidays, Blaha and his crewmates worked on Christmas Day and on New Year's Day as well.

In early January, floods caused widespread damage in the western United States, and bad weather in the eastern United States threatened to delay the launch of *Atlantis* that would bring Jerry Linenger to Mir. Undaunted, Blaha prepared for

his return to Earth, packing 15 bags of gear to be transferred to the Space Shuttle. He continued his work on the Biotechnology Systems Experiment, Svet, and other experiments; and he collected samples of microbe population from the water, air, surfaces, and crew.

Before his departure, Blaha would encounter yet one more challenge. On the evening of January 10, he heard a loud clattering noise in the Spektr module.

An investigation revealed that one of the two cooling fans was broken in the large freezer containing all of the Mir-22 life science research data. There was no spare fan on Mir. Blaha removed the front door of the freezer and affixed a temporary door to hold the temperature as long as possible. The American astronaut also removed the fan blade and reinstalled the primary freezer door with only one fan operating—a configuration adequate for only one week.

The Space Shuttle was scheduled to launch in 36 hours. The mission needed to bring a replacement fan, plus a spare, or the following Mir-23 mission life science research program would be significantly impacted.

Blaha radioed Pat McGinnis, his Flight Surgeon in the Mission Control Center-Moscow, and told him that a ham radio communication had been scheduled for 9 p.m. (noon Houston time) with Blaha's wife. He told the flight doctor to locate Matt Mueller, an engineer in Houston working with NASA, and tell him to be present at the ham radio session. Right on time, Blaha greeted his wife, and after hearing Mueller was present, spent the rest of the six-minute communication explaining the small emergency.

Mueller and Blaha had trained together for four months in Star City on all of the science experiments and equipment slated for Blaha's mission. Blaha knew this vital ground support team member could quickly understand the problem, contact the necessary people, obtain the spare fans, and have them delivered to Florida in time to be loaded on the mid-deck of *Atlantis* as it was being prepared for its launch to Mir. The plan was executed flawlessly, Blaha said later.

Atlantis (STS-81) launched on schedule on January 12, 1997. After it had docked, Blaha took special care to brief Linenger, the newest U.S. resident on Mir. During his own stay on Mir, Blaha had developed a detailed checklist to help him provide as much information as possible during the handover time. He had stressed to NASA the importance of the handover, and he had worked to ensure plenty of time had been scheduled for the two long-duration astronauts to exchange information.

When *Atlantis* landed at Kennedy Space Center 10 days later, Blaha followed some advice he had received from the Russians. He allowed Kennedy Space Center workers to carry him off the Orbiter on a stretcher so that doctors could better study the effects of microgravity on an astronaut's return to Earth's gravity. Blaha's wife and daughter greeted him with kisses and hugs.

Atlantis touches down to conclude the fifth Shuttle-Mir docking mission and returns Blaha to Earth after four months in space.

John Blaha Mails a Letter Home

December 13, 1996

This week we started preparing for the arrival of a Progress resupply vehicle. Two days before the launch we started loading up the old Progress docked to the Kvant module. We put all our dirty clothes, trash, equipment nobody wanted, 600 liters of urine, many containers of solid waste, etc., into the cargo bay.

We started sleep shifting 2 days before the launch, because we planned to undock the old Progress at 2 a.m. and dock the new Progress approximately 26 hours later. We, of course, waited until we knew the new Progress launch was successful and the spaceship was going to have a good chance of docking with us before the old Progress was undocked.

At midnight, Valeri, Sasha, and I worked with engineers on the ground to ensure we had a good seal with the hatch leading to the old Progress. When everyone was convinced we had a good seal, the Moscow Control Center sent commands to automatically undock the old Progress. Valeri installed a special control

system near the Base Block control station and was ready to fly the Progress manually, if required. He had a TV monitor, which displayed the Mir as seen from the Progress.

About 10 minutes after the Progress undocked, we could visually see it at about 100 meters through a large window in the floor of the Base Block. It was beautiful to watch this big beautiful machine with solar panels—they looked like airplane wings—pull away and finally disappear.

Twenty-four hours later we were eagerly awaiting the arrival of the new Progress. I was in the Kvant-2 module looking through one of the small windows. I finally saw the Progress at a distance of 30 kilometers.

It was a shining star rising towards us at great speed from beneath the horizon. This was an incredible sight. There we were approaching the terminator on planet Earth, and this "beaming" shining star was roaring towards us. Then all of a sudden, the light from the Progress extinguished as we passed into the shade of the Earth. Five seconds later, four lights on the Progress were turned on.

I watched the remainder of the rendezvous through a tiny window in the aft end of the Kvant module, right at the point where the docking would occur. Again, Valeri was monitoring the event with his backup control system in the Base Block of Mir.

The docking felt quite firm. Five times stronger than I remembered the Shuttle docking with Mir felt over two months ago. The Progress rendezvous approached from behind, passed the Mir radius vector, then performed an approach on the velocity vector.

We verified we had a good seal before opening the hatch at about 5:30 a.m. We were supposed to go to sleep at 6 a.m. Of course, we stayed up a few extra minutes as we searched for our crew packages.

Once we found our packages, it was like Christmas and your birthday all rolled together when you were five years old. We really had a lot of fun reading mail, laughing, opening presents, eating fresh tomatoes, cheese, etc. It was an experience I will always remember.

The Progress brought us a lot of food, fresh water, fuel for the reaction control jets, oxygen, spare parts needed to repair systems, equipment for a spacewalk, science equipment, towels, and clothes.

I thought you may be interested in reading about what it was like to have a Progress arrive at Mir. This type of event will occur many times on the International Space Station.

John

Life in Microgravity

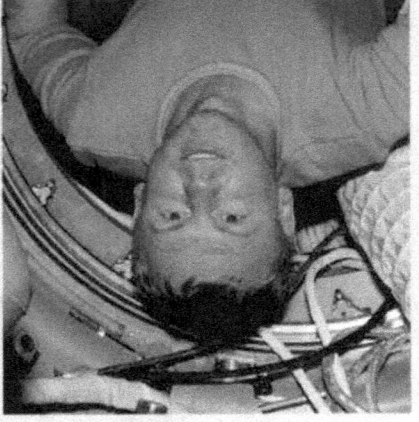

Gravity exists everywhere, but it is not felt in an orbiting spacecraft where the sensation is more like that felt by someone in an aircraft during a "free fall." Indeed, an orbiting spacecraft is, in effect, falling around the Earth, with its occupants falling at the same speed inside it.

People and objects in the spacecraft appear to be floating, yet they still possess the same mass that they did on Earth. The situation can be compared to that of a boat floating next to a dock; a person can easily move the boat while still feeling its inertia or "heft." In space, transferring heavy batteries between modules can still take hard work. A Progress supply ship can strike the Spektr module and cause crippling damage.

Living and working in microgravity presents pleasures and problems. One astronaut compared it to "flying in your dreams." She said, "You're floating. You can flail your arms. But, you don't go anywhere."

Many astronauts develop short-term "space sickness." Also, tools can float away, and simple activities—such as eating—require special tech-

niques. Furthermore, prolonged exposure to microgravity causes changes in the human body, including bone loss and muscle atrophy as well as changes in blood and fluids circulation. These can partly be countered by strenuous exercising while in orbit.

Space motion sickness is only partly understood. It is a kind of motion sickness, and different people experience it differently. Two of the Shuttle-Mir astronauts, both medical doctors, related their own observations about overcoming it.

NASA-1 Astronaut Norm Thagard had an interesting comment, resulting from his launch to Mir onboard a Soyuz spacecraft. He noted that the Russians had claimed that space motion sickness was not a problem for them. "And yet," according to Thagard, "what I found out on the Soyuz is there is a reason why they do better: You cannot move around. There's not that big volume on the Soyuz.... You just don't move around, and you certainly don't have as many head movements" during the two days it takes Soyuz to travel to Mir.

Thagard pointed out that, after a Space Shuttle launch, "you start moving your head," immediately after orbit is attained. Astronauts start unlocking lockers and getting out equipment. "You basically hit the deck running," said Thagard. "You've got so much to do and so little time in which to do it that, immediately upon clearance [and] a 'go' for orbit, you're up—just darting—and throwing big head movements...."

"There is absolutely no question that that's what causes and exacerbates space motion sickness," according to Thagard. Because he couldn't move around much onboard Soyuz, he "really never got beyond stomach awareness."

NASA-4 Astronaut Jerry Linenger had a similar observation. He wrote of the arrival at the space station of the Mir-23 crew, "It was obvious that they were all relieved and glad to get out of their capsule; but it was also obvious that the voluminous Mir nauseated them. The freedom of motion came at a price—space motion sickness. While everyone was able to smile during the press conference, the newcomers were glad when the cameras were turned off and they could remain still and feel miserable alone."

Almost every astronaut adapts to microgravity, and all of the Mir astronauts did very well on their Mir flights. NASA-7 Astronaut Andy Thomas said, "One of the things that surprised me was how quickly I could adapt to that environment. Now, true, it was my second flight, [but] I was surprised, because on this flight I adapted to being at zero-gravity psychologically very quickly, and I very quickly learned to function and to accept that environment as the normal environment, and it felt natural."

Thomas elaborated. "That was what was so strange—[that] an environment which is fundamentally so unnatural could so quickly feel natural." He added that "you become more blasé about it because you have it all the time. And, every now and again you have these little reality checks.... 'Wait a minute. I'm weightless. I can float. That's the way I move.' And, you just have to remind yourself that—yes—you are weightless.

"Of course," Thomas said, "another form of reality check was when you would go and look out the window and have this spectacular view, just to remind you of where you really were."

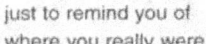

Once the basic feeling of microgravity is adapted to, astronauts must adjust to working and living with it. In one of his "Letters to My Son," Linenger wrote: "Everything floats up here. And, I do mean everything. Propulsion and suction are the keys to daily tasks of living. For example, you can

brush your teeth pretty well, as long as you keep your mouth closed. Open your mouth and breathe out just a bit: and you have foam floating away. I keep a small two-by-two gauze pad nearby, and I carefully capture the stuff in it. Then, to a plastic [Ziplock®], and remember to seal.

"I can gulp peanuts like a fish would: they float, I open my mouth and pull 'em in! … I guess that God made us all-purpose beings because, once the food goes down the pipe, it stays down. Peristalsis: who would have thought of it."

Sleeping in microgravity feels like many people assume it would feel—like sleeping on a cloud. Mir astronauts used sleeping bags that were secured so they wouldn't float off.

Bathing on any spacecraft has always been problematic. Microgravity allows water to "glob up" and drift everywhere. Thagard said that on Mir, "They give you one wet towel. And, these are just cloth towels in plastic wrapping that have some sort of antibacterial solution in it. I'd get one of the wetted towels a day and two dry towels a day. And, that works fine. If you need more, you can always just wet one of the dry towels, or put more water on the wet towel.… It worked fine for cleaning up after exercise."

Exercise on Mir had to be done against mechanical resistance; for example, by stretching cords or by running on a treadmill. In one of his "Letters to My Son," Linenger discussed some peripheral effects of treadmill running in an orbiting spacecraft. He wrote: "Sasha [Lazutkin] is running on the treadmill, medium pace…. It's definitely Sasha, and he's on the treadmill in module Kristall, and not on the tread-mill in the Base Block module. I know who it is—and what he's doing—not by sight or sound, but by feel. I can feel him. Frequency, about 1 hertz."

Linenger was working on a computer at the time. He said, "The computer and I are going up and down right now. Feels similar to being in a rowboat, near the shore, after a ski boat has gone by. Gentle, but definite swaying. The whole 13-meter 'tube' I'm in is moving. The force Sasha imparts is absorbed by the station, and it sways, resonates. If he slows down or speeds up a bit, I'll feel nothing. A peaceful float. When Shannon

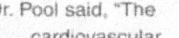

Lucid was onboard, she had to stop running at a given pace because the station would resonate at a dangerous level."

An astronaut's visual perception is also affected in microgravity. Thomas wrote in a letter from Mir: "The most frustrating thing is that you are forever losing things. You might be rummaging through a bag to find one item, while all the other contents are floating away, and before you know it, they are gone, and lost. They may even be close by to you, but as you look around you tend to focus your gaze only on surfaces, where we are accustomed to seeing things." NASA-6 Astronaut David Wolf wrote, "We just don't expect the pliers to be pointing straight at us, at eye level, one foot in front of our face."

Microgravity also affects the systems and structures within the human body. During Shuttle-Mir, Dr. Sam Pool was Assistant Director of Space Medicine at Johnson Space Center. According to Dr. Pool, "Flying in space, particularly in microgravity for long periods of time, is not very friendly to the human body." Many body functions degrade, and even countermeasures, such as exercise, have not solved the problem. Dr. Pool said, "The cardiovascular system becomes less responsive, particularly on return to gravitational field. Bone mineral is lost. Neurophysiology is definitely affected. The Russians say—and I think they're quite correct—that returning cosmonauts from long missions can't play simple games that children play because their coordination and so on is not up to it That's certainly been proven true now that we've begun to fly with them."

Several of the Shuttle-Mir science investigations studied the effects of microgravity. For more information on the Science of Shuttle-Mir, refer to the CD-ROM.

(Photos, page 80) Marsha Ivins from STS-81 (left) floats on *Atlantis'* aft flight deck while the Mir space station is visible in the overhead windows; (top right) NASA-7 Astronaut Andy Thomas moves through Kristall airlock; (bottom right) While NASA-4 Astronaut Jerry Linenger brushes his teeth, he collects his toiletries. (Above, top left) Sergei Krikalev (left) and Jan Davis (right) from STS-60 eat pieces of candy from the air while on *Discovery's* mid-deck; (top right) Elena Kondakova and Jean-François Clervoy demonstrate the sleep restraints arrangement in the SPACEHAB during STS-84; (bottom) A water bubble free-floats during STS-84.

Meanwhile on Earth
NASA-3

During John Blaha's increment onboard Mir, NASA's operations in Russia likewise experienced its first "handover" of responsibilities, from one Operations Lead to the next. Operations Lead Isaac "Cassi" Moore took over from Bill Gerstenmaier. Gerstenmaier had been so immersed in so many aspects of Shannon Lucid's flight that a good way to pass on all the knowledge was difficult to develop. By the end of Blaha's flight, the NASA operations team in Russia had expanded by several members, and the situation of a nearly "one-man show" was a thing of the past. (See Operations, page 47.)

Moore was to learn more than just Mission Control-Moscow's way of doing operations. He also was to learn how the Russian flight controllers might spend a light moment. One day Moore was in the control center, where the Russians would often play music on one of the communications channels or "loops." On this occasion, Moore could not see over his own console, but he began to hear the song "La Macarena" playing. As he later told it, "Then I realize that what I'm hearing is not coming from the voice loop … and I'm trying to figure out what's going on here. It's a little bit like the Twilight Zone. I look around, and I look over the console, and most of controllers in the room are doing the Macarena. And, it was all I could do not to fall out of my chair—because they were singing the Macarena and doing the Macarena in the control room for Mir. Now, I was trying to imagine that over here, in the control center in Houston, and I can't. That's just not a vision that I can do. But, here are these people working, … the Macarena is played, and so they're going along with it, sitting on console…. And, I tried not to make an international incident by falling out of my chair."

Operations Lead Christine Chiodo arrived in Russia toward the end of Blaha's increment to begin one of her lengthy stays in Star City. She later told the story of one Christmas she spent in Russia, when the Americans made a similarly odd impression on their Russian controller colleagues. "We had the room decorated in lights," Chiodo said. "Somebody brought in Santa hats, and we're giving out candy canes to the Russian ground team. They were looking at us like we were nuts."

These stories detail how Shuttle-Mir participants gave each other psychological support, both on orbit and on the ground. For another example, it was around 6 a.m., Houston time, when Shuttle-Mir Program Manager Frank Culbertson called the Moscow control room. Chiodo said to him something like "Jeez, you better get to bed. I'm sure your kids are going to be up early."

Culbertson replied, "Oh, I'm still wrapping gifts."

"It makes you feel like you're not so far away," Chiodo said later. "Phase 1 management was fantastic in supporting us. Every morning—usually anywhere between 6 a.m. and 7 a.m., Houston time—we knew the phone was going to ring. It was either going to be Frank Culbertson or Jim Van Laak."

The Flight Docs

They couldn't make house calls to the space station Mir but they did almost everything else, including watching over the Mir astronauts' health, training, studies, communication, and stress levels. The Shuttle-Mir "flight docs" were all-purpose physicians.

Since the Mercury Program, medical doctors have been very involved in NASA's human spaceflight efforts. These Flight Surgeons take part in astronaut selection and training, in astronaut physical health care and psychological support, and in monitoring the environment onboard spacecraft.

During Shuttle-Mir, NASA Flight Surgeons expanded their roles. Besides overseeing the astronauts' health and the life support situation onboard Mir, they traveled to Russia to assist the astronauts during training and served as important communicators and facilitators during their Mir residencies. As their duties expanded, they often spoke directly to the U.S. astronaut onboard Mir. This differs from the situation of a Space Shuttle flight, where only the Capsule Communicator generally speaks to the flight crew. In Star City, the Flight Surgeons also served as occupational safety and general medical officers, and they provided medical care for NASA's people when needed. NASA-6 Astronaut David

Wolf talked about how important the Flight Surgeons were and praised Dr. Chris Flynn, his mission's Flight Surgeon.

"The Flight Surgeons take on a new importance in a mission like this. They do more than what would traditionally be considered Flight Surgeon work. They are your alter ego. They're yourself on the ground, and they know the people you need to talk with and communicate with." He added that the Flight Surgeons were the Mir astronauts' "gateways" for all sorts of information and communication.

To accomplish this with the flight doctor, Wolf said it "takes a long relationship before the mission. He has to know you and your family, and you have to know him, and he's under a similar stress as you are on these long-duration missions."

Dr. Roger Billica was Medical Operations Working Group Leader during Shuttle-Mir. Dr. Billica said, "[It] made it so that the crewmember knew this doctor, and was comfortable with him and willing to really work with him when there were problems. It wasn't some strange voice or strange person [they heard]; it was their Flight Surgeon. I think that helped a lot when things came up, when the crewmember had questions or concerns, for them to feel comfortable, secure, confident that we were going to be able to deal with whatever the situation was."

The Flight Surgeons encountered differences between the American and Russian ways of practicing flight medicine. Johnson Space Center's Assistant Director of Space Medicine during Shuttle-Mir, Dr. Sam Pool, said that, generally, "the American system … follows the military model in that we provide health care" before, during, and after spaceflights. "We also organize emergency medical services that may be required. We provide medical certifications for selection as an astronaut and for each space mission to which [an astronaut is] assigned."

The Russians, according to Dr. Pool, have developed separate and almost autonomous systems for each phase of flight. Some physicians are involved in training the cosmonauts and preparing them for a mission, and a different group of physicians is responsible during the mission. At the time of recovery of a Soyuz spacecraft, representatives of

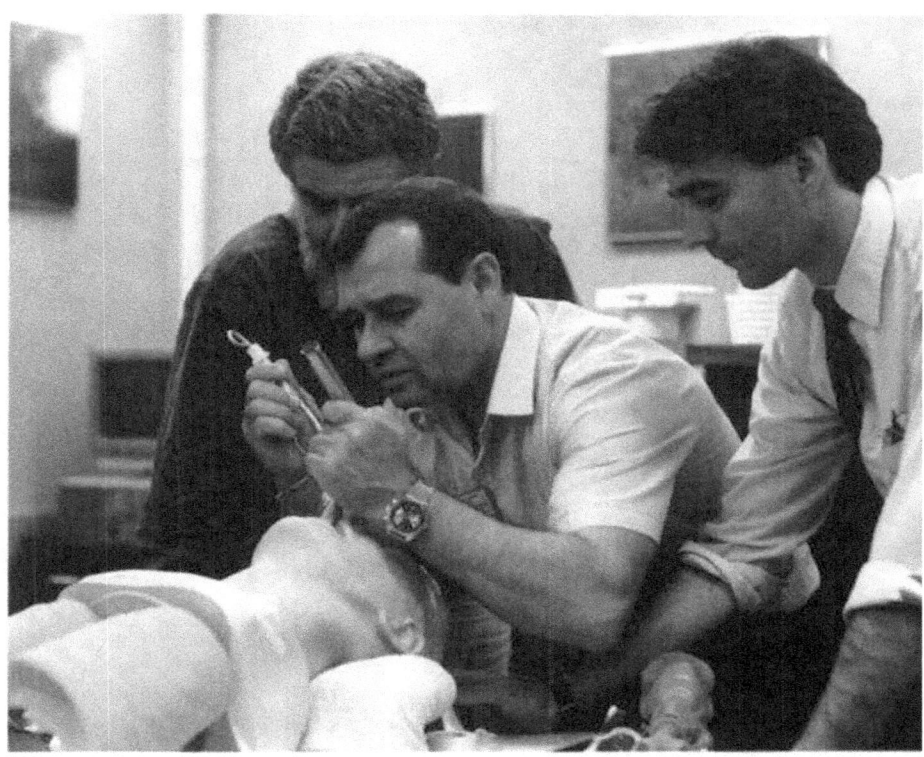

both groups are frequently at the landing site
to assist. "We found," Dr. Pool said about
working with the Russians, "that they didn't
have a single physician—a Flight Surgeon—
who had worked the entire process, from
preflight training to in-flight operations to
postflight rehabilitation. They didn't have
anyone who had that sort of experience
because their efforts were more focused and
limited. So, that is a fairly big difference in
the two systems."

Differences also existed in the ways the
American and Russian space programs certified people for training and spaceflight. Flight
Surgeon Mike Barratt said, "We often had disagreements about whose standards should be
applied to, for instance, a U.S. astronaut training in the Russian program. Each program
had a very large experience base but a very different philosophy. In particular, the Russian
philosophy was more towards functional loading—put a crewmember under a certain situation of low pressure, high temperature, sleep
deprivation, centrifuge training, whatever, and
see how they respond physiologically." NASA's
philosophy was to look more at health factors
instead of fitness. "We would differ quite a bit
on what test result would declare a person
healthy or certified for training or spaceflight,"
Barratt said.

Being a Flight Surgeon for a long-
duration mission was also different from
serving as a Flight Surgeon for a Space
Shuttle mission. It hadn't been done since
Skylab in 1973, and there was no specific
training for it other than Russian language
training. "A lot of it is an attitude and an
approach," said Flight Surgeon Tom
Marshburn, "as opposed to really step-by-
step things. Perhaps you might have to have
worked a long-duration mission to feel things
as deeply as we do, about how much you
have to stay on top of the crew schedule, and
how important it is to make sure that the
crew stays healthy. It's something we have to
be proactive [about] to make sure [it] hap-
pens. And, if we're not [proactive], it may
slip through the cracks."

The Flight Surgeon's role didn't end when
the Mir astronauts landed. Flight Surgeon Flynn
described what came after Wolf's flight. "[He]
had a lot of medical experiments that he was participating in; and, really, for the next 60 days
after landing, he was very busy. I kind of became
his chauffeur, and so I'd go pick him up in the
morning.... We were typically 15 minutes late
getting to wherever we were supposed to go. But,
I'd get to his place early enough so that we'd sort
of take time for him to recuperate" from just getting up and getting going in the morning.
"Returning to gravity is a tough thing. He was
still physically getting used to being back home."

Flynn would help Wolf emotionally while
they drove in to work. "We'd stop and get
doughnuts. Get a cup of coffee. And, we'd just
kind of sit in the car for a few minutes, thinking
about the day and what we'd accomplished the
day before." Thinking later about the roles and
responsibilities of a Flight Surgeon, Flynn said,
"That's a privilege that is kind of the best part
of this job—when you have a crewmember who
will allow you to be close to him and to really
do your job well."

Norm Thagard's Flight Surgeons were
Mike Barratt and Dave Ward. Shannon Lucid
was attended to by Gaylen Johnson. Pat McGinnis
served as physician for both John Blaha and Andy
Thomas. Marshburn worked with Jerry Linenger.
Terry Taddeo attended to Mike Foale. And, Flynn
was Wolf's Flight Surgeon.

Interestingly, three of the U.S. Mir astronauts
were also medical doctors: Thagard, Linenger,
and Wolf. Shuttle-Mir's medical support program
is discussed in depth in the Phase 1 Program
Joint Report on the CD-ROM.

Space Shuttle *Atlantis*

Launched: January 12, 1997, 4:27 a.m. EST
Kennedy Space Center, Pad 39-B

Orbit: 184 nautical miles

Inclination: 51.6 degrees

Landed: January 22, 1997, 9:23 a.m. EST
Kennedy Space Center

Mission: 10 days, 4 hours, 56 minutes

STS-81 CREW

Commander Michael A. Baker
Fourth Shuttle flight

Pilot Brent W. Jett
Second Shuttle flight

Mission Specialist John M. Grunsfeld, Ph.D.
Second Shuttle flight

Mission Specialist Marsha S. Ivins
Second Shuttle flight

Mission Specialist Peter J.K. Wisoff, Ph.D.
Third Shuttle flight

Mission Specialist Jerry M. Linenger, Ph.D., M.D.
Second Shuttle flight; remaining on Mir

Mission Specialist John E. Blaha
Sixth Shuttle flight; returning from Mir

PAYLOADS

Space Habitation (Double) Module

Shuttle Shortwave Amateur Radio Experiment

KidSat Educational Cameras

Treadmill Vibration Isolation and Stabilization System

Biorack Multipurpose Facility

Cosmic Radiation Effects and Activation Monitor

Orbiter Space Vision System

Midcourse Space Experiment

STS-81
Bringing Back the Harvest
January 12 – 22, 1997

Pictured in the Mir Base Block are (front row, left to right) STS-81 Commander Baker; Mission Specialist Grunsfeld; NASA-3 Astronaut Blaha; Mir-22 Flight Engineer Kaleri; (back row) NASA-4 Astronaut Linenger; Mir-22 Commander Korzun; Mission Specialists Ivins, Wisoff; Pilot Jett.

Onboard STS-81 was Jerry Linenger, who would replace U.S. astronaut John Blaha after Blaha's four-month stay aboard the Russian Space Station Mir. This mission, the fifth of nine dockings, transferred the most materials to date. *Atlantis* also brought back to Earth the first plants to complete a life cycle in space—a crop of wheat planted by Shannon Lucid and grown from seed to seed.

Liftoff occurred at the opening of the available launch window at 4:27 a.m. EST. At that moment, Mir was high above the Galápagos Islands, about 2,400 miles southwest of Kennedy Space Center. About 25 minutes after launch, the Mir-22 crew was notified of the Shuttle launch, which they viewed via video uplink. By the next morning, *Atlantis* trailed Mir by 6,000 miles and was catching up to it at about 600 miles with each orbit. In the SPACEHAB double module, the STS-81 crew tested the Shuttle Treadmill Vibration Isolation and Stabilization System, designed for use in the Russian service module of the International Space Station.

Rendezvous and docking occurred on Flight Day 3. At the time of docking, the two spacecraft were about 210 miles above and southeast of

Moscow. The *Atlantis* crew had sighted Mir when they were about 40 miles out. Onboard Mir and nearing the end of his 118-day flight, Blaha did not see *Atlantis* until about eight minutes before docking.

Soon after the Shuttle's docking, an informal greeting ceremony occurred in Mir's core module (Base Block), where the STS-81 crew presented the cosmonauts with new flashlights. Sections of Mir—including the Kristall module—were being kept dark to conserve power. Linenger demonstrated how to hold a flashlight in one's teeth, while using one hand to anchor oneself in microgravity and the other hand to do work.

During five days of mated-spacecraft operations, Commander Michael Baker and Pilot Brent Jett fired the Shuttle's small vernier jet thrusters to gather engineering data for the International Space Station. All nine members of the two crews floated back and forth, hauling materials between the two spacecraft. They transferred nearly 6,000 pounds of logistics to Mir, including 1,600 pounds of water; 1,100 pounds of U.S. science equipment; and 2,200 pounds of Russian logistical equipment. About 2,400 pounds of materials were moved to *Atlantis* from Mir.

Astronaut John Grunsfeld compared Mir to "exploring a cave." At first, a newcomer found it difficult to find the Priroda module, where the crews stowed much of the equipment for transfer onto Mir. Pilot Jett later described going from the docking module to the Kristall module. He said that Flight Engineer Alexander Kaleri "grabbed me and said, 'Okay. Just kind of follow this.' There was a line that went through the Kristall module, and you could kind of use it as a translation aid; but it was also very helpful because there was so much equipment, and at times the passageway got very narrow. The Kristall is kind of like their attic…. They put a lot of extra equipment there … but then, once you get … into the node and then into the Base Block, it's a lot more like what you would expect for a station."

Every evening, Blaha shared with Linenger what he had learned about living and working onboard Mir. Linenger would work with the Mir-22 crew of Commander Valeri Korzun and Flight Engineer Kaleri until the arrival in February of the Mir-23 crew of Commander Vasily Tsibliev, Flight Engineer Aleksandr Lazutkin, and German researcher Reinhold Ewald. (After a brief stay on the station, Ewald returned to Earth with the Mir-22 cosmonauts.)

Mission Specialist John Grunsfeld loads film into a camera used to record experiments.

Pictured above is the double SPACEHAB module, located in the payload bay of *Atlantis*.

SPACEHAB

The SPACEHAB Space Research Laboratory provides a Space Shuttle with extra "shirt-sleeve" space for crew-tended experiments. The single module weighs 10,584 pounds and is 9.2 feet long, 11.2 feet high, and 13.5 feet in diameter. This research facility increases pressurized experiment space in the Orbiter by 1,100 cubic feet, quadrupling the working and storage volume available. A single-module laboratory has a total payload capacity of 3,000 pounds.

SPACEHAB also provides experiments with standard services, such as power, temperature control, and systems functions. Environmental control of the laboratory's interior maintains temperatures between 65°F and 80°F.

During Shuttle-Mir, SPACEHAB single and double modules were used to carry supplies and equipment to Mir. SPACEHAB was located a little past halfway down the Orbiter's payload bay, past the docking system, and was accessed from the Orbiter's mid-deck through a tunnel adapter connected to the airlock.

SPACEHAB was used on seven Shuttle-Mir missions. Its first-ever flight was on STS-57, June 21-27, 1993. NASA leases the modules from SPACEHAB, Inc., of Arlington, Virginia.

Atlantis undocked on January 19, 1997, and performed a fly-around of Mir, 1,000 feet out from the station. Upon deorbiting, *Atlantis* reentered the atmosphere over British Columbia, Canada, to fly across the central United States and then land on a clear Florida morning at Kennedy Space Center.

Later in Houston, Jett said that Linenger "seemed real comfortable over in the Mir. I knew he was going to do a really, really great job. He's a very disciplined person, and I knew he would have a great mission. I was kind of sad to be leaving him … I knew I would see him again and I knew I'd see the cosmonauts again; but, you know … seeing Jerry on the other side of the hatch when we closed it, I was thinking that he is now part of a Russian crew; and he won't be—except for video links and audio links—he won't be really able to talk to his friends…. And, of course, I had no idea that he would go through a very critical situation like he had with the fire."

Jerry Linenger

NASA-4

January 12, 1997 – May 24, 1997

Fire and Controversy

Physician, triathlete, and astronaut Jerry Linenger flew to Mir to do world-class science and to show that a human could do more than endure in microgravity—that he could physically and mentally thrive there.

But, scarcely had his work begun when circumstances changed. A chemical fire altered the whole nature of his mission and served to change the relationship between the U.S. and the Russian space programs. Mir systems breakdowns and problems with communications exacerbated a demanding and difficult situation, both onboard Mir and on the ground. In spite of this, the Mir crews—and the American and Russian ground teams—accomplished the mission's goals, including almost all of the planned U.S. science experiments.

Launching with Space Shuttle mission STS-81 on January 12, 1997, Linenger succeeded NASA-3 Astronaut John Blaha and joined Russian Mir-22 crewmembers, Valeri Korzun and Alexander Kaleri, who would stay onboard Mir until the arrival of the Mir-23 crew. During his duration, Linenger joined his Russian crewmates and became the first American to undock in a Soyuz spacecraft and do a fly-around of Mir. He also became the first American to conduct a spacewalk from a foreign space station and in a non-American spacesuit. Linenger returned to Earth with the STS-84 crew of *Atlantis* on May 24, 1997, after a total of 132 days in orbit—the longest-duration flight of an American male to date.

Curiosity was "what got me here," he wrote in a letter from Mir to his son; but it was more than curiosity that kept Linenger going. Training and dedication played big roles. Born in Eastpointe, Michigan, in 1955, Linenger graduated with a degree in bioscience from the U.S. Naval Academy. He went on to earn a Doctorate in Medicine from Wayne State University, a Master of Science in systems management from the University of Southern California, a Master of Public Health in health policy from the University of North Carolina, and a Ph.D. in epidemiology from the University of North Carolina. His U.S. Navy duties included a stint as medical advisor to the Commander, Naval Air Forces, U.S. Pacific Fleet. Selected for astronaut training in 1993, Linenger made his first Space Shuttle flight on the STS-64 mission in 1994.

Linenger's second Shuttle flight was his STS-81 trip to Mir. While *Atlantis* was still docked, NASA-3 Astronaut Blaha gave Linenger a detailed and personalized

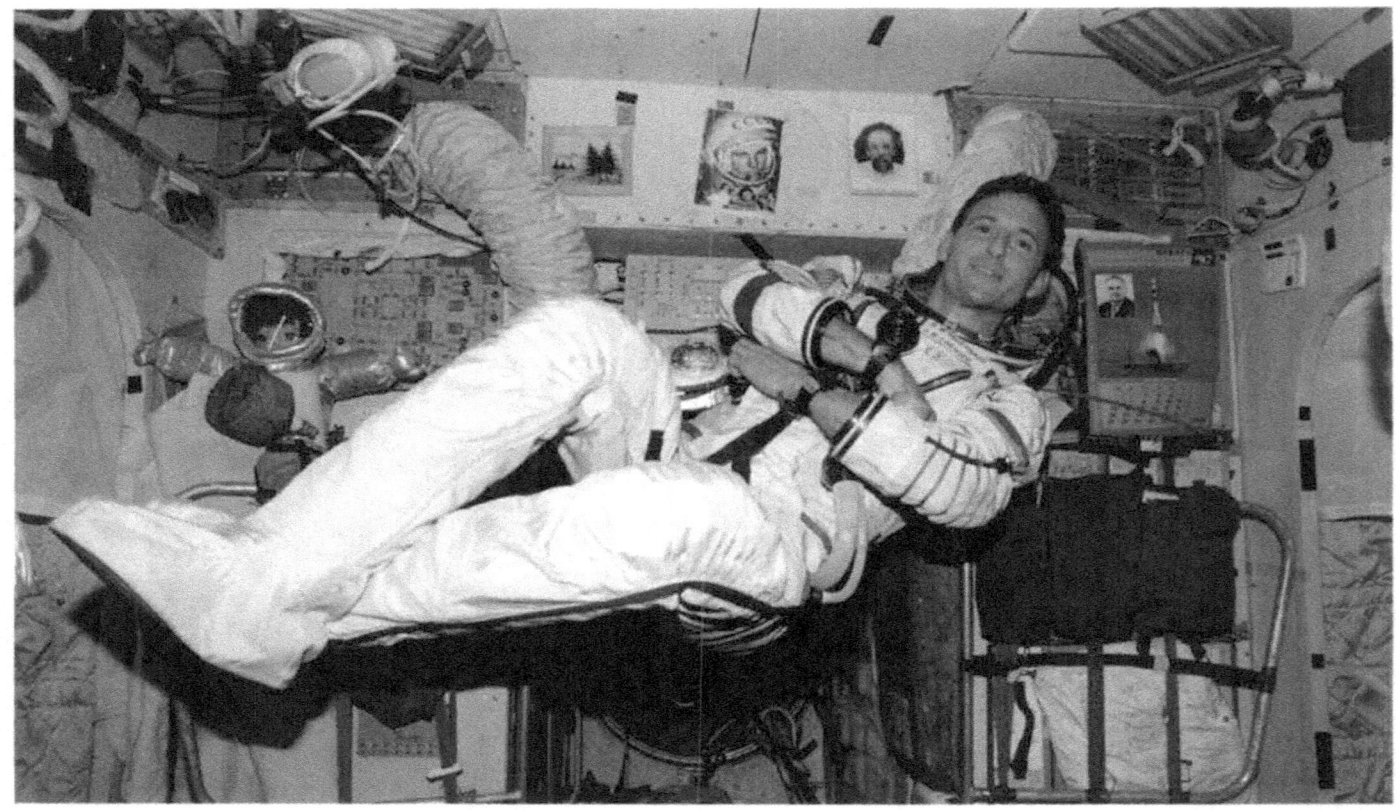

On January 23, ... [Linenger] wrote the first in a series of 72 *Letters to My Son* [that] 'Space is a frontier. And I'm out here exploring. For five months! What a privilege!' His experiences would prove to be a considerable challenge as well.

"handover" of the tough assignment. This included Blaha's insider knowledge about Mir's systems and quirks, and about the psychological stamina the job would require. The two also packed up the greenhouse experiment for return to Earth with the first plants to complete a life cycle in space—a crop of wheat planted by Shannon Lucid.

Before Linenger could become an official member of the Mir-23 crew, he had to try on and test his Sokol pressurized spacesuit, which would be needed for any emergency evacuation of Mir. STS-81 Mission Specialist John Grunsfeld described the suit-up in a dispatch he sent down from orbit: "The suit is a tight-fitting pressure bladder with a helmet, and Jerry had to squeeze and squirm, with me pushing to get him into it. Compared to our orange launch-and-entry suits, the Russian equivalent seems flimsy and delicate, but it is much lighter. We had to take care not to catch the suit on any sharp edges as we made our way to the Soyuz."

The suit-up worked, and Linenger's custom-made seat-liner was installed in the Soyuz. He signed into the space station's log as an official crewmember. His adventure was beginning. On January 23, before *Atlantis* undocked, he wrote the first in a series of 72 "Letters to My Son" in which he would chronicle his adventures onboard Mir. "Space is a frontier. And I'm out here exploring. For five months! What a privilege!" His experiences would prove to be a considerable challenge as well.

After the Space Shuttle pulled away, the Mir crew took a day off to relax and unpack some of the materials transferred from *Atlantis*. Linenger told flight controllers that unpacking his many boxes of gear was like opening Christmas presents.

He had created quarters for himself in the Spektr module; and he began his regular daily exercise regimen, which included two 1-hour exercise sessions on a treadmill and a stationary bicycle. He also started to work on some of the life sciences and medical investigations, including replacing the

Jerry M. Linenger

Jerry Linenger was born in Eastpointe, Michigan, and chose to attend the U.S. Naval Academy as much to help his family with expenses as to prepare for a career as an astronaut. He holds one bachelor's degree in bioscience, two master's degrees—in systems management and health policy—and two doctorates—in medicine and epidemiology. Among his jobs as a Navy doctor was an assignment as medical advisor to the Commander, Naval Air Forces, U.S. Pacific Fleet. Linenger became an astronaut in 1993. Two years later, he flew on the STS-64 Shuttle mission, which included the first use of lasers for environmental research.

While on Mir, Linenger wrote many "Letters to My Son." His first letter ended with:

"You know, although I am up here floating above the Earth, I am still an Earthling. I feel the pain of separation, the pride of a father, and the loneliness of a husband away from his wife like an Earthling. And maybe even a bit more acutely."

Linenger and his wife had their second son soon after Linenger returned to Earth from Mir. Leaving NASA soon after his Mir experience, Linenger and his family moved to Michigan. He has since published a book about his experiences, titled *Off the Planet: Surviving Five Perilous Months Aboard the Space Station Mir.*

Fourth U.S. astronaut to live onboard Mir • Experienced an onboard fire • First American to spacewalk in a Russian suit
First American to perform a space station "fly-around" in a Soyuz spacecraft • Set American space endurance record for men
Launched with STS-81 on January 12, 1997 • Returned with STS-84 on May 24, 1997

radiation dosimeters that had been returned to Earth on *Atlantis.*

The science laid out before Linenger entailed experiments in Earth sciences, biology, human life sciences, microgravity, space sciences, and risk mitigation experiments for the International Space Station. In an interview before his launch held at the Gagarin Cosmonaut Training Center in Star City, Linenger discussed the way he was approaching his science.

"I am a physician by background," he said, "so the life science things have come pretty easy. The more physics sorts of things—that's something that I have had to learn. It's like going back to school again and learning some of the basic science behind it, and then getting into the actual mechanics of how to carry out the experiment." From Mir, Linenger wrote to his son, "I try my best. It's important stuff, and I have to be careful to not make a mistake, like turn the wrong switch at the wrong time, or the experiment would be ruined. Two laboratories (actually 13-meter long 'tubes'—named Priroda and Spektr) full of equipment. Pretty complex. And basically a one-person show."

Linenger's first science included working in the Priroda module on a glovebox facility that would provide an airtight, contamination-free work area. He also worked with the balky Biotechnology Systems hardware that had given Blaha problems.

Linenger was settling in onboard Mir. At the time, he wrote to his son: "Let me tell you about my house. Spectacular view. Unobstructed, overlooking the oceans, the lakes, the rivers; the mountains, plains, and valleys; the city lights, the stars, the other planets. Six modules. One toilet. Dining area with two private sleep stations. Three vessel garage: Soyuz, Shuttle, and supply/garbage truck [Progress]. Each module a 13-meter tube. Lots of extras. Two modules are new additions. State-of-the-art freezers, computers, gas analyzers. Built-in treadmills and bicycles for the recreational enthusiast. Utilities: completely solar-powered. Water from tanks, urine, and condensate. Oxygen included. Radio, ham radio, and telemetry."

On February 6, three weeks into Linenger's residency, the Progress vehicle undocked, full of

garbage, to burn up in the atmosphere. After the undocking, Linenger described how it felt from inside Mir. "I felt and heard the springs pushing it away. Looked out a tiny window by the hatch and saw its three lights backing away. Stable and slow. Then the thrusters [fired] …"

The next day, Linenger and his crewmates took the Soyuz on a 27-minute "fly-around" of Mir. Later, Linenger gave an impressionistic account of how it felt from within Soyuz. "Smooth, yet firm, push-off. Spring-action. An ink pen floating forward. Then the thrusters firing. Not like an explosion, more like low growls. Short. Repeated….

"Out the window, the Earth spinning by, and the flashes of the thrusters. The space station docking port moving away. A view of module Priroda out my window. Then the whole station—all six cylinders …

"Strapped in. Crouched with my knees almost to my chest. Spacesuit on. Ventilators humming and feeling the air trickle out inside my suit. Control panel in my face. Spinning miniature globe in a glass case. Caution-and-warning lights. Operation manuals written

in Cyrillic. And feeling like we are moving, flying …

"On [Mir], you fly around inside—but you don't feel like the station is flying. Especially if you don't look outside. But in the [Soyuz], it feels like a car or [an] airplane or [a] jet—sitting in a cockpit, and flying. The Earth spins below. The space station changes position outside the window. And you feel the gentle thrust …

"The docking. Feel and hear a thud. Feel your spaceship being yanked around a bit. Glad when the pressure inside holds. Glad to open the door again."

Linenger described an interesting perception of the very odor of space that he detected in Mir's airlock. He called it "a distinct, burnt-dry smell." He added a poetic description of his home in space. "I could see a lone ray of light shining through the port window and outlining the dining table. We had left some food out for dinner. It was the only time during my stay in space that Mir looked warm, inviting, and spacious. It reminded me of opening the door to a summer cottage that had been boarded up for the winter, looking inside, and seeing familiar surroundings."

Mir was even more familiar to Linenger's crewmates. At this point, Korzun and Kaleri had been onboard for 172 days. Linenger had been in space for 27 days. The continuous American presence in space was now well past 300 days.

By February 10, company was coming. Soyuz TM-25 launched from Baikonur, carrying the Mir-23 crew of Vasily Tsibliev and Aleksandr Lazutkin, and German Astronaut Reinhold Ewald. Ewald would spend 20 days on Mir, performing experiments, and would return to Earth with the Mir-22 crew. On February 11, NASA's Space Shuttle *Discovery* (STS-82) was also in space, with seven astronauts onboard, on a mission unrelated to the Shuttle-Mir Program. The resulting 13 people in Earth orbit tied a record for the most humans in orbit at one time.

Outer space was—in a sense—full, and so was Linenger's time. He reported he had not been able to satisfy his interest in geography. "There is no free time to just go hang out and look out the window," Linenger said in an interview. "I've only done it maybe once a day. I am not sure five months is going to be long enough up here." The views he did manage to get proved

spectacular. He wrote: "Today I saw huge dust storms in the Sahara of Africa. Lake Chad drying up. Five minutes later: the Nile, the triangle of the Sinai Peninsula, and the Red Sea all in one view. Then, Elbrus and the snow-covered Caucasus." Also, his physical exercise program was now coming around. He reported: "The first couple of weeks, it was very, very difficult to run on the treadmill. It is a lot tougher than I thought it would be, but now I kind of feel I am back to my old pace and I feel real good."

The Soyuz TM-25 arrived on February 12, looking like "a stout, winged insect." The new crew brought with them a treat for the old crew—fresh fruit! Apples, bananas, lemons, and oranges. Linenger thought the fresh, earthy citrus smell meant even more to him than the taste. He also wrote to his son and described how the new crew was finding its space legs. "Sasha [Kaleri], Valeri, and I have never collided. Never even touched. I'm sure that they kept clear of my flight path early on, but now we just glide by one another with the greatest of ease. Personal space protected. Everything in place. Huge difference now. Pencils and cameras and gear a-flyin'. People propelling themselves off each other.... Cords and cables along the way getting pulled out. But already, I've observed this lessening as the new guys adapt and learn the terrain."

Mir was crowded now with six humans onboard. To supplement the Elektron oxygen-generation system, they used lithium-perchlorate canisters, which generated oxygen through a chem-

"We immediately started fighting that fire…. You had to react … to keep your head about you, …"

ical reaction. Meanwhile, Korzun and Kaleri packed their belongings for the trip home to Earth with the German Ewald. Linenger started a human life sciences investigation into immune system alterations in relation to sleep in microgravity. This required him to wear electrode sensors while he slept and provoked him to quip that he was always working, even in his dreams.

During this handover period, his waking nightmare struck. On the evening of February 24, all hands were in the Base Block finishing a dinner of Russian dishes such as jellied fish and borscht. "There was also caviar," according to Mir-23 Flight Engineer Lazutkin, "red caviar, which we brought along ourselves. You didn't get it in your rations. It's too expensive, but we treated ourselves. Let's say it was a festive evening, and all six of us were sitting around the table."

Linenger excused himself to go to the Spektr module to do some work. At about the same time, Lazutkin went into the tunnel to the Kvant-1

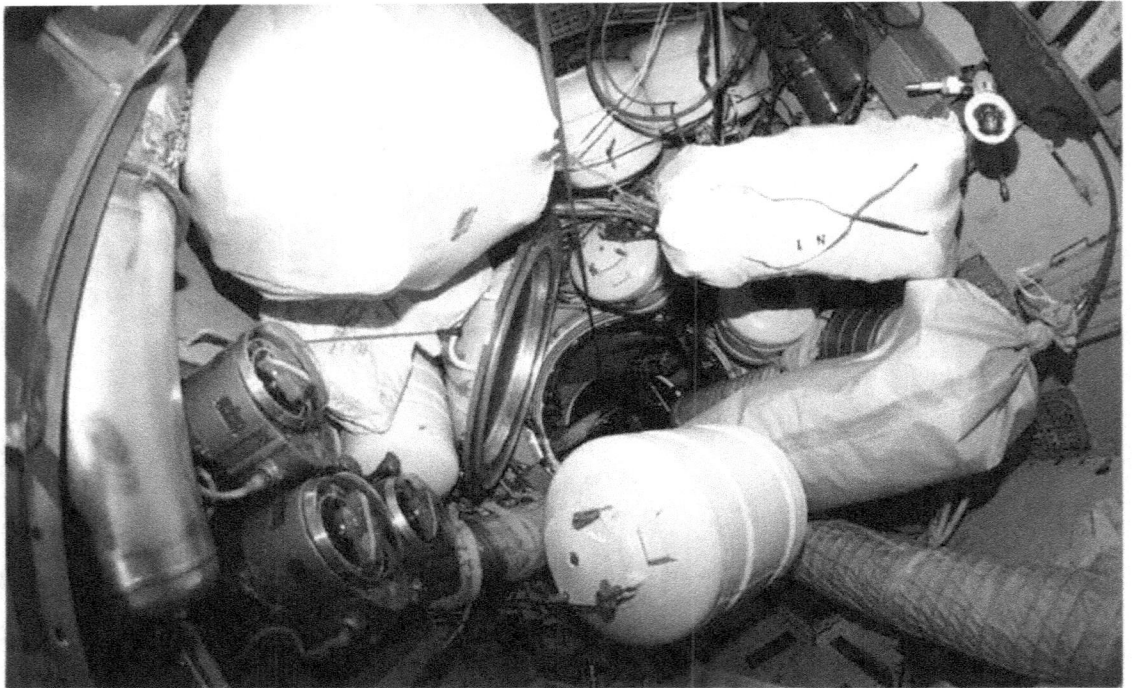

module to "burn" another oxygen-generating canister. In a complete surprise, the normally slowly combusting chemicals erupted into searing flame.

In Spektr, Linenger heard the master alarm go off. But that was not, in itself, alarming to him. The same alarm was also used to wake up the crew every morning and would sound for many reasons—both serious and mundane. Linenger's first reaction was to put on some hearing protection against the noise and to quickly save his computer data in case the power went out. He then flew toward the Base Block to see what was happening, and he noticed the first tendrils of smoke.

"We immediately started fighting that fire," he reported later from Mir. "You had to react to the situation, you had to keep your head about you, so I guess it was just a matter of survival. Going through your mind were thoughts—'We need to get that fire out.'"

The situation was critical. The fire blocked the only path between the crew and one of the two Soyuz vehicles. At the moment, there was a way of escape for only three of the six men onboard the station.

"The smoke was the most surprising thing to me," Linenger reported. "I did not expect smoke to spread so quickly." In microgravity, there are no convection currents. But, fans on Mir continually circulate the air; and this smoke "was a magnitude about ten times faster than I would expect a fire

to spread on a space station. The smoke was immediate. It was dense … I could see the five fingers on my hand, I could see a shadowy figure of the person in front of me who I was trying to monitor to make sure he was doing okay, but I really could not make him out. Where he was standing he could not see his hands in front of his face. In the distant modules at the very end of the cones, the smoke was still dense, so it was very surprising how fast and rapidly the smoke spread throughout the complex."

The fire resembled a box full of fireworks sparklers, all burning at once. The flame shot out about two to three feet in length, with bright bits of molten metal "flying across and splattering on the other bulkhead." The canister provided the fire with both oxygen and fuel. "It had everything it needed," Linenger said later.

Lazutkin said, "When I saw the ship was full of smoke, my natural reaction was to want to open a window. And then, I was truly afraid for the first time. You can't escape the smoke. You can't just open a window to ventilate the room."

The crew immediately began putting on oxygen masks. Linenger's first mask failed to activate quickly enough, so he grabbed another one. He said that in the time between trying the two masks, the space station's atmosphere became unbreathable. "I did not inhale anything, and I don't think anyone else did because the

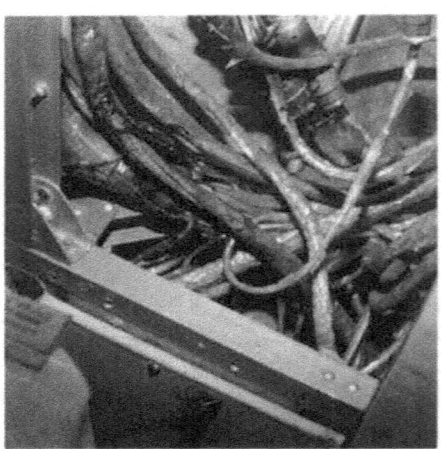

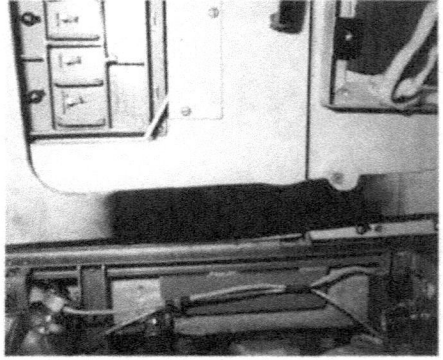

thickness of the smoke told you that you could not breathe. So, everyone immediately went to the oxygen ventilators. They worked very [well], and they protected us from inhalation injury."

Mir-22 Commander Korzun was in charge of the station. He ordered the one accessible Soyuz escape vehicle readied for evacuation. Now, they had to subdue the fire. Korzun faced the fire, using Mir fire extinguishers. He later said, "When I started spraying foam on the hot canister, the foam didn't stick and had little effect. So, I switched to water and started using that." The water turned to steam, adding to the smoke.

Throughout this ordeal, physician Linenger stayed with Korzun. He passed Korzun fresh fire extinguishers and kept monitoring his level of consciousness. Also, to stabilize Korzun in the absence of gravity, Linenger wedged his own legs into Mir's connecting tunnel and held on to Korzun's legs. "At one point," Linenger said, "I floated in front of his face, but the smoke was too dense, even at six inches [to see how he was doing]. So, I resorted to tugging at him when I could not be assured that he was still okay."

Korzun recalled later, "Jerry kept tugging my leg. 'Valeri, how do you feel?'" Linenger also kept an eye on his other crewmates. He was impressed with Kaleri's cool-headedness. During the crisis, Kaleri calmly worked at a computer, printing out reentry information for both of the Soyuz vehicles.

The oxygen canister eventually burned itself out; but smoke remained everywhere, even "in the distant modules at the very end of the cones." It was now about 100°F inside the Kvant-1 module. And, it was dark with smoke and soot. According to Lazutkin, "We even thought someone had switched the lights out in Kvant. That's how black it was."

The fire had destroyed the canister itself, as well as the panel covering the device. The crew also reported that the outer insulation on several cables was melted by the heat. Fortunately, all Mir systems continued to operate normally.

Shortly after the incident, Linenger reported from Mir, "Being a physician, I was very concerned with crew health. We set up a station for any respiratory problems that might take place. We had all the emergency gear in place. I did exams on all the crewmembers immediately following the fire, and then for 24 and 48 hours after that. I looked at oxygen saturation in the blood, checked the lungs—all the normal things you would do post-fire. From my assessment, I [didn't] see where

anyone had any serious [smoke] inhalation damage, and it was due to good action by the crew to get into the oxygen masks quickly."

The two crews went back to work. Linenger completed the first part of the sleep investigation on March 1. On March 2, the Mir-22 crew returned to Earth. Korzun and Kaleri had clocked 197 days in space. The new Mir-23 Commander, Tsibliev, had already spent 200 days on Mir during an earlier expedition. This mission was Lazutkin's first time in space. Linenger still had nearly three months left.

Trouble occurred again almost immediately. During the first week in March, an Elektron unit failed. The unit, located in Mir's Kvant-2 module, separates oxygen from the onboard wastewater and returns the oxygen to the cabin atmosphere using electrolysis. Russian ground controllers asked the crew to attempt to activate a second Elektron unit located in the Kvant-1 module. The crew succeeded in starting up that unit, but the unit was producing higher levels of hydrogen than it should. Controllers told the crew to shut down the system and to go back to the solid-fuel, oxygen-generating devices. There were still 200

Mir-22 Commander Valeri Korzun and Flight Engineer Alexander Kaleri are pictured above with the Protein Crystal Growth GN₂ Dewar freezer, Base Block module, prior to the incident. During the fire, Korzun (foreground) took charge, using the fire extinguisher while ordering the Soyuz capsule prepared for escape. Kaleri (background) calmly printed reentry information from the computer. Although six men were onboard, they had access to only one escape vehicle that had a capacity of three riders.

of these onboard—a two-month supply—and thousands had been successfully used during Mir's lifetime. Linenger and his crewmates did not object, but they made sure they always had a clear path to the Soyuz in the event of another fire.

The crew also had problems with the Progress-233 resupply vehicle, which had been undocked from the station on February 6 and placed in a stationkeeping position away from Mir. The Progress could no longer redock to the outpost as planned due to problems with the remotely operated rendezvous system. The next test of the rendezvous system was planned for the next Progress resupply vehicle, scheduled to arrive at Mir on April 8.

Being in space for more than 100 days provided Linenger opportunities to witness unique sights, including the Hale-Bopp Comet (pictured above).

> Going without power on Mir was
> "... unearthly dark.... Dark is not
> even the proper word for it."

During early March, Linenger worked on the Human Life Sciences Humoral Immunity Investigation, which involved taking blood samples to study the immune cells in the human body. He also began the Microgravity Opposed Flame Flow Spread Experiment to study flames in microgravity. He wrote, "We've got some great experiments.... We're way ahead on the power curve as far as that goes. For example, we did a flame experiment inside a glovebox, a very controlled situation looking at ventilation and how it affects flame spread." The cabin fire in late February possibly added to Linenger's interest in this. He wrote, "I realize that this is important work I'm doing up here. I am glad to be doing it and I am very preoccupied with my work."

The science was very important to Linenger. In a letter to his son, he wrote again about the flame experiment and commented on the value of having human researchers doing the science. "I'm not only getting through this experiment, but I'm

expanding it a bit, based on my real-time observations.... A programmed machine can't do that. You need a trained human observer, a scientist, to do that. Human observation and intervention give us better data, better understanding."

In another letter, he compared his situation with that of Antarctic explorers of about a century earlier. "Compared to what they endured, the space station is a five-star hotel. They ate seal and penguin meat, day after day after day. I get Russian-American freeze-dried cuisine—shrimp cocktail, veggies, [borscht]—the whole spread. They had a lightless winter. I get light and dark every 45 minutes. They wore the same clothes for over a year; I get a fresh T-shirt and pair of shorts every week. They had to trudge through uneven, unstable terrain; whereas I float effortlessly."

Still more technical problems came in mid-March as Linenger was passing the halfway point in his stay onboard Mir. The space station's orientation system broke down. A sensor in the Spektr module failed, prompting the motion control computer to switch to a backup system. During the three-minute swap-over, all attitude control was lost. The crew placed the station in what is called "free drift" and then used onboard thruster jets to stabilize its attitude. For most of that day, Mir remained in a "gravity gradient," which basically means that the most massive part of Mir naturally pointed toward Earth. Because this attitude did not keep the solar arrays pointed at

the Sun, the crew turned off the gyrodynes and other equipment to conserve power. Late in the day, flight controllers uplinked a new attitude maneuver to the motion control system computer, and the crew restarted the gyrodynes. The station's primary attitude sensor, called Omega, was still inoperable; so control was managed by a backup unit until the crew installed a spare Omega sensor and rerouted cables.

Linenger described, in a letter, the experience of going without power on Mir. "Last night it got really, really, really dark in my room, module Spektr. Lost all power. I've been in dark places before, but this was unearthly dark. Darker than any dark I've ever seen. Dark is not even the proper word for it.

"Of course, I couldn't hang out in the quiet room. No ventilators working means no air circulation. Warm air doesn't rise in space (which way is up?); there is no natural convection. No wind, no breeze; without the ventilators, only stillness."

In late March, Linenger became the fourth most experienced U.S. astronaut when his total flight time surpassed that of the 1974 Skylab-4 crew of 84 days in space. The only U.S. astronauts with more time in space were his Shuttle-Mir predecessors—Norm Thagard, Lucid, and Blaha. In a radio interview, Linenger talked about his experience so far:

From a controlled environment, Linenger studied the effects of microgravity on fire. As seen below, candle flames in a "glovebox" on the Priroda are rounded—not pointed as on Earth. (See "Combustion Experiments on Mir" and Summary at RB-RBM.)

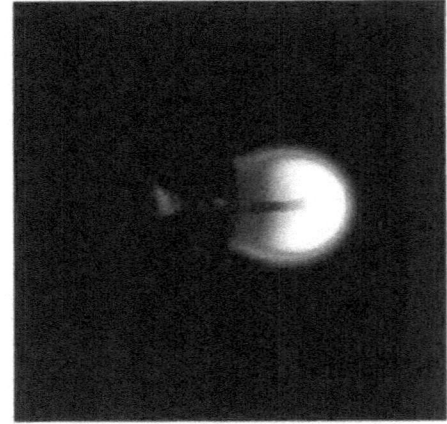

"[Mir] is a small place, but space is an amazing place to be. Just today I looked out the window ... and saw [the comet] Hale-Bopp. It looked like a flashlight in the sky, and then I looked to the north and saw the Northern Lights flickering green explosions off the northern horizon of the Earth. Then, I saw the sunrise. Moments like that lift your spirits. Loneliness and things that you might think would be very tough to bear up here get kind of mellowed out by things like that. The adventure of being in space is enough to get you through it, and I really have no difficult problem with that up here."

By early April, the Mir-23 crew had used about 70 of the oxygen-generating "candles" since the fire. They had about 130 remaining and were using about three a day. Also, Russian flight controllers had detected a leak in one of the Kvant-2 module's cooling loops. The Mir crew worked on the leak while the station's orientation was altered so that Kvant-2 was kept in the shade of other structures. Because of coolant loop problems, a Vozdukh carbon dioxide removal unit shut down. Carbon dioxide removal now had to be performed by lithium-hydroxide canisters.

The crew was now awaiting the launch of the next Progress resupply vehicle from the Baikonur Cosmodrome in Kazakhstan. The Progress was bringing repair equipment for the Mir's Elektron oxygen-generating system, additional oxygen-generating candles, and extra repair gear for the station's cooling loops as well as routine supplies of food, equipment, and personal effects for the crew. Also, new spacesuits

necessary for environmental control were Vozdukh carbon dioxide removal units (below).

for a planned spacewalk by Tsibliev and Linenger on April 29 were stored aboard the resupply ship, as were replacement lithium-hydroxide canisters, gas masks, and fire extinguishers.

The crew kept working on their science investigations. An unknown hardware failure of the orientation experiment resulted in the only cancellation of an experiment for Linenger. That experiment was to have used French equipment to examine changes to human sensory functions during spaceflight, as well as the subsequent effects experienced during a cosmonaut's readaptation to gravity.

On April 8, Progress-M 34 docked. Mir now had enough solid-fuel oxygen generators to provide several months of backup oxygen for the station, plus several weeks of gaseous oxygen stored in tanks. Mir had 14 new lithium-hydroxide canisters that could remove carbon dioxide from the air for 21 days if needed as a backup for the Vozdukh carbon dioxide scrubber.

The crew was now constantly doing repair work. Most of the science work was suspended for that week so Linenger could help his crewmates. The crew's exercise schedule was reduced to one hour per day. By week's end, they got the Vozdukh system restarted, lowering carbon dioxide levels onboard. It was not a perfect fix, but additional hardware was scheduled for delivery in May by STS-84. Tsibliev and Lazutkin also repaired the cooling loop in Kvant-2 and were able to duct cooler air into the Base Block. Lazutkin tried to repair the Elektron oxygen-generating system, but he did not succeed.

Besides causing cooling problems, the coolant loop leaks also allowed ethylene glycol to escape into the station. Linenger talked about the situation in an interview. "The ethylene glycol caused the concern that I have as crew physician: inhaling ethylene glycol. We have some respirator filter masks that we wear when we're doing the repairs in Kvant, so that lessens some of the effect. We're having some congestion—secondary I'm sure—due to some of the fumes. Temperature, of course, not only affects us but also affects some of the hardware onboard and the ability to remove the moisture from the air. You need a cooling loop for the condenser to work to gather the condensate, so the humidity is up also, which again is not the best thing for equipment. It gets very complex very quickly, and we need to fix our problems."

The environment on Mir was hot, messy, and uncomfortable by Linenger's 90th day in

space. He said he felt fine physically and also felt safe onboard despite the difficulties the crew had to deal with. He was asked if his mission to the Mir had turned out the way he expected.

"Not really," he said, "although we're out here in the frontier and I guess I expected the unexpected. We've been getting some of that. As far as the science return, it is what I expected.... We've been running more metal samples and things like that than we thought we'd be able to do. So, in spite of some of the difficulties, we've been having a very successful mission. And, some of the system

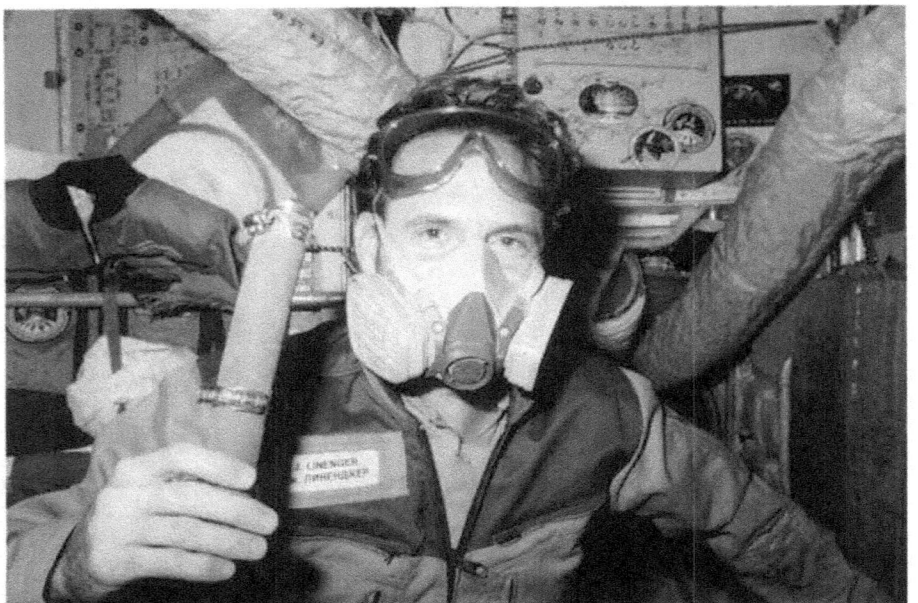

problems—I can't say that I expected them. But, ... I was trained to work on those systems and assist the crew where I could."

In mid-April, Linenger and Tsibliev began preliminary preparations for a five-hour spacewalk to retrieve micrometeorite detection packages located on Mir's docking module. This would be Linenger's first extravehicular activity, as well as the first spacewalk performed by an American from a non-American spacecraft who was wearing a non-American spacesuit.

The crew had gotten the Elektron system working, but oxygen production was insufficient so they were using supplemental oxygen from the Progress vehicle as required. They also had the Vozdukh carbon dioxide scrubber working again, but they had been unsuccessful in finding leaks in the coolant loops. And, they were now having trouble with condensation buildups. They had discovered great globs of water located behind panels in the Kvant-1 module. In microgravity, even leaks act differently than they do on Earth.

Towards the end of April, the Mir-23 crew worked at their repairs. They kept the space station in an attitude that shaded the Base Block, although that diminished power from the solar arrays. They were allowed to resume their normal two-hours-a-day exercise schedule. Linenger passed his 100th day in orbit, and the Shuttle-Mir Program marked

the 400th consecutive day in which an American had been in space. Tsibliev and Linenger prepared for their extravehicular activity.

Both spacewalkers would wear new versions of the Russian Orlan spacesuit. Their first task would be to install the optical properties monitor to the station's docking module. One of the extravehicular activity's exercises would make use

> Fearing the very real possibility of collision, instinct told me to brace for impact. I gritted my teeth, held my breath, and hoped for a miss.

of the Strela crane, which Mike Foale, future NASA-5 astronaut, described at the time. Foale said, "It's basically a long, telescoping tube just like the antenna on a radio. It can telescope in and out, and you can change the telescoping with your hands. It allows one person to clamber along it using the handholds on it to the base. Vasily will clamber down there, and he'll then tell Jerry to hold on, attach the big suitcase, and then Vasily will crank it and he'll move the whole tube over through the 90 degrees while Jerry is just floating—90 feet away from the rest of the Mir—just on the end of this pole. It

bounces around, too. It's pretty 'whippy'.... Jerry holds on—holds fast. And, he ties off the crane. At this point, Vasily makes his way along it to join Jerry; and then he and Jerry will install the optical properties monitor on the docking module."

On April 29, Tsibliev and Linenger conducted their five-hour spacewalk. They tested the new spacesuits, installed the optical properties monitor and a radiation dosimeter, and retrieved several externally mounted material-exposure panels. Linenger has said that the spacewalk, such an amazing experience, was probably the main memory he would take away from his time onboard Mir. (See "It's all downhill from here"; page 97.)

The Mir-23 crewmembers spent the day after the extravehicular activity resting and stowing their spacesuits and the articles retrieved from the exterior of the Mir. They then resumed more routine activities, including an effort to isolate a leak in a cooling loop in the Kvant-1 module. The Russian flight control team reported the Mir's oxygen-generation and carbon dioxide removal systems were operating normally. The ground controllers also asked Mir Commander Tsibliev to try out a new way of docking a resupply vehicle, by practicing on the garbage-filled Progress that had recently undocked from the station.

Up to this point, the Russians had used a Kurs ("course") automated docking system, with a manual backup remotely operated rendezvous system operated by Mir's commander. But, the Kurs system was expensive and heavy. To save weight and cost, the Russians wanted to try the manual system as the primary way to dock a Progress. However, on March 4, when ground controllers sent the Progress toward Mir, Tsibliev was never able to gain control of the vehicle. The Progress narrowly missed hitting Mir.

Linenger described the moment. "I flew to the window that faced the same general direction as the window Sasha [Lazutkin] and Vasily were using, and did so just in time to see the Progress go screaming by us. Fearing the very real possibility of collision, instinct told me to brace for impact. I gritted my teeth, held my breath, and hoped for a miss. Although the Progress had disappeared from view under the edge of the window, I quickly calculated that, having felt nothing, the Progress must have missed hitting the Base Block."

A similar, failed maneuver would indeed result in a collision—and a crisis—during the next Shuttle-Mir increment.

Linenger's increment was nearing its end. The crew spent the first half of May finishing his science investigations, tidying Mir, and making repairs. They fixed a urine processing system, sopped condensation, and repaired a condensate recovery system. To get ready for the arrival of the STS-84 crew, they checked the Elektron oxygen system and tapped into the oxygen supplies of the docked Progress vehicle. They kept looking for the small coolant leaks with no success.

On May 9, Linenger summed up his stay onboard Mir. He said, "[There are] two sets of difficulties we've had. One is the human difficulty of dealing with those things, and the other one is the space station itself. We've overcome all the difficulties. The ultimate test is we're still alive and well, we're all here exploring the frontier. On the other hand, it takes a lot of work, it takes daily attention, and it takes a lot of work from smart people on the ground looking over our shoulders and giving some guidance along the way. But, we were able to overcome about as much difficulty as you can imagine."

The crew of STS-84 on the Space Shuttle *Atlantis* arrived May 16, 1997, with NASA-5 Astronaut Foale. On May 24, *Atlantis* landed at Kennedy Space Center. Linenger had spent 132 days in orbit, the longest space mission of an American male to that date.

It's all downhill from here

April 30, 1997

All spacewalks are different; and a spacewalk on the surface of a sprawling space station has a different flavor than one conducted inside the cargo bay of the Space Shuttle, than one dangling outside a capsule attached to an umbilical, than one rambling on the Moon's surface.

Imagine this. You are in scuba gear. Your vision is restricted by the size of your underwater mask. Your fins, wetsuit, and gloves make you clumsy and heavy. The water is frigid; in fact it is thickly frozen overhead with only one entry-exit hole drilled. Your life depends on your gear functioning properly the entire time. The farther away you venture, the farther away the escape hole in the ice, and the less you can tolerate any failure whatsoever.

There is no bottom. Up and down are confused. Your path is not straight, but rather [is] around obstacles on a constantly convex, falling away, prime surface. As you round one obstacle, the next appears, and soon enough it is difficult to determine from where you came.

You are not in water, but on a cliff. Crawling, slithering, gripping, reaching. You are not falling from the cliff; instead, the whole cliff is falling and you are on it. You convince yourself that it is okay for the cliff and yourself on the cliff to be falling because when you look out you see no bottom. You just fall and fall and fall.

The Sun sets swiftly. Blackness. Not merely dark, but absolute black. You see nothing. Nothing. You [grip] the handhold ever more tightly. You convince yourself that it is okay to be falling, alone, nowhere, in the blackness. You loosen your grip.

Your eyes adjust, and you can make out forms. Another human being silhouetted against the heavens. When it first got dark, you were feet first falling. Five minutes later, as the cliff itself rotates, you feel as if you have reached the crest of the roller-coaster, and are now barreling down steeply—steeply to the point that you feel you will flip headfirst out of your seat—toward Earth. You come out of your seat, and are falling spread-eagle. Now headfirst. You want to flip back upright. You can't. You decide it is okay to be diving headfirst into nothing.

You need to work with your hands. You let go. You depend on the two tethers you placed on handholds to hold [you]. You rotate, twist, and float—all randomly and uncontrolled—still the cliff is falling and rotating. You know you are falling with it, you tell yourself surely you are falling with it because you just attached your tethers; yet it is difficult to discount the sensation that you are moving away, alone, detached. You feel as if you are at the end of a fishing pole, which gets longer and longer and thinner and thinner at the end, and you [are] the fish hooked to its flimsy end. It sways back and forth; you, being attached, sway back and forth [with it]. The pole no longer looks rigid and straight, but rather like a skinny S-curve. You are hanging to the thinnest limb of the tallest tree in the wind. The tree is falling. You convince yourself that it is a strong oak; that the limb will stay attached and not fracture, and that the forest bottom is far away.

In the midst of all of this, you carry out your work calmly, methodically. You snap a picture or two, and below notice the Straits of Gibraltar narrowly opening to the Mediterranean.

That is how it felt, best as I can describe it.

Jerry

Working Outside

Russian cosmonauts performed spacewalks or extravehicular activities (EVAs) during all of the Shuttle-Mir increments. While performing EVAs, astronauts and cosmonauts do not "walk" in space. They are actually in their own orbits around the Earth, and their spacesuits are, in effect, miniature spacecraft. EVA suits protect spacewalkers from the dangers of space—including heat, cold, and vacuum—for more than six hours at a time.

NASA's extravehicular maneuvering unit (EMU) suit is more flexible and more complicated than the Russian Orlan suit, and it is harder to get in and out of for astronauts. An astronaut first puts on the EMU pants and then rises into the top. The Orlan suit has a door in back—sometimes compared to a refrigerator door—through which a cosmonaut can easily enter. The Russian spacesuits fit roughly between the 40th and 60th percentile of the Russian population, and cosmonaut candidates must fit these size limits. NASA spacesuits fit between the 5th and 95th percentile of American adults, allowing NASA to select astronauts from a wide variety of people.

During STS-76, Astronauts Rich Clifford and Linda Godwin conducted the first American EVA outside Mir. During the Shuttle-Mir Program, four joint spacewalks occurred—NASA-4 Astronaut Jerry Linenger with Mir-23 Commander Vasily Tsibliev; NASA-5 Astronaut Mike Foale with Mir-24 Commander Anatoly Solovyev; NASA-6 Astronaut David Wolf also with Solovyev; and during STS-86, when crewmates Scott Parazynski and Vladimir Titov wore American EMU suits for their joint EVA.

(Photo left) Tsibliev performs tasks at one end of the Strela pole and is backdropped by Mir's Base Block and Kvant modules. (Page 99, bottom left photo) Linenger and Tsibliev are shown in cooling garment with headgear that serves as the suit underlayer needed to maintain appropriate body temperatures. Tsibliev and Linenger performed a five-hour spacewalk in the new Orlan-M spacesuits. (Top right photo) Linenger in the spacesuit marked with blue stripes works closely with Tsibliev (bottom left of photo) outside the Mir space station. The two Mir-23 crewmembers installed the optical properties monitor on the docking module (photo bottom right).

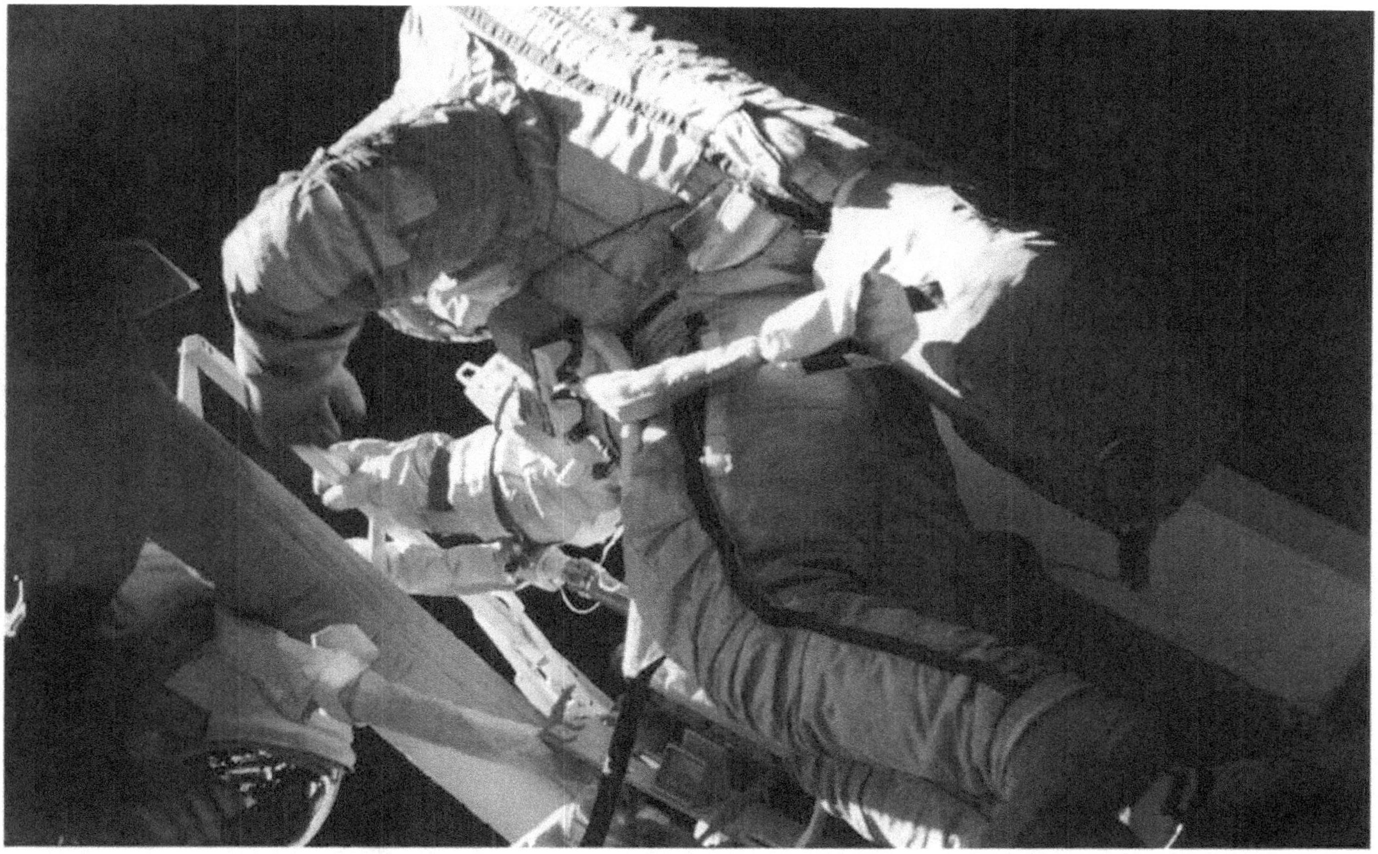

Failures to Communicate

In this new era when everyone seems to be talking on a cellular telephone, it is important to remember that instant communication isn't always as easy as it looks. During Shuttle-Mir in the 1990s, communications proved to be a major headache for NASA.

For example, NASA Space Shuttle astronauts are accustomed to having virtually constant voice communications with the support teams on the ground during missions; but Mir generally offered only one 10- to 20-minute "window" during each 90-minute orbit. As a resident on the Russian space station, an American astronaut typically participated only briefly during portions of two of these "com" passes a day. Available to the U.S. crewmember were fax, a primitive form of e-mail, and amateur radio links. None of these media worked as well as they usually do on Earth. However, ham amateur radio operators on Earth provided the Mir astronauts with important communications, passing on messages and handling "packets" of electronic mail.

On Earth, American managers and engineers were accustomed to immediate telephone communications between NASA Centers, but the phone system in Russia worked poorly. The Russian postal service was undependable. Most verbal communications—both Mir-to-ground and on the ground between managers—were carried out in Russian, which few Americans understood well. Further, the time zone difference between Texas and Russia was not conducive to creating "normal work hours" for the international teams.

NASA worked to meet the communications needs. Ground teams planned and "scripted" conferences with the Mir astronauts in advance. Written communications were used extensively. Also, the amateur ham radio conversations helped, especially with family and social communications. Efficient telephone links were installed between NASA's Russian office and its American Centers. NASA personnel often hand-carried mail when traveling to and from Russia.

But, it never got easy; and at times it was much worse than other times. During NASA-4 Astronaut Jerry Linenger's stay, communications ability became so poor that Linenger decided to forego voice contact with the ground and to do all his communicating via e-mail. In Linenger's words, "Communications were so bad that I had

to give up on them."

Linenger's Flight Surgeon, Tom Marshburn, said, "We had very limited com. It was often very ratty." At Linenger's recommendation, Marshburn said, "We went ahead and tried to do all of our communication by e-mail using the packet system. So, we had no voice communication with him for about two months. We'd hear his voice coming down on the Russian loop, and we would have a chance to say a few words to him. But, the standard com—for about two months, we didn't have it."

Marshburn thought that this arrangement was worth trying out in terms of basic communications, but he also felt that it created a psychological "disconnect between the teams." He said that Linenger "became more isolated in our minds, and probably we did in his mind as well. That didn't affect our operations so much, but there was that slight disconnect there." Also, the hardware was unable to keep up with the e-mail, and there was so much information that it could be days sometimes before information got sent up, said Marshburn. "So, we'd keep trying to plan further and further ahead. I'd say we never got more than two days ahead in the information we could get to him. So, he was pretty much flying solo a lot of the time, without any voice communication."

Shuttle-Mir Program Manager Frank Culbertson described the situation from a ground control viewpoint: "You come AOS [acquisition of signal], and you have a whole bunch of stuff you've got to tell the crew.... They've got a whole bunch of stuff they want to tell you—or maybe they don't want to talk to you at all.... The Russians call it a telephone game. So, you've got to be careful about it."

Culbertson said that some of the NASA Operations Leads and Flight Surgeons who did most of the voice communications "learned how to do it very well. They would come in very relaxed at the beginning of a com pass—not jump right into some kind of set agenda. They would come in as positive as they could [be], and then react to the crewmembers' mood to try to fit in with [it]. If they didn't get everything said or done, they'd wait 'til the next time."

NASA-3 Astronaut John Blaha thought that the voice communications were often an interference with his work. However, NASA-5 Astronaut Mike

Foale saw them as an opportunity to stay close to his Russian crewmates. Furthermore, after the Progress collision with Mir, Foale thought the Russian ground controllers felt more comfortable with his taking an active part in crucial decisions because of his previous participation in the voice communications.

And, it wasn't just the way information was exchanged that turned out to be different about the Shuttle-Mir Program. It was also how the information was expressed—a cultural difference Culbertson defined as the American "impetuosity about getting things done." Space Shuttle Commander Jim Wetherbee related a lesson he received in how to talk from Cosmonaut Vladimir Titov: "Titov one day said to me, 'I notice that the Russians talk very normally on the radio.'"

The Americans didn't. Wetherbee realized that Americans tend to talk like fighter pilots. "You use a military jargon—a lingo that is very quick and terse.... Titov said to me, 'Jim, one day you Americans will learn. It is better to talk.'"

Meanwhile on Earth
NASA-4

The Mir fire and its aftermath changed the Shuttle-Mir experience on Earth, too. Concerns for the crew's safety dominated within the Program, the world news media, and the U.S. Congress. As a result of the fire, NASA's operational involvement in Mir expanded again.

The 10:35 p.m. (Moscow time) fire surprised everyone, but it surprised American space managers later than it did the Russians since NASA officials weren't told about it until the next morning. Operations Lead Tony Sang walked into the control room about nine o'clock the morning after the fire, and he noticed many Russian life-support engineers were in the room. In Sang's words, "[Vladimir Solovyev, the Lead Flight Director,] came up to me, in his broken English, saying, 'We have a little problem.'"

According to Sang, "Ground [communications] pass was about to start, so I put on my headset and I listened. I told [Solovyev], 'Well, if there was a problem,'—and evidently it was a life-support problem—'then you can have the com.' So, I started listening and I recognized the word for 'fire.' Right then I went, 'Oh, no.'"

Sang telephoned Program Manager Frank Culbertson in Houston, where it was the middle of the night. "I told him, 'I don't know the details, but from our last ground pass … it sounds like everybody's okay.'"

Jerry Linenger had already reported from Mir. He had done pulmonary exams on the crew and found everyone to be unharmed in that respect. The immediate crisis had passed.

No NASA official had been alerted at the time of the fire. According to Sang, "I talked to Solovyev, … and he told me the lowdown of what happened. I told him, 'Why wasn't I called the night before?' He said, 'They made a terrible mistake, not calling you.'"

The Mir fire emergency was unexpected and largely unprecedented. NASA's three previous Mir astronauts had experienced much safer missions. Also, there had not been a Russian space station fire event of this magnitude ever before. There had been a smaller oxygen-generator fire onboard Mir's precursor, the Salyut, but only the generator's cloth covering had burned. This fire had involved the generator's chemical core.

Flight Surgeon Tom Marshburn later remarked that had NASA previously asked the Russians, "'So, have you had problems with this device?' they probably would have said, 'No. We've been using it for this many years and really haven't had any problems.'" In Marshburn's opinion, "You would probably have to talk to them for a while before they said what that little incident was on Salyut, because they really would feel like it was not a big deal."

After the fire, both the Flight Surgeon and Operations Lead were given cell phones, and a better procedure for alerts was developed.

Another change in the Program's management was the creation of the position of a Mir Systems Engineer, dedicated to following the health of the space station's life-supporting systems. Program Deputy Manager James Van Laak later spoke about this increasing engagement by NASA in the day-to-day operations onboard Mir. "When we have … questions, our people are able to go down the hall [at Johnson Space Center] and talk to the specialists and get good answers. I would be overstating it to say that any of our people is expert in the Russian systems, because … they're not built in the Western engineering traditions. But, pretty much without exception, their systems are intelligently designed. And, we have confidence that all the critical systems will function very well."

The fire was NASA's first big lesson since Skylab on how different the operation of a long-duration space station is, especially when compared to a shorter-duration Space Shuttle flight. Such emergencies not withstanding. Culbertson later spoke about the need for patience—most of the time. "There are a few things in a station environment that you've got to immediately jump to and do something about real quickly. Those few things are readily identified, like a fire or depressurization. Almost everything else, you go, 'Darn. It broke.' And, as long as you're still breathing and you've still got a way to escape, then there's really no need to panic or to get excited about it. You just say, 'Okay, ground. This broke. What do you want to do? Do you want to work on this? Or, do you want to leave it for now and come back to it in a few days?'

"And so, we've had to learn a more patient way of operating, in space and on the ground, than we're used to on the Shuttle because the Shuttle flight is very limited. When something breaks on a Shuttle flight … [Mission Control] brings in the cavalry, and you work on it really aggressively…. Because you have got to be able to land before you run out of consumables, and you've got to land in the same vehicle that's got the problem. That's not true on the station. You're not going to land in the same vehicle. You're not going to ever bring it back, … and [its] life is extremely long." Culbertson's patience and judgment would be tested again during the next American mission on Mir.

In April 1997, two months after the fire, the U.S. House Science Committee adopted an amendment to the NASA authorization bill, sponsored by committee Republican chairman James Sensenbrenner, Jr. and ranking Democratic Representative George Brown. The document stated that NASA must certify that Mir met or exceeded NASA safety standards. (Sensenbrenner would later ask NASA's Inspector General to conduct a review of the Program's safety.)

In May 1997, Wilbur Trafton, NASA's Associate Administrator for Space Flight, invited the Russian Space Agency to fly one cosmonaut each on STS-89 and STS-91 to continue the learning experience of STS-63, -71, -84, and -86.

During Linenger's entire mission, future NASA Mir astronauts continued to train for their durations onboard Mir. Also training in Star City was American Astronaut Bill Shepherd, who was scheduled to command the first Expedition to the International Space Station. About the safety of flying on Mir, Linenger's successor Mike Foale said at the time, "I think the Soyuz is really the biggest insurance ticket they have. And, it's the safest and the most proven they have. With the presence of the Soyuz descent capsule there the whole time, I will never feel particularly vulnerable. The fire is the worst case that you can imagine, I think."

Little did Foale know that he would experience a collision that was every bit as frightening.

Space Shuttle *Atlantis*

Launched: May 15, 1997, 4:07 a.m. EDT
Kennedy Space Center, Pad 39-A

Orbit: 184 nautical miles

Inclination: 51.6 degrees

Landed: May 24, 1997, 9:27 a.m. EDT
Kennedy Space Center

Mission: 9 days, 5 hours, 20 minutes

STS-84 CREW

Commander Charles J. Precourt
Third Shuttle flight

Pilot Eileen M. Collins
Second Shuttle flight

Mission Specialist Carlos I. Noriega
First Shuttle flight

Mission Specialist Edward T. Lu, Ph.D.
First Shuttle flight

Mission Specialist Jean-François Clervoy
European Space Agency
Second Shuttle flight

Cosmonaut Elena V. Kondakova
Russian Space Agency
Second spaceflight

Mission Specialist C. Michael Foale, Ph.D.
Third Shuttle flight; remaining on Mir

Mission Specialist Jerry M. Linenger, Ph.D., M.D.
Third Shuttle flight; returning from Mir

PAYLOADS

Space Habitation (Double) Module

Cosmic Radiation Effects and Activation Monitor

Radiation Monitoring Equipment

Shuttle Ionospheric Modification With Pulsed
Local Exhaust

Liquid Motion Experiment Protein Crystal Growth

Single-Locker Thermal Enclosure System

Midcourse Space Experiment

Electrolysis Performance Improvement Concept
Study

STS-84
Delivery and Pickup
May 15 – 24, 1997

After the sixth docking of the *Atlantis* with Mir, crews gathered in the SPACEHAB for a photo. They are (bottom left clockwise) NASA-4 Astronaut Linenger; STS-84 Pilot Collins, Mission Specialists Lu, Clervoy, Kondakova, Noriega; Commander Precourt; Mir-23 Flight Engineer Lazutkin (center); NASA-5 Astronaut Foale (right); Mir-23 Commander Tsibliev (behind Lazutkin).

STS-84 delivered U.S. Astronaut Mike Foale to Mir and brought back to Earth Astronaut Jerry Linenger. The crews transferred 249 items between the two spacecraft, including approximately 7,500 pounds of water, experiment samples, supplies, and hardware. Included in the exchange were Elektron oxygen-generating units. These were especially important after the fire that had occurred during Linenger's residency, and the Shuttle proved again that one of its main values is in bringing back from orbit faulty equipment for testing and analysis.

The May 15, 1997, predawn liftoff was perfect, after a flawless countdown and no unplanned "holds." *Atlantis* roared into orbit normally; and, as it began to pursue Mir, the STS-84 crew busied themselves with docking preparation activities. They installed the centerline camera in the Orbiter docking system, extended the capture ring on the docking system, and tested the handheld laser device used for range and closure rates with Mir.

The crew also began activities with Biorack, their main science activity.

On May 17, *Atlantis* completed the U.S.'s sixth docking with Mir. Commander Charlie Precourt greeted Mir-23 Commander Vasily Tsibliev before the Shuttle crewmembers floated into Mir's core module for a welcoming ceremony. After setting up tubes from *Atlantis* to help ventilate Mir, the ten crewmembers shared a meal before getting back to a full day of work. In addition to transferring materials, astronauts continued working on the Biorack facility and the Mir structural dynamics

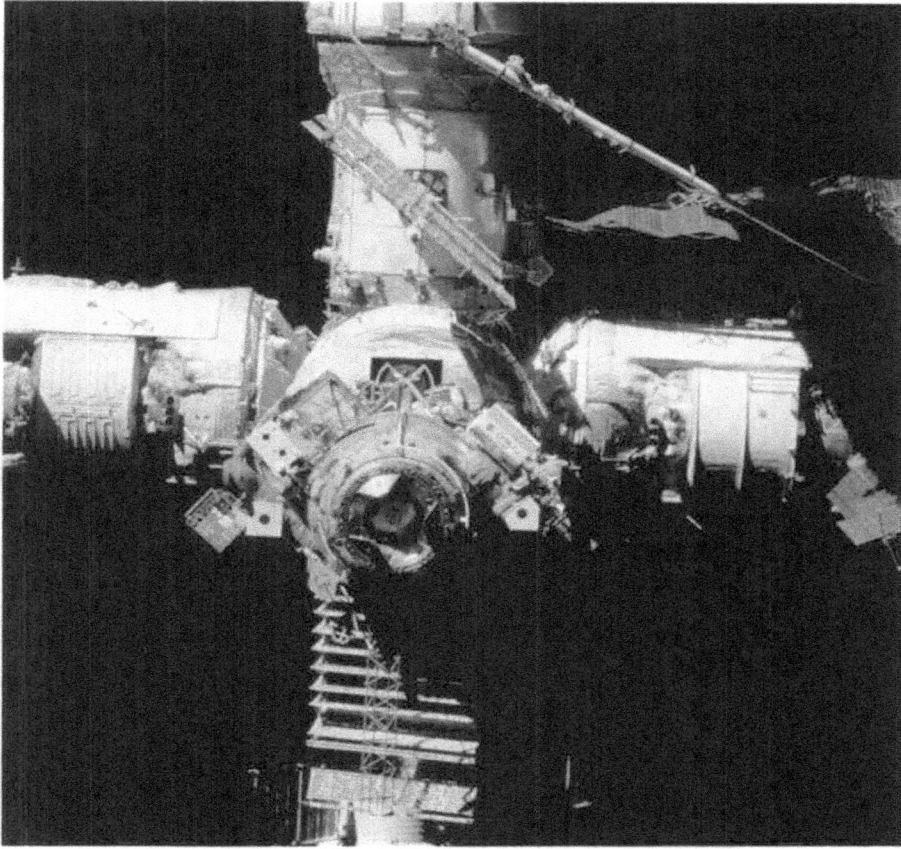

experiments, which gathered data on the effects of firing thrusters on either the Shuttle or the Mir.

The official handover from Linenger to Foale came on May 18 when Foale, Linenger, and Commander Tsibliev installed Foale's seat-liner in the Russian Soyuz capsule. Foale told Mission Control in Houston, "Your NASA-5 crewmember would like to report that he has fully switched over to the Mir side." Linenger added, "I stand relieved of duties on the Mir. It's good to be back on U.S. soil."

The crews took time to share an international dinner with foods from the homelands of crewmembers: Russia, Peru, France, Britain, China, and the United States. Commander Precourt, who speaks several languages, later said that sharing meals is rare on Shuttle flights. "So, on -84, that was one of the things I insisted [on], that we find the time in the flight plan for both the Russian flight controllers and the American flight controllers … to leave us totally alone. With this one meal, I made sure they gave us this total freedom to be together for about … two hours; and it was just a really, really memorable experience." For his own contribution, Precourt brought Texas barbecue to the meal.

Hatches between the two spacecraft were closed on May 21, setting the stage for the undocking at around 9 p.m. EDT. The initial separation was performed by springs that gently pushed the Shuttle away from the docking module, placing the Shuttle in a "free-drift" mode. Once the docking mechanism's springs pushed the Shuttle to a distance of about two feet, the Shuttle's steering jets were fired to begin slowly moving away from Mir.

Atlantis then continued to a distance of approximately 3,000 feet below Mir, where it stopped to test a European laser-docking sensor before continuing its journey back to Earth.

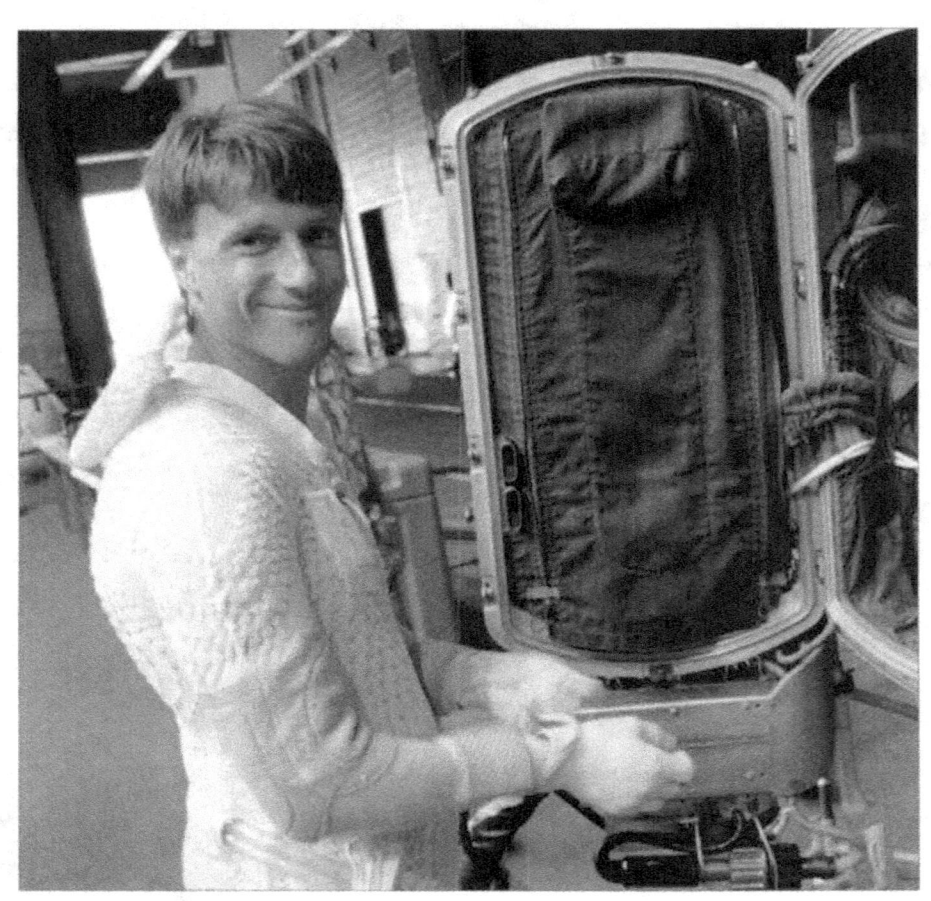

Mike Foale

NASA-5

May 15 – October 6, 1997

Collision and Recovery

Mike Foale went to Mir full of enthusiasm in spite of the fire and other problems during the NASA-4 increment. He expected hard work, some discomfort, and many challenges; and he hoped to integrate himself fully into the Mir-23 crew. The challenges became enormous when a Progress resupply vehicle accidentally rammed the space station, breaching the Spektr module and causing a dangerous depressurization. The Mir-23 crew worked quickly to save the station; and, in the troubled months that followed, Foale set an example of how to face the more dangerous possibilities of spaceflight.

Meanwhile on the ground, NASA's Mir operations were changing, too. In part because of the problems, Foale's NASA-5 increment catalyzed a broader and deeper partnership with the Russian Space Agency.

Foale's diverse cultural, educational, and family background helped him adapt to his life onboard Mir. Born in England in 1957 to a British Royal Air Force pilot father and an American mother, his early childhood included living overseas on Royal Air Force bases. An English boarding school education taught him how to get along with strangers; and as a youth, he wrote his own plan for the future of space-flight. At Cambridge University, Foale earned a Bachelor of Arts in physics and a doctorate in astrophysics. But in the midst of this progress, disaster struck. Foale was driving through Yugoslavia with his fiancee and brother when an auto accident took their lives but spared his own. This experience undoubtedly taught Foale about life's fragility and risks. Nevertheless, he continued an active and adventurous path, including diving to salvage antiquities in the Aegean Sea and the English Channel.

After university, Foale moved to America to pursue a career in the United States space program. He was working as a Space Shuttle payload officer in 1988 when he was selected for astronaut training. He flew on STS-45 and STS-56, and then became involved with the Shuttle-Mir Program as a crewmember of the "near Mir" STS-63 mission in 1994. In October 1995, while in Star City working on spacewalk issues, Foale found himself unexpectedly tapped for the NASA-5 mission. The reason? Two previously scheduled astronauts—Scott Parazynski and Wendy Lawrence—did not fit new current restrictions for the Soyuz escape capsule's seats. For a while, a modified "Three Bears" folktale circulated at NASA: "Scott is too tall; Wendy is too

docking number six, and Foale and three of his crewmates had prior experience flying a Space Shuttle to Mir. Commander Charlie Precourt had been the pilot of STS-71. Pilot Eileen Collins had been Foale's crewmate on STS-63. Cosmonaut Elena Kondakova had experienced 169 days in space as a Mir-17 crewmember in 1994-95, including six days with NASA-1 Astronaut Norm Thagard. The Mir complex had grown since any of them had seen it last. The Russian space station now included the new Priroda science module plus the new Russian-built and U.S.-delivered docking module.

Foale has described his first view of Mir, back in 1995 with STS-63 as "like seeing the great wall of China from a distance. You don't relate to it. You know you don't have to live in there." For the STS-63 crew, it was somewhat like being a tourist on a bus tour. They could see Kondakova, Aleksandr Viktorenko, and Valeri Polyakov waving at them, all excited. "We didn't understand each other very well," Foale said, "but we had Vladimir Titov onboard, who could speak with them. We lingered there.... They invited us to tea." Of course, the first Shuttle-Mir docking and a real tea with the Russians wouldn't take place for four more months, during STS-71.

On STS-84 when Foale saw Mir for the second time, the space station—besides being bigger—looked in better condition than he had imagined. "I was expecting worse and saw something better," he said later. "I saw brighter, more cheerful objects." Mir was different than the "dull, cellar-like impression I'd had in my mind." When Foale floated into the Mir living area, the atmosphere he found there cheered him. "It was a warm, welcoming, cozy place" in spite of the masses of cables and equipment and wires. Happily for Foale, "It looked like a home."

NASA-4 Astronaut Jerry Linenger greeted Foale and provided him an accurate, personal picture of what to expect. Concerning this, Linenger wrote, "To be sure, that the Mir looked and even smelled better than the astronauts had expected was not accidental. Every free moment during the weeks prior to the arrival of the Shuttle

short; but Mike is just right." Within weeks, Foale moved to Star City with his wife, Rhonda, and their family so he could train for his mission on Mir.

Like NASA-2 Astronaut Shannon Lucid, Foale saw his training in science and his self-identity as a scientist as helpful preparation to becoming a long-duration astronaut. He also valued "the willingness to undergo something very

different and foreign." He has said, "It was that trepidation—but interest nonetheless—to get through it. To go and do this strange thing. I think it comes out of a person, based on his background, culture, and family. I'm not sure it's something we could train into a person."

When Foale launched to Mir onboard *Atlantis* (STS-84) on May 15, 1997, the Shuttle-Mir Program was maturing. This mission marked

C. Michael Foale

Mike Foale was born in Louth, England, the son of a British father and an American mother. He holds a Bachelor of Arts in physics and National Sciences Tripos with first-class honors from Cambridge University, Queen's College. He also earned a Ph.D. in laboratory astrophysics at Queen's College. Before his NASA career, Foale participated in scientific scuba diving expeditions to survey Greek antiquities; and he dove on the 1543 ocean galleon, the *Mary Rose*, as a volunteer diver, learning excavation and survey techniques in low-visibility conditions.

Pursuing a career in the U.S. space program, Foale moved to Houston to work on Space Shuttle navigation problems at McDonnell Douglas Aircraft Corporation. In June 1983, Foale joined NASA Johnson Space Center and worked as payload officer for four Shuttle missions. Foale became a NASA astronaut in 1988 and served as a mission specialist on Shuttle missions STS-45, STS-56, and STS-63 before beginning training for his mission on Mir.

In his oral history, Foale made a comment about coming home: "It's also moving back to America, you see. It's not just coming back from Mir. It's coming back from being posted overseas. My post overseas was one-and-a-half years to Russia, and then it was to a Russian environment—but in space."

As the first U.S. Mir astronaut to return to space, Foale flew as a mission specialist on STS-104, the third Hubble Space Telescope servicing mission. He continues to be involved with the space program at Johnson Space Center.

Fifth NASA astronaut to live onboard Mir • Experienced a collision with a Progress resupply vehicle • Conducted a spacewalk with his Russian commander • Launched with STS-84 on May 15, 1997 • Returned with STS-86 on October 6, 1997

was spent cleaning up." Linenger personally showed Foale "how he would don the respirator, find his way into the Soyuz capsule, and activate the fire extinguisher."

According to Foale, "Jerry and I talked for a long time, maybe a total of six hours or so over three or four days.... Jerry was being very careful to tell me, 'Don't be fooled by the illusion that this is all okay while the Shuttle's here. It will change.'"

Also greeting Foale were his new Russian crewmates, Commander and Mir veteran Vasily Tsibliev and first-timer Flight Engineer Aleksandr (Sasha) Lazutkin. The schedule called for Foale to help the two cosmonauts finish their Mir-23 mission, and then to serve on the first part of the Mir-24 expedition. "I certainly enjoy their company," Foale said of his new crewmates from orbit. "We spent every meal ... chatting and talking."

Right away, Foale acted to ensure that he fit in with the crew. First, he noticed that the crew spent a lot of time dealing with e-mails and instructions from the ground, so he wrote a computer program to automate that effort, thus saving them an hour of work a day. Second, he made

sure he was present for every communications session, and he spent time with Tsibliev and Lazutkin. Foale said that it was easy to lose oneself in one's work onboard Mir, and—oddly—to lose track of the others. "It's not because the Mir is such a big space. It's because it's such a cluttered space. You could easily spend a day without talking to crewmembers." Foale's easy familiarity with his crew, and with the Russian ground team, would serve him well later when times got tough.

Foale's crewmates opened up to him about the troubles during Linenger's increment, including the fire. "Vasily talked about it quite a bit," Foale recalled. "Sasha [Lazutkin] ... took me to where the fire occurred and showed me what he was doing and how the fire happened. He gave me a long hour's description of everything that happened during the fire. It was very amusing. It was a good story with serious undertones. But, he wasn't making a big deal out of this. He was telling me a story because I wanted to learn. And other times, Vasily would talk about the near-miss of the Progress docking. That was a very close call."

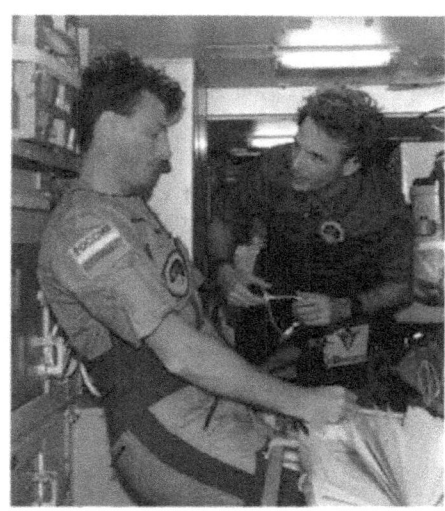

During the handover, NASA-4 Astronaut Jerry Linenger (right) provided a detailed briefing to the next long-duration American crewmember on the Russian spacecraft, Mike Foale.

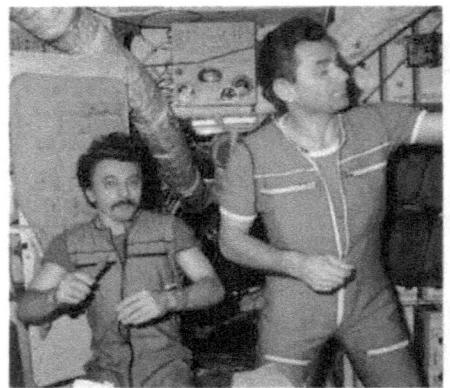
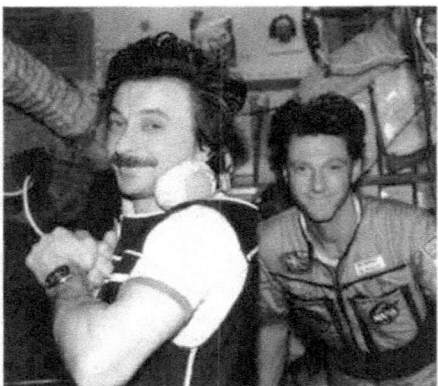

Foale settled into his work on Mir. He later said, "My whole frame of mind was, 'This is not a Shuttle flight. This is long duration. So, in the first two or three weeks, if I don't get all these things done, it's not a problem.'" He would give it what he called a "best-faith attempt … working seriously to do what I could" while building a

"I feel like I'm living in a garage," Foale reported from orbit.

good relationship with his Russian crewmates. "Sometimes in the afternoon Sasha Lazutkin would find me and say, 'Mike, you want to drink tea?' And, we'd drink tea. Then we'd go back to doing whatever we were doing. That was kind of the existence I had up to the collision."

To create his own living space, Foale tied his sleeping bag in a corner of the Spektr module, which was new but not all that comfortable. Spektr's sidewalls were skewed a disorienting 45 degrees off those of the Base Block and were crowded and noisy. "I feel like I'm living in a garage," Foale reported from orbit. Behind Spektr's panels, two drive motors slowly rotated outside solar arrays. Foale could hear them at night while he slept.

Foale got right to work on his science, setting up special containment areas where 64 black-bodied beetles would be exposed to special lighting conditions in a study of the insects' circadian rhythms. He prepared the Svet greenhouse facility for an experiment on rapeseed growth in microgravity. In addition to other investigations, he also assisted his crewmates in the continuing search for leaks in a Kvant-1 cooling loop.

Foale enjoyed working with the Mir crew "just on general maintenance tasks, where you are helping each other, just fixing tubes and using a wrench. None of it's very hard mentally, but you're working together. You're with people, and communicating and joking, and I like doing that stuff."

All in all, life onboard Mir was going very well. In terms of environmental systems, June was the quietest month in a long time. Foale said that, "compared to being on Shuttle, I feel much more healthy. I've noticed that my vision, for example, is really clear. It may be because I've managed to avoid all that paperwork I had to do on the ground in Houston. With the three or four hours of exercise that we do each day … and along with

the regular diet and all the rest, I feel very healthy. And of course, we don't get colds here. No one comes by to infect us. It's a very pleasant place. It doesn't rain!"

Figuratively, however, a storm of troubles was approaching. As June passed, the crew got ready for the arrival of a Progress resupply ship that would deliver more food and supplies. But before its arrival, Moscow ground controllers had instructed Commander Tsibliev to test the new tele-operated remote unit docking system by remotely controlling the redocking of the previous Progress, which recently had been undocked from Mir. In addition to expanding operational capabilities, the remote unit docking system was meant

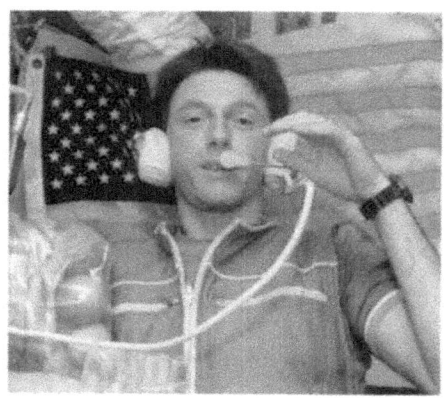

Foale (pictured above) communicates via a shortwave radio while holding a bag of fruit delivered by an unmanned Progress. Crews welcomed the spacecraft with its fresh fruits and smells that permeated the Mir as the resupply ship was opened. However, during Foale's residency, a Progress collided with the station, breaching the integrity of the complex. To stop depressurization, the Spektr module containing the leak had to be sealed off from the rest of the station. Before they could secure the hatch, Foale and crewmate Lazutkin had to clear the entryway of cables and tubes. Pictured below is the station's center node where the action occurred.

to reduce the launch weight on Progress vehicles and to eliminate the expense of the automatic Kurs equipment, which now had to be purchased from the Ukraine.

Commander Tsibliev had good reason to be concerned about the procedure. In January 1994 during a Soyuz redock, the vehicle gently bumped into Mir. During Linenger's residency, Tsibliev had attempted another remote docking; but he had lost control of the spacecraft and had narrowly missed ramming the space station. After that incident, Moscow ground controllers reasoned that perhaps a radar system had caused interference. For this next attempt, they would have the radar turned off. Foale and Lazutkin would provide the only direct measurement of the approach. They would watch for the Progress from Mir's windows, ready to use handheld lasers to help gauge the vehicle's distance and closing speed.

On June 25, 1997, Tsibliev took remote control of the Progress and fired its rockets to propel the craft toward the space station. In ways, the procedure was similar to playing a video arcade game. Tsibliev had to virtually "fly" the Progress from onboard Mir while he watched a video screen that showed an image from a camera onboard the Progress.

The Progress left its parking orbit and began moving rapidly toward Mir. But on the video screen, "it was difficult to make out the station,"

according to Tsibliev. The Mir complex "looked very similar to the clouds below it." Tsibliev's deficient perspective had a further limitation. According to Foale, "What Vasily was seeing on his screen was an image that didn't change in size very fast. That's the nature of using a TV screen to judge your speed and your distance. He couldn't determine accurately from the image that the speed was too high." By the time Tsibliev could judge the speed, the Progress was already traveling too fast. He fired the braking rockets, but it was too late.

Lazutkin finally espied the approaching Progress, and he realized the danger. "Michael, get in the escape ship!" he told Foale. Lazutkin later described the onrushing Progress as looking "full of menace, like a shark." He said, "I watched this

> Lazutkin finally espied the approaching Progress, and he realized the danger.

black body covered in spots sliding past below me. I looked closer, and at that point there was a great thump and the whole station shook."

The Progress collided with a solar array on the Spektr module. Then the spacecraft hit Spektr itself, punched a hole in a solar panel, buckled a radiator, and breached the integrity of Spektr's hull.

Foale had moved into the node at this point. He felt the impact shudder through his fingertips. He heard what seemed to be "a far-off ker-thump." Then the crew heard hissing and their ears began popping. According to Tsibliev, "The decompression alarm system immediately went off. The pressure began to fall, and the station started to spin." Precious air went rushing into the vacuum of space.

The crewmembers all realized the mortal danger of the situation. They might have to get into the Soyuz capsule and abandon the space station. Tsibliev checked a pressure meter inside the Base Block. The needle was moving down toward 600 millibars of pressure; 540 millibars was necessary for the crew to maintain consciousness.

While not knowing at the time exactly where the punctures were or even how many there might be, Lazutkin and Foale first worked in the node to seal off the Spektr from the rest of the station. In their way lay masses of tubes and cables that had been routed through the hatch and into the node since the day Spektr had been docked to the

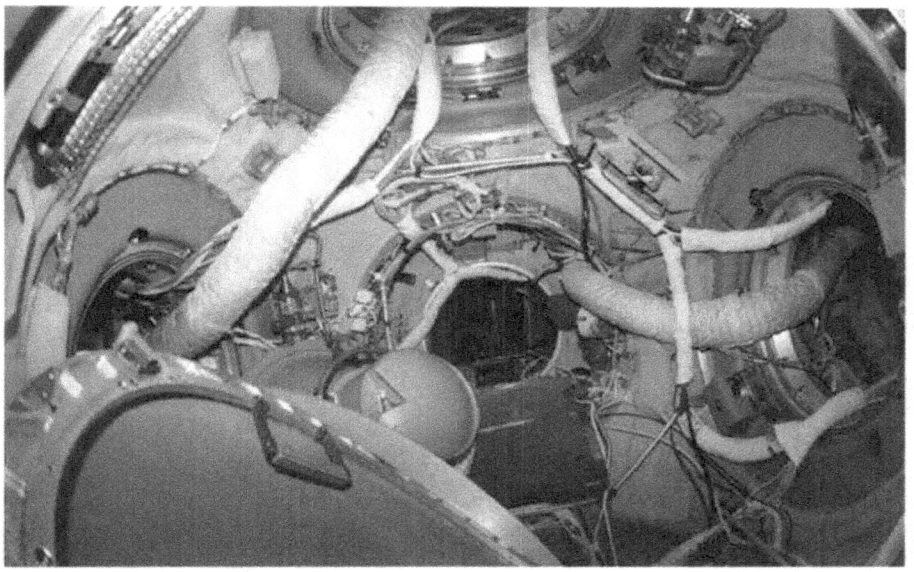

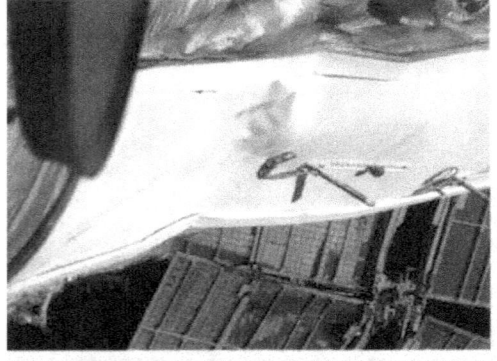

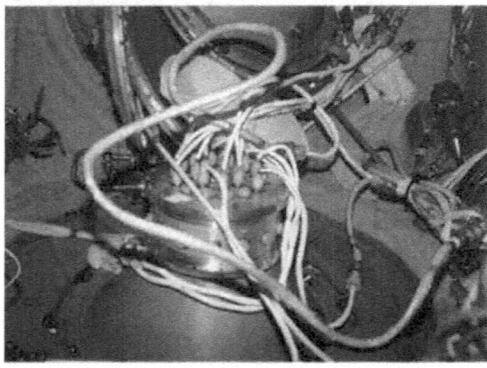

station. They worked as fast as they could. "We started pulling the cables," Foale related later. "There was a cable that burned in spots, so we had to find a way of disconnecting that one."

Once they had cleared the hatch, they needed to seal it with a cover. The node had six hatches, but all the covers had been tied out of the way. First, Foale and Lazutkin tried to free a big hatch cover that had a valve they could use later to equalize air pressures. Stubbornly, however, this one proved too difficult to untie. According to Foale, they "wasted about a minute" trying to untie that hatch. "And, the pressure's falling. The pressure's falling." Foale began thinking, "Things are getting pretty tense now."

The two crewmates had to give up on the big hatch. They found a thinner hatch cover. They untied it, and that one "popped" into place. The air pressure in the node forcefully pushed the hatch against the hatchway. "Truly," Foale thought, "there is a leak on the other side of this."

As his ears stopped popping, Foale knew they had isolated the leak and the immediate

Mir-23 crewmembers survived a crisis situation on June 25, 1997. While Commander Vasily Tsibliev conducted a test of a new tele-operated docking system, a Progress vehicle collided with the Mir space station. First, the resupply ship damaged a solar array on the Spektr module (photo top left). Then, the Progress vehicle hit and punctured the Spektr module, punched a hole in the solar panel, and buckled the radiator (middle photos). Mir began to lose air pressure and started to spin. As quickly as possible, crewmembers sealed off the leaking module from the remainder of the station; but days passed before the crew onboard and the support crews on Earth relaxed. Less than two weeks after the collision, a Progress vehicle traveled to Mir with supplies and a hermaplate designed to prevent future depressurization and to reroute power from Spektr (bottom photo). The damaged 11-year-old Mir (opposite page, top right) survived this incident and continued its journey in space until March 2001, when the space station was deorbited.

crisis had passed. As he expressed later, at this point he had thought, "Hmm. I guess I'm not going home." Then, he realized, "Well, okay. We're here for the long haul." Then, ever the optimist, he looked on the bright side. "Well, hey! We just survived a pretty big emergency!"

> ... this was the hardest time [Foale] experienced onboard Mir ... [It] was also ... [an] opportunity ... Russian controllers were now ready to let Foale take a major role ...

But a longer-term, chronic crisis had been created. Some of the cables Foale and Lazutkin had disconnected had served to provide electrical power from the Spektr's solar arrays to the rest of the station. Their disconnection, along with the station's tumbling, now caused a power loss on Mir and the shutdown of the central computer. The station fell dark and silent.

This was a novel experience for Foale, who had by this time grown used to the ever-noisy Mir. He said later, "For the first time I experienced a totally silent, still space station. There are no fans moving. There is no light on. Nothing is alive. Just our breathing is causing any sound." Lazutkin was more dramatic in his description: "The silence is deafening. You want to close your ears so you can't actually hear the sound of silence. It's painful. You experience flight in a completely different way." According to Tsibliev, for a time after the accident, "We watched the polar lights and the stars in complete silence."

The power outage lasted about a day-and-a-half. Because of the orientation of Mir's orbit at that time, the station was more often in Earth's shadow than in the Sun's light. Only when a panel happened to catch some solar energy did they have enough power to contact ground controllers in Moscow.

According to Foale, this was the hardest time he experienced onboard Mir; but that hardship was mainly because they all got so fatigued. Lazutkin went two full days without sleep.

On the other hand, this was also a time of opportunity for Foale. The Russian controllers were now ready to let Foale take a major role in the recovery of their space station. Since his arrival, Foale had been serious about talking with the ground team and had eagerly volunteered for other station work, such as sopping up condensate from the station's walls.

The collision had knocked Mir into a spin; and the power outage had shut down the gyrodynes so that the spin now went uncontrolled. To stop the spin and face the arrays toward the Sun, the crew needed to know the spin rate of Mir. However, the computer and other instruments were out of operation. So, in the dark and in the silence, Foale went to the windows in the airlock and held his thumb up to the field of stars. Combining a sailor's technique with a scientist's knowledge of physics, Foale estimated the spin rate of the space station. Then, he and Lazutkin radioed the estimates down to the Moscow Control Center. The ground controllers fired Mir's engines, and that stopped the spin—certainly not perfectly, and in no way permanently; but it showed that it could be done.

For future corrections, the crew would sometimes use the rocket engines on the Soyuz capsule. However, these engines pointed at a 45-degree angle to the axis of the main station. This and other factors created another problem in physics, as well as in onboard communications. Foale found an older, 18-inch scale model of Mir to which flashlights had been taped to approximate the newer Spektr and Priroda modules. In the microgravity of Mir, Foale set the Mir model to slowly spinning. Then he shone another flashlight onto the model, simulating the Sun. In this way, he determined how Tsibliev—who sat at the Soyuz capsule controls—should apply pulses with the Soyuz jets to set up a stable rotation and orient the solar arrays to the Sun. Over the next hours, Foale kept a star watch at his window and shouted instructions to Tsibliev in the Soyuz, many feet away.

Finally, the crew took turns catching a few hours of sleep. Then, they got back to their hard work in the dark. According to Foale, "We basically hunkered down and had to deal with a station that had all power removed from all modules except for the front two." The crew started moving batteries from the darkened modules to the Base Block to charge them up. They kept a supply of charged-up batteries ready to power the Base Block if the power went out again. This chore and other work occupied them for the week after the collision.

The now-out-of-reach Spektr had been Foale's bedroom. All of his private articles and

many of his experiments were sealed behind the hatch in the vacuum of space. Tsibliev searched the Russian supplies and found a toothbrush for the American. Foale worked with NASA officials on the ground, who put together a care package to be sent up on the next Progress ship, which arrived automatically on July 7.

Also onboard the Progress was the hermaplate, a modified node hatch cover that had been hurriedly built to allow the reattachment of Spektr's cables. Tsibliev and Lazutkin, to try to effect these reattachments, were scheduled to don spacewalking suits and perform what amounted to an internal spacewalk into the vacuum of Spektr.

As the crisis continued, Foale actually learned to enjoy rotating the station, in spite of the fact that this was a tricky procedure. Tsibliev was understandably worried about wasting fuel that might be needed for a trip home; but on Foale's instructions, he pulsed the Soyuz jets to cause complicated movements "up and to the right, or down and to the left." Lazutkin came up with an ingenious way to use a normally Earth-observing periscope to track the Sun and trace the station's motion.

Another painful twist of fate occurred on July 13. While Tsibliev exercised on the stationary bicycle, he recorded an irregular heartbeat.

Medically, this disqualified him from conducting the internal spacewalk.

NASA and Russian officials agreed that Foale could participate in the spacewalk instead of Tsibliev, and Foale started training. He was already one of NASA's extravehicular activity experts, and the Russians now trusted his judgment and skill.

But then, an event occurred that might make anyone feel jinxed. On July 17, during one of the training exercises, Lazutkin mistakenly disconnected an important power cable. The cable was one of hundreds, and the action was an easy mistake for a fatigued crewman; but it caused another severe power outage and a computer collapse. "Oh, my feelings!" Lazutkin said later. "Shooting yourself would be easier! It was terrible. The station emergency alarm went off. I realized instantly that I'd made a mistake."

The station fell into another period of tumbling without power, in which it was discovered that the Soyuz escape capsule's power could not be switched on unless the main station's power was also working. This fact had serious implications for the availability of the Soyuz during a crisis. Because of this and all the problems of the Mir-23 crew, Moscow ground controllers delayed the internal spacewalk in the Spektr module until the Mir-24 crew arrived in early August.

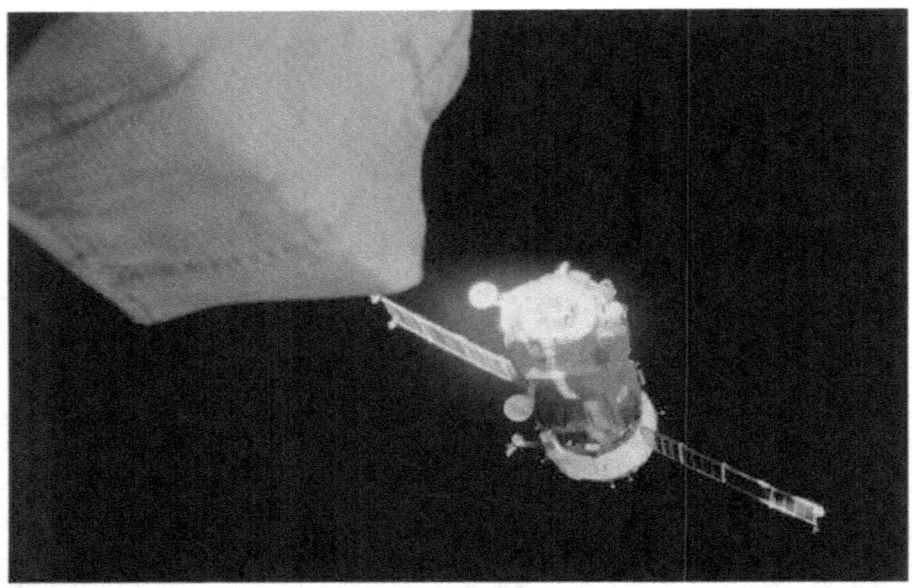

After a one-week handover, Tsibliev and Lazutkin boarded the Soyuz capsule for their trip back to Earth (above). The two had endured one of the most challenging long-duration missions in the history of spaceflight — one that included a fire and collision. Remaining onboard, Foale (bottom photo, right) was joined by the Mir-24 crew, Flight Engineer Pavel Vinogradov (left) and Commander Anatoly Solovyev.

accomplished a one-week handover. With the fire, the collision, and the other challenges, Tsibliev and Lazutkin had weathered one of the most challenging long-duration missions in the history of spaceflight. Regardless, leaving Mir was for them a moment of great nostalgia. According to Tsibliev, "No one who has been there thinks of Mir simply as a pile of metal. It's as if it touches you inside, and you feel as though you're a part of the station." Lazutkin later said, "I didn't want to leave the station because I felt it was like a living creature."

Before reporting to Mir, Foale had gone through some training with Mir-24 Commander Solovyev and Flight Engineer Vinogradov. He was familiar and comfortable with them, which was good because Moscow ground controllers had planned a lot of work for the whole crew. The crew would start by moving the Soyuz to another docking port. Later, Solovyev and Vinogradov would perform an intravehicular activity (IVA) to attempt to restore power through the damaged Spektr. Then in about three weeks' time, Solovyev and Foale would perform an extravehicular activity to inspect the Spektr module's hull for damage. This was good news to Foale, who now looked forward to participating in this spacewalk.

Repositioning the Soyuz included a fly-around to inspect the Spektr module. Foale got the job of taking photographs during the flight.

Foale worked to salvage all he could of his mission. Needless to say, the Progress collision dealt a great blow to his scientific investigations. He was, however, able to continue with several of the experiments. For example, the beetles had survived. The greenhouse was working. Many of the second set of broccoli plant seeds sprouted, including some seeds that were generated from the first set of plants. This was the first time that a second generation of space-borne plants had ever been grown. Foale was also able to do Earth observations with a Hasselblad camera. And, he and his crewmates prepared the space station for the next crew.

With all the power outages, a lot of condensate had built up on Mir's interior surfaces. According to Foale, "Fifty percent of my time

was spent just mopping up water. It was like cave diving, going into a dark module with a full-length suit on." Foale mopped up the water, either with old underwear or used clothes, or with a device that sucked the water into an airtight bag.

The Mir-24 crew arrived in their Soyuz spacecraft on August 7, and the two crews

> Foale worked to salvage all he could of his mission.

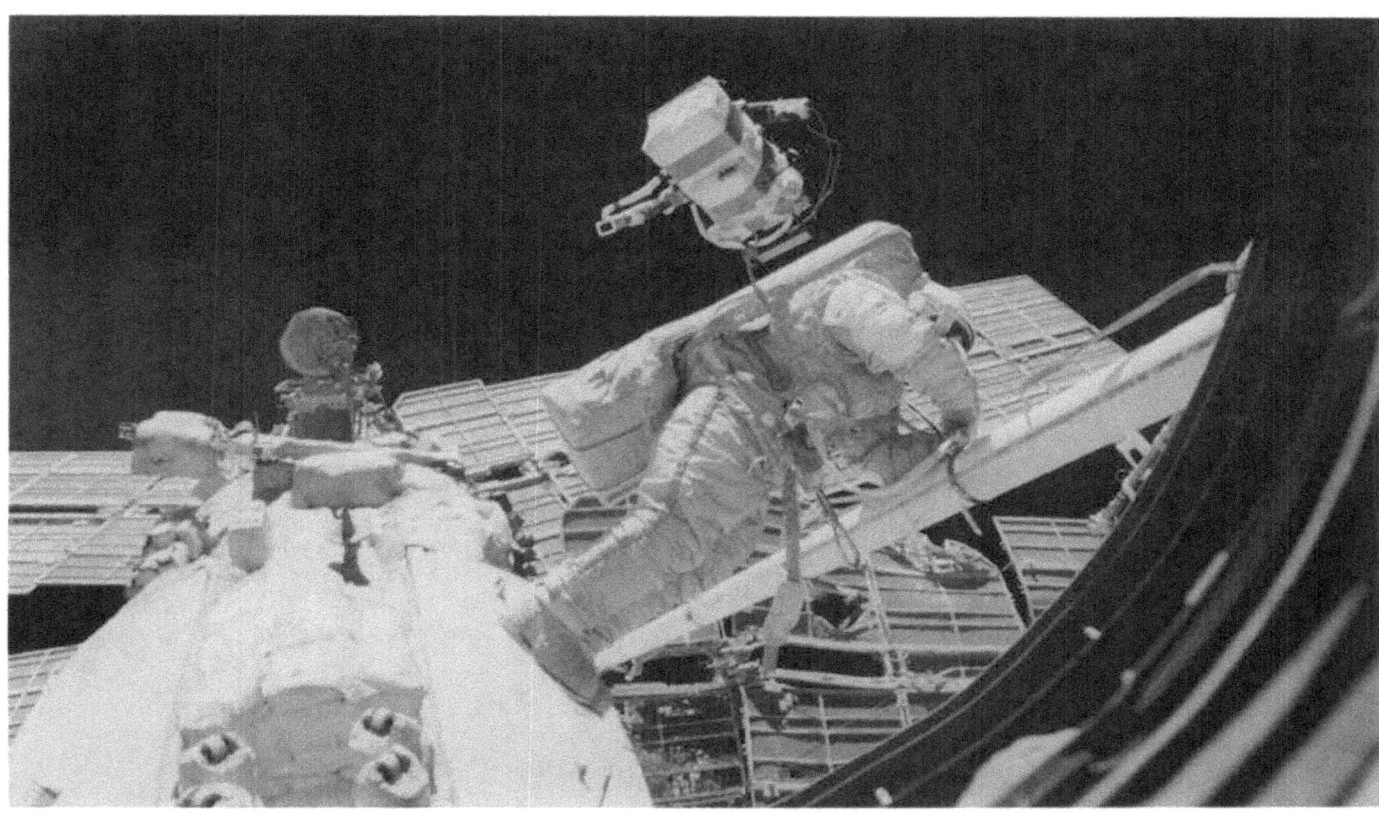

Not only did this assignment provide him with fantastic views of the station, but the fly-around got him "out of the house"—his first time outside the Mir space station in months.

The next day, the crew docked a new Progress vehicle. The docking proceeded normally with the automated Kurs system, until the Kurs failed at about 200 feet out from Mir. Controllers told Solovyev to go ahead and use the tele-operated remote unit docking system—the same system Tsibliev had been using at the time of the collision. This time the remote docking system worked, except for a short dropout of the video image, and Solovyev docked the Progress.

When the time came for his crewmates' IVA in the Spektr module, Foale stationed himself inside the Soyuz escape module. In case problems with hatches and depressurization occurred and the entire crew had to make a quick escape, Foale would be ready to assist in an evacuation.

During the IVA as Solovyev and Vinogradov opened Spektr's hatch, there was a fairly rapid depressurization of the airlock they were in. This indicated the puncture hole in Spektr's hull was large—perhaps a half an inch. At this point, Vinogradov noted that one of his extravehicular activity (EVA) suit gloves was not sealing well, and they had to repressurize the airlock and get a new glove.

> Although they observed a lot of damage to Spektr and its attachments, the two space-walkers could find no actual hull breaches.

The two cosmonauts then went to work. Vinogradov floated feet first into the darkened module to begin the job of connecting the power cables to the special hatch plate. Solovyev joined Vinogradov a short time later, helping him inspect several areas behind panels for leaks. They found

no obvious signs of puncture inside the module. Vinogradov described Spektr as being in generally good shape, with a few "white crystals" floating around, possibly from soap or shampoo, and a thin layer of frost on experiment counters that had been exposed for two months to the vacuum of space. They were able to reconnect power from two solar arrays and from part of a third. They also retrieved one of Foale's laptop computers and some photographs.

After the IVA, the crew prepared for the EVA to inspect Spektr. They also repaired systems and mopped up more condensate. Foale enjoyed assembling a truss structure to be used during the EVA. In another example of on-orbit problem solving, Foale and Solovyev came up with a way to get the balky structure out through the airlocks.

About three weeks into the Mir-24 mission, Foale and Solovyev performed their six-hour extravehicular activity. For most of this spacewalk, Foale was positioned at the base of the 60-foot Strela crane from which he moved Solovyev, at the other end, to the places Solovyev needed to work. Once or twice, Foale himself moved to the end of the Strela to hold Solovyev's feet while the Russian worked, digging with a raisin knife under Spektr's insulation and searching for holes. Although they observed a lot of damage to Spektr and its attachments, the two spacewalkers could find no actual hull breaches.

Solovyev rotated some solar arrays to provide more power that would help the station substantially. However, twice more during the mission, computer failures caused a loss of electrical power for about 24 hours each time.

The crew continued working on the condensation problems. They found "balls of water a cubic meter in size" in two modules, Kristall and Priroda, where temperatures had dropped into the 40s. However, at the same time, the Base Block's temperature had risen into the 90s; so the crew ducted its warmer air into the cold module. This initially caused an increase in condensation as the warm moist air hit the cold metal. But slowly, the modules began drying out. According to Foale, "We first powered up the Kristall, and it dried out fairly nicely; and then the last week before docking with STS-86, we finally dried up Priroda enough" to turn on its power. "And, that was an amazing thing for me," to see all the modules we had access to "all powered-up, finally."

Foale's eventful mission to Mir was coming to an end. Before NASA-6 Astronaut David Wolf arrived, Foale said in a radio dispatch to the ground, "[Of] course, I am getting excited because STS-86, the Space Shuttle *Atlantis*, will be coming, I hope, to pick me up and switch me out with [Dave].

"During the next ten days, I will be extremely busy packing up the 140 or so items that have been sent to me for return to Earth, as well as conducting the last pieces of research that we were unable to do when we had less electric power."

Foale referred to a growing controversy on the ground—over whether Wolf should succeed him on Mir—by saying, "I'd like to summarize really why I think Dave Wolf should stay onboard space station Mir when I leave. (See After Weighing the Risks, page 128.) Really, I think it comes down to the fact that, even though this flight has been one

of the hardest things I have ever attempted in my life, I have to remember what John F. Kennedy said when I was about four years old. Forgive me if I get it wrong. He said, 'We do not attempt things because they are easy, but because they are hard, and in that way we achieve greatness.'

"I believe out of this cooperation of America with Russia, which is not always easy,

we are achieving some extremely great things. And for these reasons, I think I've really valued my time onboard space station Mir. I will always remember the last three or four months with great, great alacrity and nostalgia, I'm sure. I really count all that we are doing together, America and Russia, to be extremely valuable to future cooperation on the Earth …"

Meanwhile on Earth
NASA-5

Just as the collision changed everything about Mike Foale's experience onboard Mir, it affected all parts of the Shuttle-Mir Program on the ground. Program Manager Frank Culbertson was again wakened with news of a dangerous accident onboard Mir. He and other NASA managers now had to determine what had happened, what the risks were to future U.S. Mir astronauts, and whether to continue the program.

On June 25, 1997, at Mission Control-Moscow, things were going along like most other Shuttle-Mir days. According to Operations Lead Keith Zimmerman, "It was just a regular communications pass coming up, so I went down into the main room to sit on our console. By pure coincidence, I happened to take an interpreter with me that day …"

Zimmerman plugged his headset into the console, expecting a normal day and an average Progress docking. But when the communications pass started, he noticed the crewmember on Mir was talking very fast. "My Russian's okay," Zimmerman said, "but I could only catch a few words." The words he heard were "Progress" and "Spektr." And then, Zimmerman said, "The interpreter got a really funny look on his face, and he said, 'I think they hit something.' It was just the very curious way he said it. I was thinking they hit their hand or something like that."

Zimmerman asked his interpreter to explain, and he was told, "Progress hit the Spektr module." Now, Zimmerman could hear Mir's alarms sounding in the background. He looked at the display and saw the space station's air pressure was starting to drop. "At that point, everyone … realized, 'Uh-oh, we've got a really serious problem here.'"

Within minutes, people were pouring into the control room.

Vladimir Solovyev, the senior Russian Flight Director, had been over in the Progress control room because, during dockings with Mir, the Progress is the "dynamic"—or the moving—vehicle. Solovyev now came into the Mir control room, and he began managing both control centers. According to Zimmerman, Solovyev "just started issuing orders," saying things like,

> … Frank Culbertson was again wakened with the news of … [an] accident …

"'Okay, you guys do this.… You guys do this.… Commander [Tsibliev], go do this.… Mike [Foale], do this.'" Solovyev directed the Mir crew to get out some of the air tanks that were used to repressurize Mir after spacewalks. "He told them, 'Get one of those and start opening it now, to kind of keep the pressure up while the leak's still going on—to give you more time to close the hatch.'"

The Mir crew had closed the hatch to the Spektr module by the end of the communications pass. According to Zimmerman, "The hatch appeared to be holding, but there was no guarantee." Then, the communications pass ended. On the ground, an anxious discussion began: "Was the crew coming home? Was the hatch going to hold? Are we going to come up [again] in an hour and give them a call, and the crew won't be there because the hatch didn't hold and they're in the Soyuz now?"

Needed On Mir

Shortly after the Progress resupply vehicle collided with Mir, Keith Zimmerman, NASA's Operations Lead in Moscow, talked to Mike Foale about his situation and his experiments onboard Mir. The following edited excerpts are from that radio discussion.

Zimmerman: Michael, this is Keith.

Foale: Okay. What do you need to know?

Zimmerman: First, how are you doing?

Foale: Great. As great as you can be without all your stuff.

Zimmerman: That's the next question. What personal stuff did you lose that needs to be replaced? And, what medical stuff did you lose that needs to be replaced?

Foale: I propose that we send up a complete Shuttle medical kit. And then, if you have room, the MSMK [Mir Shuttle medical kit]. Things like aspirin and some of the more common items out of the MSMK.

Zimmerman: Okay. Copy that. What about personal stuff? Hygiene kit, exercise shoes, personal items, clothes?

Foale: Exercise shoes for sure. I don't have those now. I also need my harness for the treadmill, and I need the expanders.

Zimmerman: Copy all.

Foale: I would really like to have a shaver. And, I'd like to have toothbrushes and toothpaste. About two or three tubes of toothpaste. That would do it.

Zimmerman: Copy.

Foale: Greenhouse is almost complete, except it hasn't got the leaf bags. The leaf bags are in Spektr. If you can get leaf bags on to the [Progress], it would be great. We've lost all of the [life sciences investigations], the [microbial sampling kits], the [bar code data logger], the [universal battery charger], and all the sleep equipment in Spektr.

Zimmerman: What about [Mir's computer interface to payloads system]?

Foale: [Crew onboard support system laptop] and all my hard drives are in Spektr. And, I have no printer. A printer would really help. If someone made an image of my hard drive, that would be great if they could send that up. I lost everything in terms of computer stuff.

Zimmerman: I guess the only other real question right now is, on Priroda, do you know what things were powered on when the power got cut, which switches may still be on, that sort of thing?

Foale: I turned off all [payload utility panels] and I turned off all equipment in an orderly fashion.

Zimmerman: Excellent.

Foale: The only thing that I'm worried about now is the beetles. The beetles are living on batteries.

Zimmerman: Right. The batteries will last 30 days; but they won't be able to do a light pulse, which is on Tuesday, so we will try to get one amp by then. Where were you during the event?

Foale: In Base Block watching Vasily, getting ready to take laser marks with the laser range finder, when Sasha said to get into the Soyuz quickly. I was actually in the node at impact.

Zimmerman: Do you have any words for [your wife]?

Foale: Just that things are fine. I felt the same as when I landed in the water [in an airplane without power]. It's all over now, and I'm glad we're getting it all together again.

Zimmerman phoned Culbertson, waking him at about three o'clock in the morning, Houston time, to tell him all that was known at that point. Zimmerman and Flight Surgeon Terry Taddeo then put in a call to Mike Foale's wife, and they set aside one of their telephones to be used solely for Culbertson and for Foale's family.

About an hour passed before the next communications pass. But it seemed to arrive, as Zimmerman said, "before you could even really blink. It seemed like the hour was over and it was time … to see: Was it still holding together? And sure enough, the first thing [the controllers] said to the crew was, 'Are you there? Okay, you're there. Good. What's the pressure? You're not on the Soyuz? That's the first thing. Okay.'" The Mir crew reported the pressure, and it appeared to be holding.

There followed several days of Mir attitude recovery and power problems, but the main, immediate crisis had passed. On the ground, the Shuttle-Mir teams hustled to save what they could of Foale's science program and to gather replacement supplies to be sent up to him on the next Progress vehicle.

The collision would shuffle the assignments of at least three NASA astronauts. Before the mishap, Wendy Lawrence, Lawrence's backup David Wolf, and Wolf's backup Andy Thomas had been continuing their training at several locations in Russia, including the Gagarin Cosmonaut Training Center in Star City and in water survival simulations in the Black Sea. The accident newly demonstrated the need for the U.S. Mir astronaut to be able to perform spacewalks wearing a Russian Orlan extravehicular activity suit. This, in effect, denied Lawrence her upcoming Mir opportunity because she did not fit the suit. (Lawrence had earlier been disqualified because she didn't fit the Soyuz escape capsule. But new size restrictions requalified her until the Orlan suit became an issue.) Wolf was moved up into her place, although he had to undergo even more intensive training to be ready for his flight. Thomas, who as Wolf's backup had not been scheduled for a Mir residence of his own, would soon be moved into Wolf's former slot as the last NASA-7 astronaut.

In the United States and Russia, several teams investigated the safety aspects of continuing the Shuttle-Mir Program, and all found that the Program should continue. Johnson Space Center Director George Abbey appointed a Mir-Progress Mishap Investigation Team, chaired by Astronaut Michael Baker, who had commanded the Mir docking mission the previous January. Program Manager Culbertson led a comprehensive Flight Readiness Review. Former Astronaut Fred Gregory,

NASA's Associate Administrator for Safety and Mission Assurance, led a separate internal review. General Thomas Stafford led an independent Safety Review Panel, which included Dr. Ralph Jacobson, President Emeritus of the Charles Draper Laboratories. Finally, Thomas Young, member of the National Academy of Engineering, led yet another independent group that looked at the Inspector General's Report to Congress about the safety of the Shuttle-Mir Program and at NASA's handling of it.

Shuttle-Mir Program Manager Frank Culbertson testifies before the U.S. Congress, September 1997.

Inspector General Roberta L. Gross cited several areas of concern about Mir and its operations. These included the Soyuz capsule's viability as an escape vehicle; Mir fire hazards; oxygen and carbon dioxide management problems; crew fatigue and stress; ethylene glycol exposure from leaking coolant systems; astronaut training and systems-knowledge limitations; and U.S.-Russian communications problems. The letter even cited the Russian cosmonaut pay system as a possible contributor to safety problems, because the cosmonauts were paid bonuses whenever they performed higher-risk procedures, such as the manual dockings of Progress vehicles.

On September 18, 1997, Program Manager Culbertson testified before the House Science Committee. (Read Culbertson's testimony from the Congressional Mir Safety Hearing on the CD-ROM.) Chairman James Sensenbrenner, Jr., remarked, "There has been sufficient evidence put before this hearing to raise doubts about the safety of continued American long-term presence on the Mir." In his response, Culbertson focused on "two broad questions." First, "Is there suffi-

cient value and benefit to be gained from continuing the missions aboard Mir?" Second, "Can we conduct those missions safely?" He addressed the safety question, point by point, and he returned to the question of value by saying, "There is much more to be learned. When you are exploring new territory or preparing yourself to take a major step into the unknown, who can say when you have learned enough?"

For its part, the Stafford group concluded, "Not only is the Mir [Space] Station deemed to be a satisfactory life support platform at this time, but it is anticipated that significant operational and scientific experience is still to be gained through continued joint operations." The Young Panel found, "The safety issues cited in the Inspector General's report have been analyzed and assessed by [the] NASA Phase 1 Team. NASA has an adequate safety assessment process that is complete and thorough. We found no safety concerns that were not being considered by NASA safety assessment processes." The Young Panel concluded that the corrective action already taken by NASA had made both the future use of the oxygen-generator canisters and the probability of a repeat of the collision and decompression incident "to be acceptable risk." However, the Young Panel did call for additional failure analysis of the oxygen canisters, more timely corrective actions for problems that might occur in the future, and the inclusion of "safety and mission assurance" inputs by U.S. Mir astronauts as a formal part of the review process for critical Shuttle-Mir functions.

Describing his ordeal in deciding whether to fly the next astronaut, David Wolf, to Mir, NASA Administrator Dan Goldin said, "I wouldn't want to inflict this pain" he had been going through "on any human being. Believe me, I don't sleep nights. And, there's only one thing that has been on my mind for weeks now—the safety of our American astronauts."

Wolf later discussed the decision process from his perspective. "It put our leadership—Mr. [George] Abbey, Dan Goldin, and Frank Culbertson particularly—in a tough spot, because here they had to say [that] something which is inherently never fully safe is safe enough. So, I applaud their leadership in this. I also applaud the good questions that people like [House Committee Chairman] Mr. Sensenbrenner and other critics ... brought forward, because if we couldn't accurately ... address those questions, ... then we really didn't have any business flying. History showed that it

was the right decision. But, those were good questions and we had an obligation to answer them."

Wolf went on, "Now, on a more philosophical level, it's easy to be good partners ... with the Russians when things are going easy. But, it's when things are difficult that we really can show what good partners we will be."

Philosophical questions aside, the decision to fly Wolf instead of Lawrence was handled professionally by all involved. During a press announcement at the time, Culbertson described what had gone into it. "Making a change in crew assignments or the flight plan for a particular crew is one of the most difficult decisions we have to deal with in a program like this ...

"As events occurred on the Mir, it became clear ... that having Mike [Foale] as a qualified EVA crewmember was a critical component of the mission. We started thinking ... whether that should be a factor in future missions. My first preference, of course, would have been to have Wendy Lawrence qualified as an EVA crewmember. We looked at her qualifications and what 'fit-checks' she had experienced [for the Russian EVA suit], and it became clear fairly quickly that she was not qualified to operate in the Orlan suit just because of size."

The other alternative was to evaluate her backup, Wolf. Wolf had had almost 150 hours of extravehicular activity training in the United States. Culbertson said, "It became apparent that we would be wiser to respond to the reality of the situation, and maximize our resources and participation, and see if there was some way that we could have David [Wolf] trained in the Orlan suit, ... and whether there was enough time to do that." Both Lawrence and Wolf were trained on the science program and the Mir systems. Either one could have operated as a full crewmember. The only deciding factor, Culbertson said, "was whether we thought it was important to have the potential for EVA participation during this upcoming mission."

Lawrence would later fly to Mir along with Wolf as STS-86 crewmates. Wolf later related, "Wendy was still a critical part of the mission [i.e., Wolf's mission on Mir]. In fact, I could not have succeeded at this without her. She knew more about the planning of the transfer, and placing gear, and what all the gear was than I did, of course. So, she went up and just made things work fast—up onboard—and got me organized and off on a great start."

The damaged Russian space station survived the June 1997 collision to continue its journey in history.

Space Shuttle *Atlantis*

Launched: September 25, 1997, 10:34 a.m. EDT
Kennedy Space Center, Pad 39-A

Orbit: 184 nautical miles

Inclination: 51.6 degrees

Landed: October 6, 1997 at 5:55 p.m. EDT
Kennedy Space Center

Mission: 10 days, 19 hours, 22 minutes

STS-86 CREW

Commander James D. Wetherbee
Fourth Shuttle flight

Pilot Michael J. Bloomfield
First Shuttle flight

Mission Specialist Scott E. Parazynski, M.D.
Second Shuttle flight

Mission Specialist Wendy B. Lawrence
Second Shuttle flight

Cosmonaut Vladimir G. Titov
Russian Space Agency
Fifth spaceflight

Mission Specialist Jean-Loup Chrétien
French Space Agency
Third spaceflight

Mission Specialist David A. Wolf, M.D.
Second Shuttle flight; remaining on Mir

Mission Specialist C. Michael Foale, Ph.D.
Fifth Shuttle flight; returning from Mir

PAYLOADS

Space Habitation (Double) Module

Mir Environmental Effects Payload

Extravehicular Activity Development Flight Test

Seeds in Space Education Activity

KidSat Educational Activity

Cosmic Radiation Effects and Activation Monitor

Cell Culture Module-A Experiment

Shuttle Ionospheric Modification with Pulsed
Local Exhaust

STS-86
Loaded With Experience
September 25 – October 6, 1997

By STS-86, the Space Shuttle had proven itself as a reliable "space truck" that was capable of heavy-hauling supplies to a distant, rapidly moving address and then precision-parking to make a delivery in orbit.

However, after Jerry Linenger's fire experience aboard Mir and Mike Foale's close call with the Progress collision, most of the public's attention was not on the Shuttle's capabilities but, rather, on whether it was safe to send astronaut David Wolf to take Foale's place. Congressman James Sensenbrenner, Jr. wrote that he did not expect that the next mission would succeed "without a potentially life-threatening situation." After a review, NASA Administrator Dan Goldin decided to press on—and the STS-86 crew delivered Wolf to begin a successful mission on Mir.

Atlantis was literally loaded with Shuttle-Mir experience. Commander Jim Wetherbee had led the "near Mir" mission of STS-63. Mission Specialist—and Mir veteran—Vladimir Titov had been the backup to Sergei Krikalev for STS-60 and had been aboard with Wetherbee on STS-63. French astronaut Jean-Loup Chrétien had already

David Wolf (above with stars-and-stripes cap) flew to Mir with the STS-86 crew to become the next U.S. astronaut to reside on the station. Pictured are (clockwise from Wolf) Mission Specialist Chrétien; Pilot Bloomfield; NASA-5 Astronaut Foale; Mission Specialist Lawrence; Commander Wetherbee; Mir-24 Flight Engineer Pavel Vinogradov; Mission Specialist Parazynski; Mir-24 Commander Anatoly Solovyev; Mission Specialist Titov.

flown on Mir. Wolf had been trained for Mir by the Russians, as had Scott Parazynski and Wendy Lawrence. Earlier in the program, Parazynski had been found to be too tall to fit safely in the Soyuz escape capsule and Lawrence was deemed too short. Lawrence was later deemed too short for the Russian Orlan EVA suit. So, this flight was especially poignant for her as she had originally been slated to replace Foale and Wolf had been her backup. (Lawrence would get to fly to Mir again to help close out the program with STS-91.)

STS-86 was *Atlantis'* last trip to Mir. Scheduled as another nighttime launch, the mission experienced a normal countdown; but about one minute into flight, a fuel cell behaved erratically. After Shuttle separation from the external tank, a thruster malfunctioned. In all, however, *Atlantis* and its systems performed very well throughout the flight; the chase, rendezvous, and docking proceeded normally.

Since the Mir space station had been having trouble—losing attitude control and drifting into slow tumble—the STS-86 crew had trained until almost the last minute, practicing simulations of various docking situations. The crew was also bringing to Mir a new attitude control computer. The timeline called for giving it to Mir Commander Anatoly Solovyev on the second day of docked operations. But, according to Wetherbee, "Anatoly told us, 'No, I don't want to wait 'til the second day. I want you to give me that

computer as soon as we have the handshake ceremony.' So, I decided to go him one better ... to give it to him during the handshake ceremony. Titov went down and drew a happy face on the outside of this box, and [then] we opened up the hatch. I shook hands with [Solovyev] with one hand, and with the other hand I gave him his attitude control computer.... Then, each of the other crewmembers gave him this big, huge bucket of water that we bring up for them to use."

Besides ferrying supplies and equipment, and bringing Wolf to replace Foale, one main activity of STS-86 was a five-hour spacewalk by Parazynski and Titov to test tools and techniques for future construction of the International Space Station. Their plans included retrieving experiment packages that had been mounted on the docking module during STS-76.

After undocking, Pilot Mike Bloomfield flew *Atlantis* to a point 600 feet beneath and in front

of the Russian station to gather data from a European Space Agency navigation sensor in the Shuttle's cargo bay. Bloomfield then brought *Atlantis* back to a point just 240 feet from Mir where he began a fly-around of the station.

Onboard Mir, Commander Solovyev opened a pressure valve in the station's node, blowing air into the depressurized Spektr module. Mission Specialist Titov, aboard *Atlantis*, and Flight Engineer Pavel Vinogradov, aboard Mir, both reported seeing particles seeping from the base of the damaged solar array on Spektr, but not enough to identify the exact location for a hull breach resulting from the June 25, 1997, Progress collision.

After two "wave-offs" due to weather conditions, the Space Shuttle *Atlantis* glided to a smooth landing at Kennedy Space Center and returned Astronaut Foale to Earth after his 145 days in space.

Joint Effort

For the third time in 1997, a U.S. astronaut and a Russian cosmonaut walked and worked together in space while the *Atlantis*-Mir space complex orbited the Earth in flawless fashion.

Astronaut Scott Parazynski and his *Atlantis* crewmate, Russian Cosmonaut Vladimir Titov, spent five hours and one minute in the Shuttle's cargo bay and at the docking module itself, collecting four suitcase-size packages called Mir Environmental Effects Payloads (MEEPs) that Astronauts Linda Godwin and Rich Clifford had left outside Mir during STS-76, 18 months earlier.

Parazynski (left) and Titov (right) also attached a 121-pound instrument called a solar array cap to the docking module for future use by Russian cosmonauts to seal off a suspected breach in the hull of the Spektr module. The two spacewalkers wrapped up their work outside *Atlantis* by testing the extravehicular activity rescue jetpacks. These jetpacks were designed to propel astronauts back to safety in the Shuttle's payload bay if the astronauts become untethered while performing an extravehicular activity.

This was Parazynski's first spacewalk and the fourth spacewalk for the veteran Titov, who had conducted three other spacewalks as a Mir commander in 1988. Also, it was the first time a non-American conducted an extravehicular activity from a U.S. spacecraft.

David Wolf

NASA-6
September 25, 1997 – January 31, 1998

Recommitment to Mir

Before David Wolf got his chance to fly to Mir, the risks had become more than statistical probabilities. Alarming events had happened on the Russian space station. Wolf's predecessors, Jerry Linenger and Mike Foale, had faced fire and collision. The U.S. Congress, the world's news media, and even former astronauts now questioned the wisdom of sending up another American to Mir. Wolf should not go, some said, because Mir was too old and too dangerous. NASA and the Congress ordered extensive reviews.

The buck finally stopped at Wolf's two bosses: Shuttle-Mir Program Manager Frank Culbertson and NASA Administrator Dan Goldin. Each considered the situation from his own perspective. Culbertson drew on his experience as an astronaut and his sensibilities as Wolf's friend. Goldin focused on safety and the mission's importance to the nation. Culbertson and Goldin used every minute of the time they had, which was all the time they needed. Goldin announced the recommitment to fly Wolf on the morning before the launch of *Atlantis*.

On September 25, 1997, when Wolf rose spaceward to begin his 128 days in orbit, he looked forward to doing the hard, sometimes grimy duty onboard Mir. He later noted that his predecessors had "done a great job fighting the alligators and it was now time to drain the swamps." He also hoped to make an exciting spacewalk alongside a Russian extravehicular activity master.

Wolf flew—and he flew with gusto and affability. In his own mind, he was not flying in spite of the risks. Nor was he flying cavalierly because of them. Wolf flew to Mir because he had carefully weighed the mission's risks. He had considered its value to humanity's progress. And, like all astronauts, he knew that he wanted those intangible, personal rewards that would come from an historic, out-of-this-world opportunity. He said later, "I was sure I was going the whole time, and I never had one moment of second thought…. I carefully went through the issues, and I was absolutely comfortable stepping through the various scenarios and responses…. In fact, I became more convinced that we should continue the more familiar I became with the details." He was confident that others would ultimately feel the same, and he stressed that it was exactly at this time—when things were difficult—that "we should demonstrate the strength of our partnership with the Russians."

Astronauts David Wolf, left, and Wendy Lawrence, center, and Cosmonaut Sergey V. Zaletin participate in Soyuz training activities in Russia.

Wolf examines petri dish (page 120) while conducting research in the Priroda module.

David Wolf's background and personality were suited to long-duration spaceflight. He is an astronaut, a physician, a researcher, an inventor, and an aerobatics pilot. Born in Indianapolis, Indiana, in 1956, Wolf had grown up to become a gregarious Jack-of-all-trades who enjoyed both theoretical thinking and skinned-knuckle mechanics. While attending medical school at Indiana University, he worked at the Indianapolis Center for Advanced Research, specializing in medical ultrasonic signal processing. He completed his internship there and, in 1983, joined NASA's Medical Sciences Division at the Johnson Space Center in Houston. He directed the development of the space bioreactor that was to see service on Mir. He also worked as chief engineer in charge of the design for the space station medical facility. Wolf became a NASA astronaut in 1991 and flew as a crewmember onboard STS-58, a medical research mission.

When tapped for the NASA-6 increment on Mir, Wolf was already in Russia, training as the backup to Wendy Lawrence. He inherited her slot when she was disqualified due to the new stipulation that all Mir astronauts must be extravehicular activity-ready and fit within the size limits of the Russian Orlan suits. For Lawrence, this was yet another "too short" situation, similar to the one that had bumped Foale into the role of flying on NASA-5. For Wolf, it hurried his trip to the space station. He had been planning on flying last, after Lawrence's increment.

Besides delivering Wolf, the STS-86 mission included the delivery of a new motion control computer and a solar array cap. On the final day of docked operations, Wolf's crewmates—Mir-24 Commander Anatoly Solovyev and Flight Engineer Pavel Vinogradov—installed the new computer and, shortly before Atlantis docked, activated Mir's gyrodynes to help keep the space station stable.

During his residency aboard Mir, Wolf wrote a series of "Letters Home." In one of them, he related the undocking: "The Space Shuttle Atlantis' hatch shut, its docking hooks released, and its translational thruster jets fired. I could clearly see Jim's [Wetherbee] and Mike's [Bloomfield] faces peering through the Shuttle's overhead window. We waved. Mike's other hand moved, and another minus Z-axis translational pulse increased our rate of separation. Another

volley of bright orange rocket engines flashed against the dimly moonlit Earth."

Atlantis pilot Mike Bloomfield held position for nearly an hour until the two spacecraft moved into orbital daylight. Then, Bloomfield did a slow fly-around of Mir while his crewmates took video and still photos of the entire station, concentrating on the damaged Spektr module. In an attempt to find the hole in Spektr's hull, Commander Solovyev opened a valve to pump air into the module, hoping that the Atlantis crew would spot debris being vented out to space. Mir Flight Engineer Vinogradov saw a particle floating away from the module, but said he couldn't detect the exact location where it came from.

As Atlantis left Mir on its journey back to Earth, Wolf already felt at ease in his new orbital home. STS-86 Commander Jim Wetherbee later recalled how Wolf had ended every telecommunications session while the Shuttle was still docked. According to Wetherbee, Wolf told the ground controllers, "Now, be careful down there. You're awfully close to the ground. You don't want to get hurt." Wetherbee thought this "was a great way to think about risk." Wolf was "up there, floating

NASA-6 Astronaut David Wolf (center) poses with crewmates Mir-24 Flight Engineer Pavel Vinogradov (left) and Commander Anatoly Solovyev.

around in what astronauts tend to think of as a relatively risk-free environment—although there are some risks—but he's telling the people on the ground, 'Don't worry about me. You take care of yourselves, and I'll be okay up here.'"

Along with his new crewmates, Wolf started work on a variety of projects, including the three-dimensional biotechnology tissue culture, optical properties monitor, and Canadian protein crystallization experiment—the latter designed to analyze the crystalline structure of 32 proteins in an effort to improve drug development and design.

Wolf also explored his new world-above-the-world; and he wrote that, in ways, it hearkened back to the age of classic science fiction. "The central command post has keys that look like worn ivory. Leather shrouds serve where plastic would now be chosen. The metal machining is recognizably Russian, and of the highest quality. Its overall character brings forth the image of the 'time machine' from H.G. Wells' classic."

> [Wolf ended telecommunication sessions with], "Now, be careful down there. You're awfully close to the ground."

Wolf mentioned "tables with things on both sides," "a bicycle with no seat," and "a set of heavy tools held in place by rubber bands." He also described the Soyuz escape capsule as "an amazing vehicle which comes straight out of Jules Verne. An absolutely beautiful piece of handcraftsmanship. Very Уютный (OO-yoot-niy) [cozy] with wood-grained control handles and those beautiful ivory keys again. As I write, it waits, fueled and ready to bring us back to Earth."

Mir was, in ways, "cozy," according to Wolf; but in other ways, it was a mess. Wolf later related how he immediately offered to take full part in the housekeeping on Mir. "The first day, after the Shuttle left, I noticed Pavel cleaning up with rags … a large amount of condensate on the heat exchangers to the Elektron unit.… We had troubles—in fact, a complete failure of the condensation-removal system." Behind panels, "large condensate globs, bowling-ball size or beach-ball size sometimes … of gooey, slimy, ice-cold fluid [were] … starting to track down the structure and into the wiring."

Wolf went up to Vinogradov and said, "You never have to do this again." According to Wolf, the flight engineer "looked at me kind of funny. I said, 'I'm doing it. Don't worry. You've got better things to do up here.' I didn't realize what I was getting into, because it took anywhere from four to eight hours a day, the rest of the mission, every single day except a few. Nevertheless, I think that went a ways to their putting me on the team…. There's no small or unimportant job on the space station. All of it has to get done, and that was the best thing I could come up with to free up their time." And, to become part of the team.

On October 8, a new Progress resupply ship successfully docked to the station's Kvant-1 module. The spacecraft carried 1.7 tons of supplies for the station, including science equipment, fresh food, 100 liters of water, and clothing and other personal items for the crewmembers. The Progress also brought up a motion control system computer to serve as a backup to the one that *Atlantis* had just delivered.

In a "Letter Home," Wolf described the Progress docking with Mir. "It was almost eerie to see the robot ship loom in out of the darkness. The view from Anatoly's tele-operated pilot station was as seen by the cargo ship, closing in on this amazing space station. Its computer mind correcting for errors in the cross-hairs on the docking

During the Shuttle-Mir Program, *Atlantis* (above) docked with the Russian space station seven times.

(Below) Mir-24 Flight Engineer Pavel Vinogradov works on equipment panels in Mir.

Maintaining an active space station for more than a decade involved using a variety of tools (top photo) that were strapped down until needed. Other items needed on a regular basis were stowed ready-for-use, such as those located near the control console in the Base Block (middle photo). Always available to the crews was a docked Soyuz spacecraft (bottom photo) that could serve as their emergency escape vehicle.

target, just as Anatoly would have done himself. It behaved almost human. Anatoly's hands were lightly poised on the remote control sticks, ready to manually take over at the first sign of bad decision-making by the computer pilot. He and Pavel checked approach speeds and positions from the console. In their minds they had transported themselves and were sitting in the cargo ship.... As I watched their moves and words [and saw] how confidently they worked together from training and experience, my few thoughts of what happened to Mike Foale a few months ago were quenched." Then, "*Thunk,*" the Progress docked. "It hit pretty firmly—which is normal. No pressure sensations in my ears. Docking mechanism properly engaged. The silence of tuned nerves was broken by laughter and handshakes. Supplies had arrived."

Wolf collected data during the undocking of the old Progress and the docking of the new Progress for the Mir structural dynamics experiment. This experiment measured forces exerted on the Mir by events such as vehicle dockings and thruster firings, and even the crew exercising. The crew performed routine maintenance on various Mir systems. They replaced batteries in the Base Block and in the Kvant-2 module with new batteries brought up on *Atlantis*; and they rearranged other batteries to ensure that the batteries were fully charged. They worked on the urine recovery system that provided water for systems. And, Solovyev and Vinogradov began their preparations for a second planned internal spacewalk into the Spektr module. This intravehicular activity (IVA) was designed to recover additional power from Spektr's functioning solar arrays by restoring the array's ability to swivel and track the Sun's light.

The crew installed a new control unit on the Elektron oxygen-generating unit in the Kvant-2 module. The Elektrons in both the Kvant-1 and the Kvant-2 modules had been running simultaneously to increase the oxygen in the station in preparation for the IVA. In the Kvant-2 module, the crew installed a new drive unit on one of the gyrodynes. On Mir, eight to nine gyrodynes were needed to maintain attitude control; Mir was now operating on 11 gyrodynes.

After three weeks onboard Mir, Wolf reported life as "quite good. The air is extremely fresh and clean. Anybody who wants to work hard can find a lot to do up here. You don't need to be a rocket scientist to stay very busy." He said he

David A. Wolf

David Wolf was born in Indianapolis, Indiana. He earned a bachelor's degree in electrical engineering at Purdue University and an M.D. at Indiana University. He served as a Flight Surgeon in the U.S. Air Force.

Wolf joined NASA's Medical Sciences Division at the Johnson Space Center in 1983. There he directed the development of a space bioreactor that resulted in state-of-the-art rotating tissue culture systems. He was also assigned as chief engineer in charge of the design for the International Space Station medical facility. Wolf became a NASA astronaut in 1991. He flew with John Blaha and Shannon Lucid on STS-58's 14-day science mission in 1993.

While on Mir, Wolf completed an almost four-hour spacewalk and felt honored to gain this experience with veteran Russian cosmonaut, Anatoly Solovyev. He said: "When I was nine years old, I saw Ed White do the first American spacewalk, and it was [at] that moment that I decided I'd like to be an astronaut and, in fact, I'd like to do a spacewalk as an astronaut. It was 31 years later that I did it. It was worth every minute of the wait.

"But I never dreamed it would be from a Russian spacecraft, in a Russian spacesuit, speaking Russian with a Russian who had been out 16 times, the most experienced spacewalker in the world, and that's what Anatoly Solovyev is. So, it was a real first-hand lesson from the number-one guy in the field, and that was a privilege."

Wolf remains a NASA astronaut. He is preparing for future flight assignments and working in the International Space Station Program.

Sixth NASA astronaut to live onboard Mir • Conducted a spacewalk with his Russian commander
Launched with STS-86 on September 25, 1997 • Returned with STS-89 on January 31, 1998

---

had taken on the personal goal of keeping the air filters clean. "I plan to leave this station a bit better and cleaner than when I came."

Of course, the August collision with the Progress vehicle had severely damaged some of that equipment; and on October 20, Solovyev and Vinogradov conducted an internal spacewalk, suiting up to enter the vacuum of the Spektr module. During the grueling six-hour, 38-minute intravehicular activity, the cosmonauts had to be careful not to get tangled up in the disorder that was now filling Spektr. They were able to connect two power cables that increased solar power to Mir by 15 to 30 percent. Wolf stayed in the Soyuz capsule during the internal spacewalk, monitoring Soyuz systems and conducting Earth observation photography. The crew spent the rest of the week powering up Mir systems and modules.

American democracy was powering up on Mir as well. In late October, Wolf cast the first American election ballot from orbit, using an electronic mail system developed among Johnson Space Center, Mission Control-Moscow, and the

County Clerk's Office in Harris County, Texas. In an interview from orbit, Wolf said about his voting, "It's important. It makes me feel attached to the ground like I didn't feel before. I feel it's more important here in space even than I did on the ground. Voting's important to all of us. It's what puts the people in charge." Ironically,

> Then, "*Thunk*," the Progress docked…. "The silence of tuned nerves was broken by laughter and handshakes. Supplies had arrived."

Russian democracy had beaten America into space. In June 1996, Shannon Lucid's crewmates Yury Usachev and Yuri Onufriyenko had voted in a Russian presidential election.

In early November, Solovyev and Vinogradov conducted two more extravehicular activities. They replaced an aging solar array on the Kvant-1 module. The new array had been stored in a

compartment on the Mir's docking module since its delivery on the STS-74 Shuttle-Mir docking mission two years before. Solovyev and Vinogradov also installed a device on the outside of the module that would enable the crew to hook up an additional Vozdukh carbon dioxide removal system for the station.

During the first extravehicular activity, Vinogradov commemorated the 40th anniversary of the launching of the first Sputnik satellite by manually deploying a replica of Sputnik. This effort was part of a joint project between Russian and French high school students. Wolf videotaped the event. He said later, "The first thing that went through my mind was, 'God, I've got to get this video camera on it and get the picture right.' But after that, when I reflected on it— 40 years is how long it's been. I believe it was October 4, 1957, when the first satellite was launched by the Russians. And, you just look at where we've come in that amount of time. It was not much more than 40 years before that [that] the first airplane flew. And, I thought a lot about

the pace of technology and how our lives are changing, and how NASA needs to be a part of designing our future way of life just as it was a large part in creating our current quality of life."

The extravehicular activity's final task for Solovyev and Vinogradov was to retrieve a panel from an old and disconnected solar array on the side of the Base Block. The panel would be returned on the STS-89 mission in January to be analyzed for micrometeoroid damage.

The cosmonauts had a problem repressurizing the Kvant-2 module's exterior airlock after the first spacewalk. For the second extravehicular activity, they used an interior compartment as a backup airlock. Then, they tightened clamps and latches around the circumference of the leaky hatch to hold pressure; the problem thus appeared at least partly solved. This was another example of the differences between short- and long-duration spaceflights. On space stations, many repairs have to be done "on the fly" instead of being brought back to Earth, as they are on the Space Shuttle.

Wolf later wrote about his duties during one of the extravehicular activities. "My job was issuing the computer commands to the new array's deployment mechanism, and something didn't work. Now, we were 'off nominal' and 'out of the checklist,' [and] going fast, in Russian, and with [only a] short time left on the spacesuit carbon dioxide scrubbers." In coordination with Mission Control-Moscow, the Mir crew improvised manual procedures to command the solar array deployment. As Wolf wrote, "Retract two steps. Disable motion quick. It's jammed. Try to re-extend by one step. What are the motor power indications? Is there a center-section deployed indication? We need to reinitialize the sequencer." And, all of this was done, Wolf wrote, "as fast as my fingers could press buttons and in Russian." After the experience, Wolf said he now had an answer for that repeating question, "What has been your toughest moment so far in the mission?"

Mir-24 Commander Anatoly Solovyev (above) displays a water dispenser onboard Mir's Base Block

Joined by Flight Engineer Pavel Vinogradov (left), Solovyev attempts to repair a solar array on the Mir space station. (Photo, right)

Solovyev (pictured opposite page) is regarded as the most experienced spacewalker in the world.

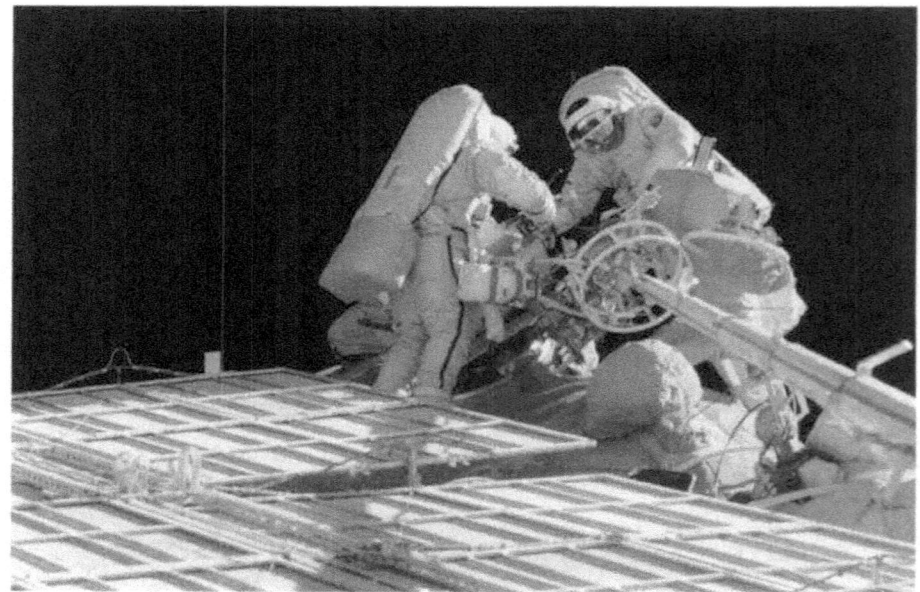

The week's EVAs were Solovyev's 13th and 14th spacewalks in his five tours of duty on the Russian outpost. He had now conducted five spacewalks during Mir-24. Vinogradov, on his first spaceflight, had now made four spacewalks. With the two EVAs behind them, the cosmonauts were given the weekend off to relax before resuming scientific research and routine maintenance activities.

Wolf was beginning his seventh week as a Mir crewmember. In an interview, he said, "I love space, and I'm getting a good dose of it here. It's fun — everything from floating and flying to handling floating equipment. It's a beautiful view. You never get tired of looking at the Earth." Wolf reported that when sleeping, "I dream that I'm in space. Last night, in fact, I dreamed I was with my friends and we were playing volleyball, and it was a wild game." While awake, Wolf said, "I'm surprised every time. I look up and I wonder how I'll reach that—and then I realize I can just float up and get it. It takes a moment for that to register, so I'm not fully used to it. But, I get better at it every day."

Wolf also realized that he would have to re-accustom himself to a changed Earth when he went home. He said that, after a Shuttle mission, "The likelihood of things changing in a drastic or significant way is small. Heck, you barely miss much of your mail. But, when you're gone for months, ... you really start to lose touch a little bit with the Earth." He was starting to feel as if Mir was his world, and the Earth was beginning to feel a little "dreamlike." It was important "to be able to keep in touch with it—visually and through any media and means that you can." Video movies were particularly effective. They had more emotional impact under those remote conditions.

On November 13, Mir suffered a temporary power loss during a test of the newly installed solar array on Kvant-1. This caused the shutdown of the motion control system computer and interrupted Wolf's scientific investigations. The crew transferred fully charged batteries from the Kristall module to the Base Block, and they restored power to five of the 11 gyrodynes that provided attitude control to Mir. However, the batteries alone could not support the many hours required to stabilize the station. So throughout the weekend, the three crewmates alternated shifts to monitor systems; and whenever Mir drifted into an attitude favorable for solar energy collection, they temporarily powered up the battery chargers.

After Weighing the Risks

On September 25, 1997, the morning before liftoff of STS-86, NASA Administrator Dan Goldin finally announced the firm decision to send David Wolf to Mir. Goldin's statement underscores the reasons to fly and outlines the decision path that NASA followed.

Ever since becoming NASA Administrator in 1992, I have worked hard to make the Agency operate faster, better, and cheaper. The dedicated people at NASA have accomplished that, but never at the expense of our highest priority: safety. Today, I continue that commitment.

It is only after carefully reviewing the facts, thoroughly assessing the input from independent evaluators, and measuring the weighty responsibility NASA bears with putting any American in space that I approve the decision to continue the next phase of the Shuttle/Mir mission.

Tonight, the Shuttle *Atlantis* will launch, sending David Wolf to replace Michael Foale and to continue an American presence on Mir.

This is a decision that all of us at NASA do not take lightly. We share our fellow Americans' deep concern for our astronauts' safety. And, we have heard the calls of some who say it is time to abandon the Mir. We at NASA, especially Michael Foale, are deeply touched by this outpouring of emotion. However, we know the decision to continue our joint participation aboard Mir should not be based on emotions or politics. It should not be based on fear. Our decision should be based, and is based, on a scientific and technical assessment of the mission's safety and the Agency's ability to gain additional experience and knowledge that cannot be gained elsewhere.

In a status report from Michael Foale, he urged our continued participation aboard space station Mir and that David Wolf join the Mir crew. I have also spoken to David Wolf. I asked him if he is confident in NASA's safety review and if he thought we should go ahead. He answered with a resounding "yes."

As the person who bears the ultimate responsibility for America's space program, I have been diligently reviewing the independent and internal safety assessments. I have concluded Shuttle-Mir has a thorough review process that ensures continued American participation aboard Mir does not put human life in unnecessary peril. Briefly, I would like to share that review process with you.

The first step in the review consists of a comprehensive, all-systems analysis to conduct the mission safely and successfully. Each major system and component critical to the crew's safety and the mission's success is reviewed and determined to be ready for flight. This review is led by Shuttle-Mir Program Manager Frank Culbertson. This step concluded last week with the final Shuttle Flight Readiness Review, a separate comprehensive review of all aspects of Shuttle mission readiness. This review resulted in unanimous approval to proceed with launch of Shuttle *Atlantis* to Mir.

The second step is an internal review led by Fred Gregory, NASA's Associate Administrator for Safety and Mission Assurance. At the Flight Readiness Review, Col. Gregory gave his certification of Shuttle-Mir flight safety.

The third step is an external independent review by a NASA Advisory Council Task Force, known as the Stafford Commission. This step, as with the two previous, is part of each and every Shuttle-Mir mission review. Led by former Gemini and Apollo astronaut Lt. Gen. Thomas Stafford, the review includes eight other non-NASA members. To address recent Mir problems, General Stafford took the unprecedented step of asking Dr. Ralph Jacobson, President Emeritus of The Charles Draper Laboratories, to head a smaller team and take a fresh look at this particular mission's safety and operational readiness.

Yesterday, I was briefed on General Stafford's and Dr. Jacobson's reviews of Mir systems. They concluded "not only is the Mir station deemed to be a satisfactory life support platform at this time, but it is anticipated that significant operational and scientific experience is still to be gained through continued joint operations."

Lastly, when the Inspector General raised concerns about the NASA safety review, we added even another step to our Shuttle-Mir review process. I asked Thomas Young, member of the National Academy of Engineering, to lead yet another independent group of some of our country's best to look at the Inspector General's report. This group set out to find if there was integrity to NASA's safety review process and to ensure no stone was left unturned. Mr. Young stated in his report to me, "The safety issues cited in the Inspector General's report have been analyzed and assessed by the NASA Phase 1 team. NASA has an adequate safety assessment process that is complete and thorough. We found no safety concerns that were not being considered by the NASA safety assessment process."

In light of increased scrutiny and heightened emotion, I can assure you, this intensely rigorous internal and external review of the Shuttle-Mir analyzed—thoroughly—risk, readiness, and, foremost, safety.

I will not trivialize the risks involved in human space exploration. Like all Americans, I know every time an astronaut travels to space there is risk. When we build the International Space Station we will encounter similar problems and there will be danger.

But, NASA is ready.

We are ready because the reviews assure us. But, we're also ready because it's the right thing to do. Americans press forward.

We overcome the unexpected. We discover the unknown. That has been our history. That's America's destiny.

I love this country very much, and I feel privileged to serve with so many dedicated, talented, and courageous individuals. Today, more than ever, I am proud of everyone at NASA for their commitment to America's future and for their service to humankind. And, to David Wolf, Michael Foale, and the rest of the Shuttle *Atlantis* crew … Godspeed. We'll see you when you get home.

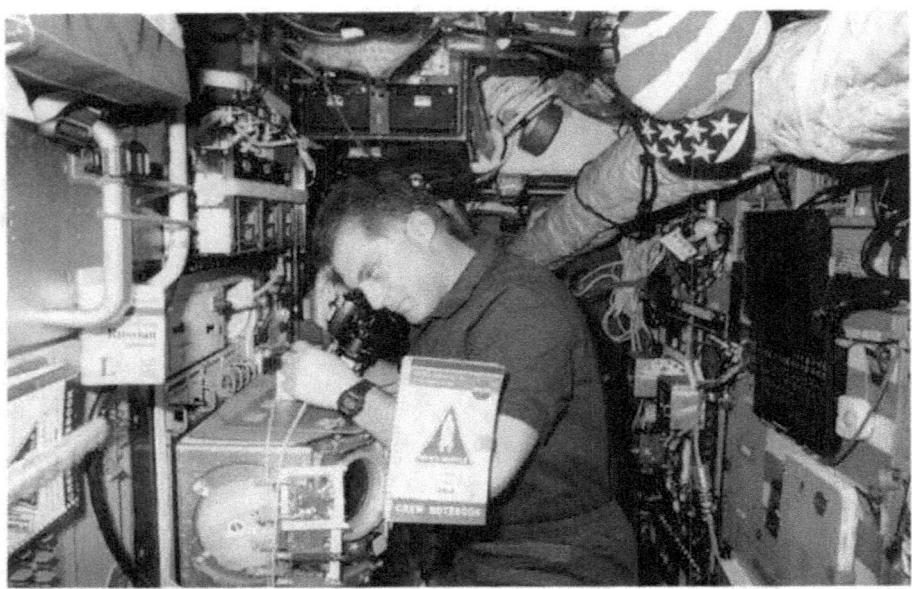

NASA-6 Astronaut David Wolf sorts samples and takes photo documentation of microbial investigations of Mir. His crew notebook, where he records notes and charts data from experiments, floats nearby.

Meanwhile on Kvant-2, a slow air leak persisted in spite of the efforts by Solovyev and Vinogradov during the second spacewalk. The situation posed no danger to the station, as the hatch door on the instrumentation compartment behind the airlock provided an air-tight seal.

Wolf was now midway through his four-month mission. In an interview, he said that while the Russian space station was like a "fixer-upper" car in which things broke down, "you can trust it to take a long trip." He said that Mir was clean and in good shape. Then, he joked, "It could use some new carpet." Wolf also compared doing science on the Space Shuttle and on a space station. "One difference is in the laboratory," he said. "Right now, for instance, on the computer display, we are watching a laser image of a crystal as it grows. You can see the sides of the crystal as microgravity helps it grow, and we get to send these images down to the scientists on Earth. We then listen to what they say and we make decisions and change what we do in response to the results. It's more like a laboratory would be run on Earth; whereas on the Shuttle, it's kind of a sprint race. You get up there [in orbit] and, in eight, ten, 12,

14 days, just do all you can. Here, we can plan and change the plan during the mission."

The American astronaut aboard Mir sometimes worked as a scientist and wrote as a novelist. In a letter, he described a view of Earth, "Ghostly outlines of continents just illuminated by the half Moon. At an unfelt five miles per second, we blow out of the Earth's shadow and into the harsh unattenuated sunlight. Solar arrays alertly take notice and rotate precisely into position to capture a bit of this fortuitous energy. We blaze over that moving line on the Earth that separates night from day. The dominant features on the planet below are two tectonic plates. One holding the Tibetan Plateau and the other India. The plates are clearly smashing together, incidentally elevating the Great Himalayan Mountain Range. Eyes now adjusting, looking real close, there, snow-covered Mt. Everest and Katmandu. It's a rare clear day over France, England, and Italy. Hazy, even smoky, into China and southern Siberia. Some large smoke plumes, a lot of forest clearing going on there. Just ahead, to the east, the incredible blue Lake Baikal, perhaps the biggest lake in the world. Set like a gemstone into the Earth's crust."

During the week of November 21, the crew experienced another power-down, this one resulting from a malfunction of the refurbished motion control system computer that had been brought up on STS-86. They replaced the computer with a new one that had arrived on a Progress. Wolf later remarked that the two periods without power, of roughly 48 hours each, "offered a unique perspective of spaceflight." With no systems operating, with no fans and no pumps going, "the incredible quiet of space was experienced. From the dark, quiet ship, the surreal experience of space was ever more intense."

Commander Solovyev was intent on restoring Mir to top condition; and soon after power was regained, the cosmonauts installed a new Vozdukh carbon dioxide removal system in the Mir Base Block to serve as a backup to the unit that was currently functioning. An extensive amount of drilling, sawing, and wiring was required to fit this large system behind the station's panels.

Wolf helped his crewmates with the systems activities and repairs, and he continued his science program. His favorite investigation was the Three-Dimensional Biotechnology Tissue Engineering Experiment, perhaps because of his involvement in its early development before becoming an astronaut. He later echoed other Mir residents when he said, "It felt good being back in the laboratory." The combination of activities led to very long workdays and, Wolf noted, "boredom was not a factor for this crew."

During December, Solovyev and Vinogradov spent time troubleshooting leaks in a backup cooling loop aboard Mir. They also released additional oxygen into the station from the tanks aboard the Progress, and they loaded refuse into Progress for its planned jettison from Mir before a new Progress resupply ship arrived. Christmas was approaching. The new Progress brought gifts from home as well as a small Christmas tree and traditional candy from the Red October candy factory near Moscow. The objects assumed that characteristic intensity typical of long-duration spaceflight.

The NASA-6 resident worked at measuring bone loss during long-term spaceflight. Studies had shown that long-duration exposure to microgravity causes a gradual loss in total bone mineral, a condition that mimics osteoporosis, which afflicts many older people. Wolf's work with this involved critically timed injections of isotopes, many blood draws, careful dietary logs, and urine collections. By learning more about bone mineral

loss and recovery in space veterans, researchers hoped to develop better treatments for those people who suffer bone disorders on Earth. Wolf also conducted an investigation to study the human body's ability to produce antibodies in microgravity in response to vaccination. Previous research had indicated that some of the human body's immune responses appear to be suppressed during long-duration spaceflight. In this case, the vaccination against pneumonia demonstrated normal antibody production.

In early December, Wolf's flight received a five-day extension with the adjustment of the launch date of *Endeavour* (STS-89). The joint decision by U.S. and Russian officials to delay the launch of STS-89 enabled the Mir-24 cosmonauts to complete three spacewalks planned for late December and early January. The delay also allowed Wolf additional time to complete his science program on Mir.

In a pre-holidays interview on December 12, Wolf discussed how he was feeling onboard the space station. "After about a month, I was feeling extremely good; and after two months, I realized just how good you could feel in space. And, I'm feeling better and better every day, enjoying

Pictured above are the Himalayan Range and Mt. Everest of Nepal; the Tibetan Plateau appears in the foreground. This is one of the views of Earth observed by the astronauts during the Shuttle-Mir Program. For more Earth observation photos, see pages 150-153.

Getting Settled in on Mir

October 6, 1997

It's a bigger job than one might think when every item you touch just floats off if you don't Velcro it, or strap it down, or bungee it in place. Great—my long-lost and invaluable electric shaver just floated by. The first place to look for lost items is in one of the air filters. A bowling ball would find its way there in zero gravity. Unfortunately, it's an obstacle course on the way and a lot of items don't make it all the way. One really learns, by the school of hard knocks, to work in little sequential steps. I keep finding myself with too many things in my hands and no way to put them down. Velcro is the lifesaver for organization—but what about 150 film cans, 25 cassette tapes, 25 CDs, 40 sets of clothes, 7 cameras, 20 lenses, over 1,000 components of scientific gear, 10 hard drives, 100 optical discs, 50 floppies, 2 critical PCMCIA memory cards (find them in all this), 4 watches, 6 computers (not counting the one we delivered for Mir—which is working flawlessly—knock on wood), 4 months of food, 30 packs of no-rinse shampoo, 60 more of body soap, razor blades, bottle of whiskey (just checking if you are still reading), and literally 6 tons more. The organizational/inventory task alone is daunting. Then come the radiograms instructing us to begin using all this. I just found my razor a minute ago. Darn, where did that radiogram float off to?

My cubicle (really the airlock) has a view that is out of this world. I share it with three spacesuits. More later. Taking a tour of the air filters.

Dave

working in space more and more, learning to handle the difficulties of working in space better and better … although I miss home more and more." His hopes for the New Year included a world at peace, in which all the borders—invisible from space—would "mean less in terms of wars and the problems we have, and more in terms of helping each other." Commander Solovyev spoke of meeting the New Year in a way "that cannot be duplicated on Earth … with the Australians first," because Australia was just west of the International Dateline.

He said, "A lot of people come out and want to talk to us on the [ham] radio and to [extend] us the best of wishes. This is a rather difficult time for us in that regard, but in general it's also very pleasant. There is, apparently, no champagne here for us to celebrate the New Year, although there is, of course, the desire."

Wolf was asked about the possibility of his making a spacewalk with Solovyev. He answered, "You know that since I was nine years old … I've wanted to do a spacewalk…. We trained intensely before the mission. I continue to train onboard. The spacesuit is in itself a spaceship. It has all the fundamental systems of a spacecraft …, and in a mission this long you need to study on orbit…. I'll look forward to taking a lesson from this fine spacewalker."

On December 17, the Progress resupply vehicle, filled with trash for disposal, undocked from Mir and was commanded to stop a short distance from the station. A small German-built robot camera, called Inspektor, was deployed from the Progress for its first flight test. Inspektor was supposed to first circle the Progress to test its maneuvering system and navigational capability. Then, on computer command from the Mir cosmonauts inside the station, Inspektor was supposed to approach Mir and place itself in an elliptical orbit around the orbital outpost—to become, in effect, a satellite of a satellite. But, Inspektor's star tracker guidance system overheated; and Mir's crew reported that, through binoculars, they could see it was pointed in the wrong direction. The robot camera on Mir was abandoned after a malfunction. Russian flight controllers had to stop the experiment. Inspektor was allowed to drift away from the space station to later burn up in Earth's atmosphere.

That disappointment didn't affect the Mir crew's overall sense of accomplishment. Wolf wrote, "The whole team is really hitting its stride right now. This is what the space station era is all about. Taken together, the little efficiencies turn a 16-hour day into, say, [an 11-hour day]. That sure means a lot at midnight when you want to float back and appreciate the adventure. Tolya [Solovyev] and Pasha [Vinogradov] are master craftsmen as they handle this ship. Occasionally Tolya, flashlight in teeth, will disappear behind a wall panel, tools and parts in tow. Hours later, as the sounds of drilling and wrenching subside, he emerges."

The holidays passed happily, but New Year's Day reminded everyone that things could always change onboard Mir. The motion control system computer failed again; but, because the station was in a good attitude at the time of the incident, no Soyuz jet firings were required to stabilize it. The batteries were also in good shape, and no damage was caused to any systems due to shutdown. However, during the recovery of the computer, all but the Base Block and the Kvant-1 modules were powered off as a conservation measure. The new Vozdukh carbon dioxide removal system scrubber in the Base Block was also shut down briefly. The crew turned off both Elektron oxygen units, and they used oxygen-generating candles and oxygen from the Progress vehicle until the Elektrons were reactivated. The entire crew also had to work on replacing a cooling system pump to keep temperatures in Priroda and Kvant-2 at comfortable levels. Temperatures throughout the mission ran well into the 90s. Indeed, throughout the entire Shuttle-Mir Program, conditions onboard Mir were hot and humid. However, they were well-tolerated by the crews, who dressed mainly in shorts and T-shirts.

On January 9, Solovyev and Vinogradov conducted a five-hour spacewalk. They examined and photographed the leaking airlock hatch on the Kvant-2 module, and they retrieved NASA's Optical Properties Monitor Experiment. Mir Astronaut Jerry Linenger had installed this experiment on the docking module during his extravehicular activity nine months earlier. Wolf monitored the spacewalk from the Base Block, and he photographed and videotaped the work conducted outside by his crewmates.

Later that week, Wolf finally had his own turn to try spacewalking. He ventured outside Mir for four hours with Solovyev, the world record holder. Among other duties, Wolf sampled areas of Mir with a spectrometer that

gauged the impact of the space environment on the surface of the space station. Wolf later called his spacewalking experience "spectacular."

However, during the spacewalk Wolf's spectrometer had a failure in its display, requiring intense coordination with Vinogradov inside Mir. As Wolf and Solovyev attempted to reenter Mir, the external airlock refused to make a totally airtight seal; and the atmospheric pressure, about 20mm of mercury, was too high for the spacesuits' cooling units to operate but far too low to allow de-suiting. The two spacewalkers were forced to retreat into the emergency backup airlock, and the high workload in the absence of cooling led to a rapid increase in their suits' internal temperatures. The airlock's shape, combined with the fact that the suits remained stiff in the fully pressurized state, required the two partners to depend on each other to connect the backup cooling umbilicals. This "off-nominal" spacewalk served as further proof that multinational crews could work together in real time, in multiple languages, and on complex operational tasks.

Wolf's four months in space came to a close in January 1998 when *Endeavour* (STS-89) launched on January 22 and traveled with Andy Thomas, the final NASA Mir astronaut, to the space station.

Right to the end of Wolf's increment, challenges persisted for the Mir-24 crew, and the crew persevered in meeting them. Although Wolf and his Russian crewmates had essentially met for the first time in space—and had not trained together—a theme of teamwork and mutual respect characterized the mission. During the final two weeks, the Russian e-mail system went down altogether. The crew depended on this link for all written information from the Mission Control Center-Moscow. The crew now

Meanwhile on Earth
NASA-6

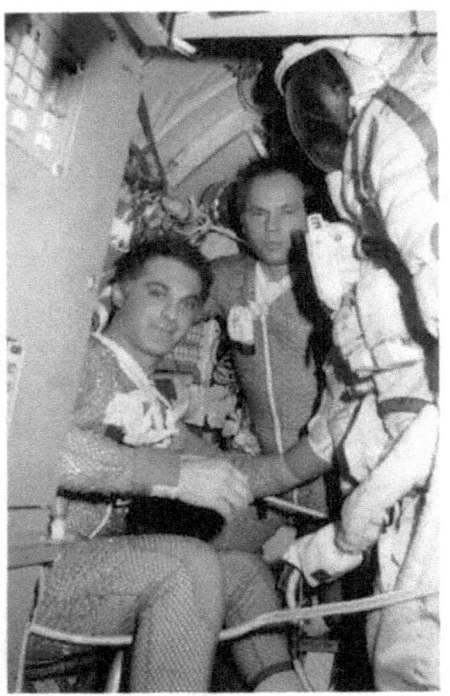

Wolf (left) and Mir-24 Commander
Solovyev in the Kvant-2 module prepare
for their four-hour joint spacewalk.

relied completely on verbal communications
passes, which at this point occurred about five
times a day for about eight minutes each time.
The large amount of technical information trans-
mitted in these passes severely curtailed all other
forms of communication. As a result, Wolf did
not hear about Earthly events, such as space pioneer John Glenn's assignment to a Shuttle flight,
until after *Endeavour* had docked when a reporter
asked him for his reaction to it during a press conference. Surprised, Wolf turned to Shuttle
Commander Terry Wilcutt for confirmation.

Wolf returned to Earth on January 31,
1998. During his time on Mir, he completed 36
scientific investigations in six disciplines, including biotechnology, fundamental biology, human
life sciences, microgravity sciences, Earth science,
and advanced technology research. In effect, a
backup crew had been called up for duty and
had successfully completed the flight program.
The risk of persevering through difficult times
was now paying off.

NASA announced in early October 1997
that Astronaut Andy Thomas would
serve the seventh and final tour of duty
by an American onboard Mir. Thomas had been
in Star City since early in the year, training as
David Wolf's backup. Astronaut James Voss,
who had previously trained as backup to Mike
Foale, was assigned as Thomas' backup.

In mid-October, Cosmonaut Salizhan
Sharipov, a Russian Air Force officer, was named
as an additional crewmember for the *Endeavour*
STS-89 mission that would take Thomas to Mir.
Sharipov was a good illustration of the "new civics"
in the territory of the former Soviet Union. He
was an ethnic Uzbek who was born in Kirghizia,
and he was now a citizen of Russia. His main role,
during his first spaceflight, would be to assist the
Endeavour and Mir crews in the transfer operations
of the eighth Shuttle-Mir docking mission.

While Sharipov was training at Johnson Space
Center in Houston, Thomas and Voss trained in
Star City outside of Moscow. Thomas spent much
of his time focusing on Mir systems and spacewalk
training. He was also fitted for his Sokol suit, the
spacesuit worn while in the Soyuz capsule during
Mir fly-arounds and possible evacuations.

During November, Thomas continued spacewalk training in the Russian Hydrolab facility.
He also took part in several hands-on sessions
with science experiments slated for his stay on Mir;
and he took a class to familiarize himself with the
science program of French cosmonaut Leopold
Eyharts, who was scheduled to accompany the
Mir-25 crew early in Thomas' mission.

On November 17, 1997, the United States
and Russia announced the first four crews to
inhabit the International Space Station (ISS).
Their expertise was heavily drawn from Shuttle-
Mir Program experience. Astronaut Bill Shepherd
would command the first Expedition, and would
be joined by Cosmonaut Yuri Gidzenko and by
Cosmonaut Sergei Krikalev, who had been the first
Russian Shuttle crewmember on the STS-60 mission in 1994. Shepherd and Krikalev had been
named to the mission in 1996, and both had
already been training in the other's country's space
program. This first crew would inaugurate Phase 3
of the International Space Station Program (Phase

1 being the Shuttle-Mir Program and Phase 2
being the early construction Space Shuttle
missions). Cosmonaut Yury Usachev, one of
Shannon Lucid's crewmates on Mir, would
command the second ISS Expedition. He would
be joined by American Astronaut Susan Helms
and by Astronaut Voss, the Shuttle-Mir backup
astronaut for Foale and Thomas.

Astronaut Kenneth Bowersox was scheduled
to command the third ISS Expedition. He would
be joined by Cosmonaut Mikhail Turin and by
Cosmonaut Vladimir Dezhurov, who had been
Norm Thagard's Mir-18 commander in 1995.
(Although remaining as backup to Expedition
One Commander Shepherd, Bowersox would be
replaced in 1999 as the third Expedition com-
mander by former Shuttle-Mir Program Manager
Frank Culbertson.) The fourth ISS Expedition
crew would be commanded by Cosmonaut Yuri
Onufriyenko, Shannon Lucid's commander dur-
ing Mir-21. He would be joined by Astronauts
Daniel Bursch and Carl Walz. Walz had flown to
Mir on the STS-79 mission.

At the end of November 1997, Thomas successfully completed his training at the Gagarin
Cosmonaut Training Center and was certified by
the Russian Chief Medical Commission for his
planned four-month Mir flight. Thomas and Voss
then returned to the United States to conclude
their training for *Endeavour's* STS-89 launch.

As Wolf spent his holidays onboard Mir,
many of NASA's people in Russia gathered to
celebrate. Wolf's Operations Lead Patti Moore
later recalled it as a personal "high point" for
her in Russia. She had dreaded being away from
her family in America, but Flight Surgeon Chris
Flynn's wife, Alice, cooked Christmas dinner for
the Americans in Russia. Moore recalled that
another Operations Lead, Christine Chiodo,
had "even made a little stocking for every single
person there and had a little gift for everybody.
We had a Christmas tree with presents under it.
Even the cats got presents from Christine. All of
that was very special, because you were away
from home at a tough time. But, I don't think
anybody was really depressed. I think everybody
called home and was happy to talk to their
family…. But, we were all there together."

Who Went to Mir and Who Didn't

I n the words of the STS-89 Commander Terry Wilcutt, the Shuttle-Mir astronauts comprised an "all-volunteer force." Spending months onboard a former adversary's orbital outpost did not appeal to all of NASA's astronauts—nor did working as a "guest researcher" while speaking a difficult foreign language.

To everyone, it was clear that the assignment would require prodigious amounts of stamina, strength, and sacrifice. Also, some astronauts thought that a Mir assignment might take them out of consideration for future Shuttle flights.

Furthermore, those who did apply for Mir could not be sure that they would get there. As with all spaceflight assignments, Mir astronauts faced many potential obstacles— including health restrictions, program needs, and operational constraints. (As it turned out, all who were selected, but who were then later disqualified, got assigned as crewmembers on Shuttle missions to Mir.)

Norm Thagard's backup, Astronaut Bonnie Dunbar, remained active in the Shuttle-Mir Program; she flew to Mir as a crewmember of STS-71 and STS-89.

Two other NASA astronauts actually scheduled to spend time on Mir lost their opportunities because of Russian size requirements. Scott Parazynksi was found to be "too tall" for revised limitations on the

seats for the Soyuz escape capsule. He was able to fly later as a crewmember on STS-86, during which he made a spacewalk on the Mir-Shuttle complex.

Astronaut Wendy Lawrence was twice affected by size restrictions. First, she was found to be "too short" for the Soyuz seats.

Scott Parazynski and Wendy Lawrence look out from the Soyuz spacecraft docked to the Mir. The two STS-86 mission specialists had both trained to be Mir residents but were declared as "too tall" and "too short."

Then after new measurements, she again became a "prime" candidate for Mir; so she trained for the NASA-6 mission. However, one of the outcomes of the Progress vehicle collision was a new requirement that all Mir residents be qualified for extravehicular activity. Again, Lawrence was deemed to be too small—this time for the Russian spacewalking suits. David Wolf, her backup, bumped up into her slot. Like Dunbar, Lawrence continued to contribute to the Shuttle-Mir Program. She served as NASA's Director of Operations-Russia and flew as a

crewmember onboard STS-86 and STS-91.

The chronology shows Thagard and Dunbar were announced in February 1994. Shannon Lucid, Jerry Linenger, John Blaha, and Parazynski were announced during Thagard's flight in March 1995. Lawrence was announced in September 1995.

A month later, both Lawrence and Parazynksi were disqualified due to the size limitations.

Mike Foale learned about his selection in October 1995 and began training in January 1996. Lawrence was re-announced for a Mir residency in August 1996, but she was replaced by Wolf in July 1997. The last Shuttle-Mir astronaut, Andy Thomas, was announced in October 1997.

Although Thomas' backup, Astronaut James Voss, did not get a Mir assignment of his own, but he was later selected for the second Expedition crew to the International Space Station.

Public attention is almost always on the astronauts, but Wilcutt's term "all-volunteer" could also be applied to nearly all the NASA employees who worked in the Shuttle-Mir Program. The same circumstances that required sacrifices of the astronauts required many on the Shuttle-Mir team to donate extra time and effort, both in Houston and in Russia.

Space Shuttle *Endeavour*

Launched: January 22, 1998, 9:48 p.m. EST
Kennedy Space Center, Pad 39-A

Orbit: 160 nautical miles

Inclination: 51.6 degrees

Landed: January 31, 1998, 5:36 p.m. EST
Kennedy Space Center

Mission: 8 days, 19 hours, 48 minutes

STS-89 CREW

Commander Terrence W. Wilcutt
Third Shuttle flight

Pilot Joe F. Edwards, Jr.
First Shuttle flight

Mission Specialist Bonnie J. Dunbar, Ph.D.
Fifth Shuttle flight

Mission Specialist Michael P. Anderson
First Shuttle flight

Mission Specialist James F. Reilly, Ph.D.
First Shuttle flight

Cosmonaut Salizhan S. Sharipov
Russian Space Agency
First Shuttle flight

Mission Specialist Andrew S.W. Thomas, Ph.D.
Second Shuttle flight; remaining on Mir

Mission Specialist David A. Wolf, M.D.
Third Shuttle flight; returning from Mir

PAYLOADS

Space Habitation (Double) Module

Shuttle Ionospheric Modification with Pulsed
Local Exhaust

Closed Equilibrated Biological Aquatic System

TeleMedicine Instrumentation Pack

Global Positioning System Development Test
Objective

Mechanics of Granular Materials Experiment

Space Acceleration Measurement System

Astroculture Biological Experiments

STS-89
Last NASA Astronaut to Mir
January 22 – 31, 1998

Posing for a group portrait on the Base Block of the Mir space station are, lower left, clockwise: NASA-6 Astronaut Wolf, Mir-24 Flight Engineer Pavel Vinogradov, STS-89 Mission Specialists Sharipov, Reilly, Pilot Edwards, NASA-7 Astronaut Thomas, Mission Specialists Anderson, Dunbar, Mir-24 Commander Solovyev, STS-89 Commander Wilcutt

STS-89 initiated the last NASA Mir residency by delivering Andy Thomas and bringing David Wolf home. The crew, which included cosmonaut Salizhan Sharipov, helped strengthen NASA's partnership with Russia in preparation for the International Space Station (ISS).

The Space Shuttle *Endeavour* became the first Orbiter after *Atlantis* to dock with Mir. This was *Endeavour*'s first flight since STS-77 in May 1996, after which it went to Palmdale, California, for new main engines and an external airlock for future ISS duties. A few unexpected delays occurred in the week before launch. The Russians wanted to wait until some specific work would be finished onboard Mir, and NASA needed to repair some of the Orbiter's heat-shielding tiles.

Endeavour roared skyward in another spectacular nighttime liftoff to Mir after a prospect of stormy weather. One of the few problems associated with the launch was that the solid

rocket boosters parachuted into rough seas. They could not be towed back to Cape Canaveral for three days, and two of the four parachutes were lost because of the delayed recovery. Repairing booster cracks and other damage would result in a cost of at least $7 million. However, high in space, the STS-89 crew effected a textbook rendezvous and docked with Mir over Russia at an altitude of 244 nautical miles. A significant flight modification for STS-89 was the way Commander Terry Wilcutt approached Mir for docking. He brought the

Endeavour in nose-forward to try out techniques needed for several ISS dockings.

The handover from Wolf to Thomas had only a few hitches. Clearly delighted to see *Endeavour*, Wolf waved from a Mir window and did a slow somersault. Crewmembers aboard the Shuttle teased him on the radio. Bonnie Dunbar said, "We're just discussing the fact that maybe Andy forgot his suitcases and we might have to take him back." However, it wasn't Thomas' suitcases but his Sokol pressure suit that caused the problem. The Russian custom-made suit would be needed in the event of an emergency evacuation of Mir. When Thomas tried on the suit in orbit, the torso was a little too short, making it difficult to pull it over his shoulders. Thomas speculated he may have "grown" a little, a common effect of microgravity. He tried on Wolf's suit and found that its sleeves were too long. Mir Commander Anatoly Solovyev helped Thomas alter his suit by detaching some internal straps. Later, Thomas said that it fit "like a glove." If necessary, he could safely descend to Earth in the Soyuz.

Endeavour continued its career as a logistics workhorse when 7,400 pounds of gear were transferred between the two ships. In addition to transfer activities, the STS-89 mission carried the SPACEHAB double module, which contained many scientific experiments.

By STS-89, Payload Commander Dunbar had developed a deep professional and personal

attachment to the Shuttle-Mir Program. This was her second Mir mission. Involved since the program's beginnings, Dunbar had trained as backup to NASA-1 Astronaut Norm Thagard. She then flew to Mir on the STS-71 flight to pick up Thagard and drop off the Mir-19 crew of Solovyev and Nikolai Budarin. With STS-89, she met up with Solovyev again, now a part of Mir-24. "When we opened the hatch," Dunbar related later, "it was really fun to see Anatoly." Nearing the mission's end, she and her colleague had a little fun. According to Dunbar, "One of the things we did—before we closed the hatch and left—was [that] I went in on the other side of the hatch and they took a picture of Anatoly and [me] inside the Mir.... When Mission Control in Moscow called us after undocking, they asked me what side of the hatch I was on. I told them [that], unfortunately, I was on the Shuttle side."

The conclusion of the STS-89 mission saw 13 people in three different spacecraft in orbit. *Endeavour* pulled away from Mir soon after a Soyuz had launched from Baikonur with two Russian cosmonauts and a French researcher. Talgat Musabayev and Budarin were on their way to Mir with Leopold Eyharts. Eyharts conducted experiments onboard Mir for three weeks before returning to Earth with Wolf's Mir-24 crewmates. Musabayev and Budarin remained on Mir with Thomas.

David Wolf (above right) reviews procedures with NASA-7 Astronaut Andy Thomas during the crew handover. (Left) Bonnie Dunbar holds the roster signed by crews during the Shuttle-Mir Program, and an STS-89/Mir-24 plaque.

Andy Thomas

NASA-7

January 22 – June 12, 1998

Smoother Sailing

As the last American to live onboard Mir, Andy Thomas benefited from the maturing U.S.-Russian partnership. The Mir space station was now safer, more versatile, and more robust. Communications had improved both between ground and orbit, and between the two space programs. Yet at the same time, life on the outpost still posed many risks. Mir was aging, sailing past its planned lifespan into a time when things could be expected to break down with some regularity. Further, there remained the human challenges with languages, cultures, and politics.

But, Thomas had his own strengths and the experiences of his six NASA predecessors. He knew, better than they had known, just what to expect. He went to Mir to conduct an intense science program and to make the most of his 140 days in orbit. He succeeded. And when STS-91 came to fetch him home, the Shuttle-Mir Program came to a successful end.

A naturalized U.S. citizen, Andrew S. W. Thomas had exploration in his blood. More than a century before his service on Mir, his great-great-grandfather had served on the first expedition to cross Australia from south to north. Born in Adelaide, Australia, in 1951, Thomas received a bachelor's degree in mechanical engineering and a doctorate in mechanical engineering, both from the University of Adelaide. He began his professional career as a research scientist with the Lockheed Aeronautical Systems Company, researching fluid dynamic instabilities and aircraft drag. He became the head of the Flight Sciences Department and managed a research laboratory and wind tunnel facility.

In 1987, Thomas was named manager of Lockheed's Flight Sciences Division. He joined the Jet Propulsion Laboratory in 1989 and soon was appointed leader of their program for microgravity materials processing. NASA selected Thomas for astronaut training in 1992. In May 1996, he flew on *Endeavour* (STS-77) as Payload Commander during the time that NASA-2 Astronaut Shannon Lucid was entering her third month onboard Mir.

On STS-77, Thomas carried with him an old steel flint, a memento of his great-great-grandfather's explorations in the Australian Outback. But for all the adventure of his Space Shuttle and space station flights, Thomas said later that he didn't see himself "in the same league" with people such as Lewis and Clark or other early explorers,

We heard Pilot Joe Edwards call out. 'Three at 104,' signifying all three were running at rated power. But we were still firmly bolted to the ground with eight very large explosive bolts, so the engine thrust made us lurch over, giving us the eerie sense of falling forward. Suddenly, with the six seconds counted away, there was a thundering roar with massive vibration ... as the solid rockets ignited, the hold-down bolts exploded, and we were driven off the launch pad and upwards into the sky. You did not need a window to know what was happening."

When *Endeavour* rendezvoused with Mir, Thomas caught his first glimpse of his new home. "We could see the station out the overhead windows—first as a point of light off in the distance that slowly grew brighter as we approached. Soon the characteristic shape of Mir could be made out, with its cruciform layout of modules and their protruding solar arrays. These panels are very winglike in their shape, and indeed Mir has often been likened to a giant insect in its appearance. We slowly approached Mir from below toward the Kristall module that carries the docking fixture."

After docking and the opening of the hatches, Thomas wrote, "My first views of the station were a little daunting. And it was very confining as we floated down the Kristall module to the Base Block. There was a lot of equipment stowed on all the panels and in every available location. But it did open out at the Base Block, which is more spacious by comparison."

He later said, "It was a bit of a shock just how crowded it was and how much stuff was in there. And, that took some getting used to. But, you can get used to it.... At no time did I feel claustrophobic up there.... There's enough room

(Previous page) Andy Thomas floats through the docking module hatch into the Shuttle airlock. During his residency, Thomas found the stars and the distant city lights particularly peaceful (above). He captured many of his Earth observations on film and kept the camera readily on hand (below).

such as his great-great-grandfather. He allowed that the early explorers had "planned everything, and had a lot of support." However, "Once they'd gone, they were gone. They were alone." Thomas said, "When we do these flights, it's true we're alone up there. But, we have radio communications to a huge group of people [on Earth] who have a lot of resources to provide assistance in the event something goes wrong and a lot of guidance about what should be done next. So, in that sense we're not alone."

When Thomas launched to Mir on STS-89 on January 22, 1998, the experience was already a different one for him. On his first Shuttle mission, he had sat on the flight deck with its

many windows and its view of a receding Earth. For STS-89, Thomas sat on the mid-deck without a view. However, according to Thomas, "Not having an outside view lets your imagination provide the imagery, and this can give you an emotional rush, possibly even more than seeing."

In the first of his "Letters from the Outpost," Thomas related his liftoff to Mir. "The weather had been questionable that day and there was still some uncertainty as to whether or not we would actually go. But ... the Control Center then called us to start the auxiliary power units that provide the steering hydraulics, and we could hear the units spinning up to speed, deep below us. Then came the call to close and lock our visors, and to initiate our oxygen flow—a protection in the event of a depressurization during the climb-out.... It was clearly getting serious as we waited those long last few minutes and seconds until liftoff. The three of us on the mid-deck shook hands together and wished ourselves well for the flight. Then the cabin became quiet.

"At six seconds before launch, a deep rumble started, shuddering the Orbiter as its three engines were ignited and run up to full speed.

Andy Thomas served the final U.S. residency on the Russian Space Station Mir, shown here in its 1998 configuration.

The custom-made Russian Sokol suit worn by Thomas (right) had to be adjusted after his arrival to Mir due to his "growth" in height—an effect of microgravity. Assisting with the alterations was Mir-24 Commander Anatoly Solovyev.

| [Endeavour] was a breathtaking sight as it pulled away... In Earth-shadow, plumes of flame from the maneuvering jets lit up the solar arrays of the station.

that you don't feel claustrophobic, but you are aware that it's a confined environment—that you don't have a lot of options of places to go."

Although Thomas was now safely onboard Mir, a potential showstopper occurred on the second day of docked operations. Before he could officially become a Mir-24 crewmember, joining Commander Anatoly Solovyev and Flight Engineer Pavel Vinogradov, Thomas' custom-made seat-liner had to be installed in the Soyuz escape capsule and his Russian Sokol pressure suit had to be checked for fit. Surprisingly, Thomas and Commander Solovyev were unable

to get Thomas into his suit. Microgravity had allowed Thomas' spine to expand; he was now too tall for the suit. Two days of discussions on the ground and some re-tailoring of the suit by Solovyev took place before the Shuttle-Mir managers, and Thomas himself, became satisfied that all was safe to proceed.

After four days docked to Mir, *Endeavour* readied for its return to Earth. For Thomas, "This was a moment of mixed emotions.... On the one hand, I was sorry to see my colleagues leave; but on the other [hand], it meant that I was now able to get on with the mission." The Space Shuttle was a breathtaking sight as it pulled away and flew around the station. In sunlight, *Endeavour* shone brilliant white. In Earth-shadow, plumes of flame from the maneuvering jets lit up the solar arrays of the station.

Thomas made his home in the Priroda module. He set up a computer with access to the informational and recreational CDs that NASA had provided. He found the bag containing his books, music recordings, stationery and art supplies, and personal hygiene items.

He wanted to dive deeply into his scientific experiments. But on January 31, only two days after the Shuttle departed, a Soyuz capsule arrived with his future crewmates: Mir-25 Commander Talgat Musabayev and Flight Engineer Nikolai Budarin. Accompanying the two cosmonauts was French researcher Leopold Eyharts, who would

stay for the three weeks of handover and then return to Earth with the Mir-24 crew.

Thomas watched their approaching capsule. "Their Soyuz appeared over the horizon, first as a small point of light that slowly grew to its identifiable shape with its attached habitation module and protruding solar panels." He and his Mir-24 crewmates watched the final approach on a video monitor, and they felt a slight bump as the Soyuz docked to the station. After checking the integrity of all the seals between the vehicles, they opened the hatch and welcomed the new crew aboard. Thomas later wrote, "It was strange to see them all again here in orbit," because he had not seen them since he left Star City for Houston early in the previous December.

This mission was Musabayev's second Soyuz flight to Mir. Budarin had flown to Mir with his Mir-19 crewmate, the same Solovyev he was now meeting again in space. That was onboard STS-71 in 1995, when *Atlantis* performed the first Shuttle-Mir docking and brought U.S. Mir Astronaut Norm Thagard and his Russian crewmates back to Earth.

While the six crewmembers were onboard, Mir showed its erratic side when a software glitch in an onboard computer placed the station into free drift. This time, however, Mir's motion control computer never shut down, and Mir's briefly

French researcher Leopold Eyharts (below, left) traveled with the Mir-25 crew to the station and stayed onboard for three weeks. He is pictured with Mir-25 Flight Engineer Nikolai Budarin and Mir-24 Commander Anatoly Solovyev.

Andrew S. W. Thomas

Andy Thomas was born in Adelaide, South Australia. He earned a bachelor's degree and a Ph.D. in mechanical engineering at the University of Adelaide. After coming to the United States, Thomas worked for Lockheed Aeronautical Systems Company where he led a research department engaged in experimental and computational studies in fluid dynamics, aerodynamics, and aero-acoustics before becoming manager of Lockheed's Flight Sciences Division. He later joined NASA's Jet Propulsion Laboratory, investigating areas such as aerodynamics and microgravity materials processing.

Thomas was chosen for astronaut training in 1992. He provided technical support to the Space Shuttle Main Engine Project, the Solid Rocket Motor Project, and the External Tank Project at the Marshall Space Flight Center. He flew as payload commander onboard STS-77 in May 1996 and performed more

microgravity research.

In his oral history, Thomas talked about exploration. Even after his remarkable achievement onboard Mir, Thomas did not see himself "in the same league as ... Lewis and Clark ... or the people who explored the Antarctic and the Arctic regions. Because when they went off on those very courageous journeys, they went by themselves.... Once they'd gone, they were gone. They were alone. When we do these flights, it's true we're alone up there. But, we have radio communications to a huge group of people [on Earth] who have a lot of resources to provide assistance in the event something goes wrong ..."

Thomas continues with the NASA astronaut program, preparing for future flight assignments and working in the International Space Station Program.

Seventh NASA astronaut to live onboard Mir • Launched with STS-89 on January 22, 1998 • Returned with STS-91 on June 12, 1998
Concluded more than two years of continuous U.S. presence on Mir

unpowered gyrodynes continued to spin while the crew corrected the problem. In other systems activities, the cosmonauts worked on one of Mir's 11 operational gyrodynes, replacing some electronics with spare parts brought up on STS-89.

Having two crews onboard Mir for three weeks limited workspace and stowage space. Thomas would later say that this was the hardest part of his stay on Mir, "when we had a lot of people aboard and it was very crowded." Spirits were high, however; and a Dutch observer, who recorded Mir radio traffic, reported, "The mood among the four cosmonauts, one astronaut, and one spationaute is excellent. There is a lot of joy and they do not complain about their modest housing."

The two cosmonauts and the French spationaute departed Mir in their Soyuz capsule, and made a safe landing in Kazakhstan during a blizzard. Thomas later said from orbit that after the Mir-24 crew left, "it then became a lot easier—getting into the work routine, the

recreation routine. I became very comfortable onboard the station, and at no point did I feel like I needed to leave. It became a very sort of comfortable, easy lifestyle in some ways."

Thomas began his complement of science activities, which would focus on 27 studies in the areas of advanced technology, Earth sciences, human life sciences, microgravity research, and International Space Station risk mitigation. His investigations would conclude some experiments started on the six previous U.S.-Mir missions, as well as begin some new research. One of the first experiments he activated was an X-ray detector device, designed to gather information on the background cosmic radiation aboard the station. He started several experiments, including the astroculture unit that provided a controlled environment chamber to support plant growth in space. He spent much time ensuring that the Biotechnology System Co-culture Experiment was rotating as expected and that the proper doses of media and nutrients were reaching the reactor chamber.

And, he soon began collecting urine samples to support the Renal Stone Risk Assessment Experiment, which studied the risk of kidney stone formation due to a sudden absorption of calcium by the body during spaceflight.

During this time, among their other duties, Thomas' crewmates replaced some hardware systems on the station, including two different water reclamation systems—one that recycled water from urine, and one that recycled water condensed out of Mir's atmosphere.

In a media interview in early February 1998, Thomas spoke about his life on Mir. There had been questions in the press about his Russian language abilities, especially pertaining to social conversations. Thomas agreed on the importance of socializing, "because we spend a lot of time together in a confined space, not just working as professionals but around the dinner table." Talking about things in general and "sharing experiences of the day" were important. Thomas said, "I've been talking with Talgat a lot, and

A masked Thomas sets up the Biotechnology System Co-culture Experiment conducted to better understand the biochemical and morphological interaction between breast cancer and endothelial blood vessel cells—an understanding that could lead to the development of new treatments for breast cancer.

Tools and personal equipment. Your toothbrush. Your comb. Those are big adjustments in the lifestyle that you have to make when you're in this kind of environment."

Thomas confirmed what the previous NASA Mir astronauts had said—that life on Mir was difficult in several ways. "It's hard because you're isolated. I mean, I have a very stimulating work-day every day, with a lot of challenging activities. Of course, the view is always there, and it's an amazing view. But, each day tends to roll into the next and there comes a certain monotony. You have to use your own resources to make the life interesting, to keep your motivation going. It's undeniably a challenge because you're in a confined space. It's crowded, and you have some difficult objectives. So, there are great challenges of taking on a mission like this. There's no doubt about it."

For Thomas, the biggest issue on Mir was the shortage of stowage space. "We're always fighting this problem of storage, of where to put things in order to do work." Still, his biggest surprise was how quickly he had physically adapted to living in microgravity. To Thomas, weightlessness felt like a "perfectly natural thing."

On February 20, the Mir crew celebrated the 12th anniversary of the launch of the Mir's Base Block, as well as 700 days of continuous American presence in space. In 1986, the Base Block, perched atop a Russian Proton rocket, had launched from the Baikonur Cosmodrome in Kazakhstan to begin what was planned to be five years in Earth orbit. Now in 1998, the Mir space station was composed of eight permanent modules. All but the punctured Spektr were habitable.

> ... the crew celebrated the 12th anniversary of [Mir], as well as 700 days of continuous American presence in space.

Also on February 20, the crew boarded the Soyuz capsule and backed away from Mir to free the Kvant-1 port for the redocking of the Progress capsule, which had been filled with trash and placed in a parking orbit on January 30. Instead of doing a fly-around as other crews had done, they held steady while ground controllers rotated Mir. Musabayev then manually flew the Soyuz back to a smooth docking at the transfer node recently vacated by the previous crew's

we've been working together in the Priroda module. We're having a good time together. We joke and kid around. I'm sort of cueing him on English, and he's cueing me on Russian.... We're telling a lot of war stories together and talking a lot about music and things, and having a good time." His conversational Russian would continue to improve.

Thomas also said that the Russian space station was "proving to be a very interesting place to live and work [in].... If you want to have fun, zero gravity is a great place to do it. But, I would have to admit that if you want to do very careful, detailed work, zero gravity is tough because you'd be amazed how easily you lose things. You take something and you just let it go for a minute, and you turn your back and you come back and it's gone somewhere. You won't find it again. I've had a terrible time just losing things, putting things down and forgetting about them, and they come loose and go flying off somewhere.

Australian native Andy Thomas brought some reminders from his homeland to his temporary home on Mir.

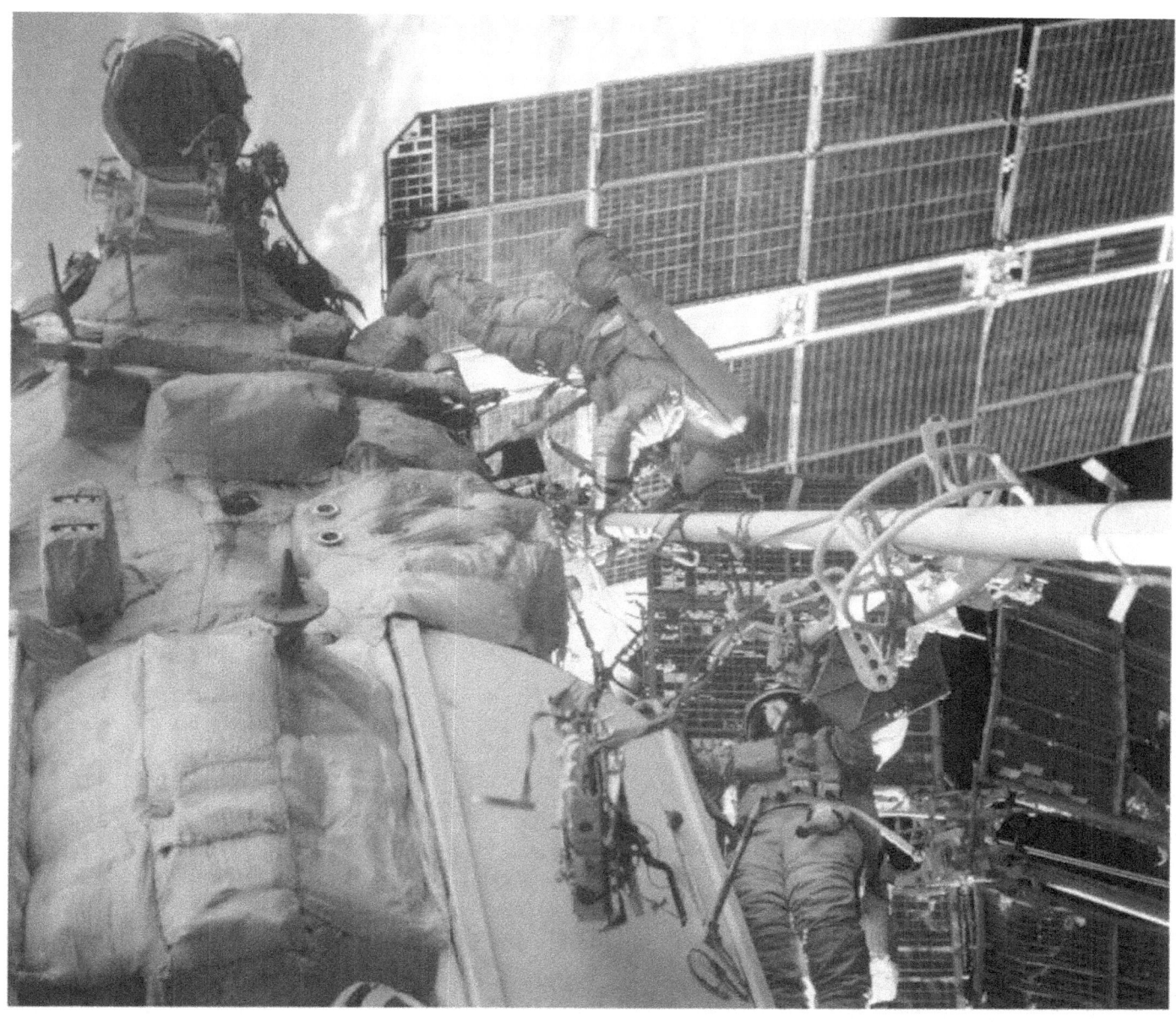

Soyuz. Back onboard Mir, the crew worked on life support systems, mainly atmospheric systems such as oxygen generators, a carbon dioxide scrubber, and a pressure valve. On February 23, the Progress vehicle was redocked, mainly using the Kurs automated system but also testing the manual tele-operated remote unit docking system that had posed serious problems to recent expeditions. When the crew reopened the re-docked Progress, a bad smell issued through the hatch—probably some overripe garbage.

Three days later, the crew got a much worse surprise. Smoke started issuing from a device that removed contaminants from the air in the Kvant-1 module. Thomas was exercising on the treadmill in Kristall at the time. When he finished, he floated past the Base Block and was alarmed to see thick smoke drifting throughout the cabin. Evidently, switches on the device had been misconfigured. It had overheated, causing a fire within the unit, and fumes were being blown into the cabin. When Commander Musabayev

Mir-25 cosmonauts Talgat Musabayev (left) and Nikolai Budarin conduct an extravehicular activity to install handrails and foot restraints near the radiator on the damaged Spektr module, and to stabilize the module's solar array with a special brace.

finally noticed the smoke, he quickly turned off the apparatus. Fortunately, this contained the fire within the unit. The fire was allowed to burn itself out, and extinguishers were not needed. Regardless, the cloud of smoke soon spread throughout the entire space station, and smoke and odor could be noticed in all the modules. Of further concern was the fact that the fire alarm system had failed to generate an alarm despite the obvious and thick smoke.

Over subsequent days, the air cleaners slowly removed the fumes; but the crew continued to feel their effects for some time. The crew took contamination readings and reported them to the ground, but these were thought to be unrealistically high and were attributed to instrument errors. However, analysis of air samples that were later returned to the ground showed that the readings were, in fact, accurate. Carbon monoxide levels had reached 20 times the recommended safety levels, and had remained very high for a couple of days. Fire had again produced one of spaceflight's foremost perils—poisoned air in a small, confined place.

On March 4, Musabayev and Budarin attempted the first of five scheduled spacewalks of the Mir-25 expedition. Their plan was to brace the Spektr module's damaged solar array. However, the cosmonauts could not open all of the ten latches on the airlock hatch. The last latch was so stuck that Budarin broke or bent three of

the wrenches available in the airlock trying to release it. Ground controllers called off the EVA and rescheduled it for later in the mission, after the next Progress resupply vehicle could deliver new equipment.

During the extravehicular activity attempt, Thomas maintained voice contact with the Mission Control Center-Moscow. The Dutch radio listener on the ground reported, "He did

> ... the air cleaners slowly removed the fumes; but the crew continued to feel their effects for some time.

this in reasonable and certainly comprehensible Russian." Both Musabayev and Thomas had earlier told reporters that their spoken communications had greatly improved. (Besides Russian and English, they could also use German. The Dutch listener complimented Thomas' "excellent German" and reported that "it was clear that Musabayev had picked up a lot of the German language from the community of former Volga-Germans in his native country" of Kazakhstan.)

The cosmonauts began work to replace Mir's air conditioner, which had not been working since the last month of NASA-6 Astronaut David Wolf's increment. Although Wolf had made the comment that he liked the heat because it "gave him more energy," temperatures

in the Base Block had been in the 90s. The crew also had to rely on the dehumidifier on the Soyuz capsule and the Vozdukh carbon dioxide removal system to lower humidities.

In science activities, Thomas continued to work with the Biotechnology System Co-culture Experiment. He was having trouble with air bubbles forming in the facility's rotating chamber, so researchers on the ground instructed him to reduce the rate at which media and nutrients rotated around the reactor chamber.

With 48 days onboard Mir under his belt, Thomas was asked in a television interview how he was dealing with the isolation and confinement. He said, "You need to be able to psychologically remove yourself from it. And for that, you use recreational aids, much like you would on Earth actually. We have music, CDs, and tapes. I've got a good repertoire of movies on videotape that I can play. I brought some paperback books to read. So, these are the kinds of things that I use in order to just relax and unwind and relieve tension and get away from things."

Recreational aids may have been available, but at this point in the mission there was little time for relaxation. Musabayev reported to ground controllers that the work burden was too high and procedures often took longer than the time scheduled for them. Thomas was having problems doing all of his scientific work while fitting in the physical exercise necessary to counter the effects of microgravity.

On March 15, the older Progress separated from Mir. The vehicle burned up in the Earth's atmosphere, east of New Zealand over the Pacific Ocean. A new Progress resupply spacecraft arrived on March 25; but in the last moments of approach to Mir, its alignment wasn't perfect. Commander Musabayev switched over to the manual teleoperated remote unit docking system for the last 20 meters and docked it to Mir. In an interview later, Thomas called the docking "seamless, a beautiful piece of work.... We felt a slight nudge and a shudder in the station as the docking took place and as the systems latched together. It was

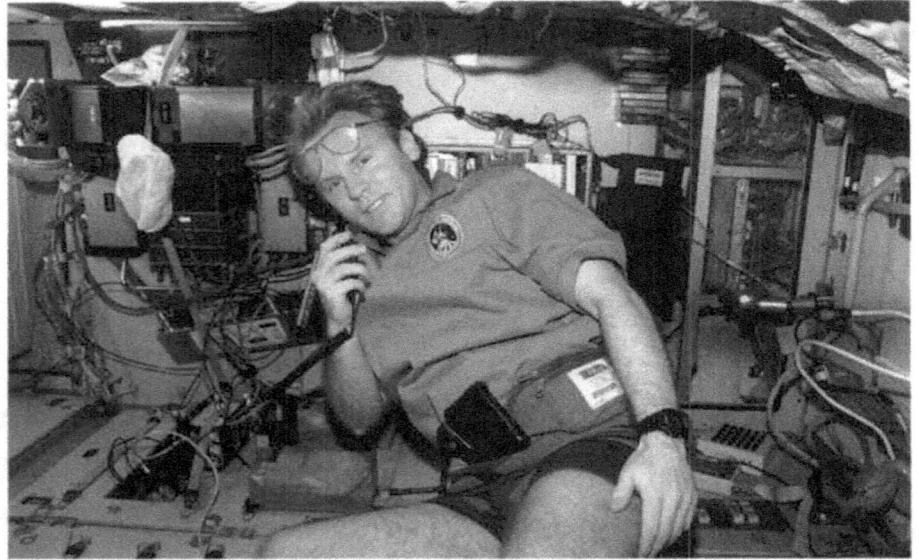

Thomas conducted research in 27 different areas of scientific study in the Priroda module.

all very benign, really, and went very well; and the commander executed the work flawlessly."

This time, when the crew opened the hatch, they were greeted with a clean aroma of fresh apples. Other welcome items onboard the Progress included letters from home, a computerized photo album for Thomas, new latches for the airlock hatch, new tools, fresh food, a CD player, and three 2-volume sets of Beatles music. On March 22, Thomas' 60th day in orbit, the Mir crew and the Shuttle-Mir Program marked two years of a continuous American presence on Mir.

Musabayev and Budarin began preparing for a series of what would be five extravehicular activities in the month of April. Thomas worked at his science investigations, including an immunity experiment for which he periodically took blood and saliva samples. This investigation compared the human body's ability to produce antibodies in microgravity with its ability to produce antibodies on Earth. Thomas also processed samples for the material science experiment called the Queen's University Experiment in Liquid Diffusion (QUELD). This joint U.S. and Canadian experiment used a special furnace to analyze the phenomenon of diffusion, which is the slow mixing of materials by the random movement of molecules of one substance into another.

On April 3, Musabayev and Budarin performed a 6½-hour spacewalk to install handrails and foot restraints, install a new workstation, and brace a solar array. All of this work was performed on the damaged Spektr module. Thomas monitored their progress and recorded their efforts on video. The work took longer than expected, so the cosmonauts had to postpone the work station and bracing jobs. Also, in a short but somewhat chilling event, contact with Musabayev was lost for several seconds. When it was restored, he reported that he had accidentally turned off his spacesuit's power supply, stopping many functions, such as communications, cooling, and ventilation.

Mir-25 crewmembers Nikolai Budarin (left) and Talgat Musabayev wave from the Kvant-2 hatch as they prepare for one of their EVAs.

Three days later, the cosmonauts conducted a fatiguing 4½-hour spacewalk. They completed the installation of handrails and foot restraints on Spektr and stabilized Spektr's solar array with the special brace. But, ground controllers instructed them to return to the station before working on the Kvant-1 boom propulsion system. While they were outside the station, a problem had developed with Mir's attitude toward the Sun, and the cosmonauts were needed inside Mir to direct firings of Priroda's thrusters.

In an April 10 interview, Thomas talked about his life on Mir. He said that the view was always gorgeous. "You see texture. You see color. The mountains you can see as folds in the land. And, you can see things like mountain ranges as a collection of mountain ranges.... You can see the way they're all folded together, a bit like a rumpled carpet, and they're all connected.... The areas that are farmed stand out, as opposed to the natural areas with different shades of green. Of course, it depends on the season."

Thomas praised the Russian food. "The soups are outstanding and the juices are just marvelous,

and there's plenty of it. I also have an abundance of American food at my disposal. We exercise regularly. I'm on a treadmill running 2.5 to three kilometers ... every day, something that I'm not perhaps as disciplined [to do] on Earth as I should be." He said his mission was going well. "I feel good. That's one of the amazing things ... You can feel very good in this environment, which—if you think about it—is a very alien environment to us."

He had lost some weight, said Thomas, which "was probably not a bad thing." He had experienced a few aches and pains that he thought were a consequence of spinal extension due to microgravity. But, he had had no ear or stomach problems. "I feel very normal. ... very healthy and comfortable. So, I have no complaints about it at all."

On April 11, the two Russian cosmonauts performed their third spacewalk. Thomas continued to document their work and provide ground controllers with systems data. This extravehicular activity, which took 6½ hours to complete, went very smoothly. The spacewalkers achieved their planned goal of detaching the old thruster and pushing it away from the space station. Thomas videotaped their progress from inside Mir. On April 17, Musabayev and Budarin went outside again to work on the boom jet assembly. This time, something went wrong with Budarin's communications; but he was able to use a backup system. On April 24, they performed yet another extravehicular activity, and finished their boom jet work. The total spacewalk time for their five April excursions was just over 30 hours.

Toward the end of April, Thomas learned that the launch of *Discovery* (STS-91) in June would be delayed five days to accommodate launch preparations at Kennedy Space Center. In an interview, Thomas said, "I wasn't too surprised. I've done a lot of support work at the Cape. I know they're under a very demanding schedule with the processing flow. So, it didn't surprise me too much, and it's only five days." Thomas was then nearing his 100th day in orbit. He said that what he was really looking forward to, upon his return to Earth, was a period of not having "schedules in my life, and just being able to be free to do what I want [to do]. To take a walk or go to the store. Or go to visit friends.

(Above) Musabayev is shown at the end of the Strela outside the Mir space station.

(Opposite page) Thomas and the crew of STS-91 take a final look at the Mir as *Discovery* departs from the Russian space station on the last mission of the Shuttle-Mir Program.

was sweeping out of Mexico and darkening skies as far away as Houston. Thomas said the Yucatan was so covered by smoke that he could not make out the coastline.

In mid-May, Thomas began packing his belongings and scientific hardware, and conducting an inventory of the U.S. equipment aboard the station. His Russian crewmates performed maintenance on thermal loops and the Kvant-1 Elektron oxygen-generating system. They found a small leak in the condensate recovery system, which they fixed by replacing a separator unit and resetting a valve. Also, the primary cooling loop for the Kvant-1 module shut down automatically. The cosmonauts checked for leaks but found none, so the loop was brought back online.

The crew loaded refuse into the Progress docked to Mir, and launched the vehicle away from the station to prepare for the arrival of a new resupply ship that carried a surprise for the crew—a guitar for Musabayev.

Thomas' scientific research program was wrapping up with the focus of his attention on the biotechnology experiment. The air bubbles in the chamber had not hampered the growth of the three-dimensional cells, and the experiment would remain powered on until the arrival of the Space Shuttle *Discovery*.

But before Thomas could depart the station, the attitude control computer failed again. The crew had to use the Soyuz jets to face Mir's arrays toward the Sun and regain solar power. They were able to activate the gyrodynes in time for STS-91's docking on June 4.

Back on Earth, the American astronaut described his feelings of being on the American Space Shuttle, pulling away from the Russian Space Station Mir. "Perhaps one of the most moving moments, though, was as we drew further and further away. We went into the night side of the planet, and I could see stars, and the running lights of the station were on. You couldn't see the station. All you could see was lights flashing, and they were just going off into the distance, these flashing points of light fading out slowly. That was kind of an emotional moment, because I knew that would be the last time I would see it—ever."

I think that's going to be the best part about being back."

By early May, Thomas was finishing up some of his scientific investigations. He processed the final pair of samples for the QUELD and completed the second of three experiment sessions of the study into the risks of renal stones in microgravity. He had photographed the Arenal volcano in the Philippines and a large dust storm that swept from the Sahara to the Mediterranean Sea. He also talked about large fires in Honduras and the Yucatan Peninsula, and the smoke from these fires that

A View from Space

May 2, 1998

As I have orbited around the Earth, I have spoken to many amateur radio operators as well as [to] television journalists conducting interviews. The questions perhaps most frequently asked are, "What is the view like from space?" and "What can you see?" Over the course of the four months that I have been on Mir, I have taken many opportunities to look out the window and take photographs, and the view is captivating, both day and night.

When you first look down on the Earth you see its obvious curvature, and the thin layer of atmosphere on the horizon with the dark blackness of space above it. It is striking to see the abundance of clouds carpeting the planet. Very seldom do we see extended areas that are free from cloud cover, particularly in the tropics. We can see these clouds building to thunderstorms during the day, and then collapsing at night back down to Earth and spreading out in huge circles as if they had been poured down onto the planet.

As you continue to watch the Earth, you begin to recognize land forms and can see that some countries have broad features allowing them to be recognized at a glance; northern Africa has its desert regions, South America has its forested regions, and Australia [has] its redness. Then there are the characteristic coastlines that we are so accustomed to seeing on a map that stand out very clearly from space; the boot shape of Italy, the Red Sea, the Mediterranean, the Florida peninsula, the Gulf of California, and so on. Finally, there are readily identifiable geographic features that only occur in certain places: the huge expanse of Lake Baikal, the Namib Desert, the Himalayas bounding the plains of Tibet and the fertile areas of India, and the Andes separating the rain forests from the western deserts of South America. After even a short time in orbit, we learn to recognize these and can quickly know our approximate position above the Earth from a glance out the window.

Evidence of human habitation is visible from low Earth orbit. Cities can be seen, although, surprisingly, they do not stand out readily. But we can make out their grid-like patterns of streets. In remote areas, certain roads and railway lines can be seen as faint lines across the Earth, such as the road through the rain forests of Brazil, and the long straight railway line crossing southwestern Australia, but generally these are too small to make out clearly. The fencing off of farmland into individual fields can also be made out, particularly in the Midwest of the U.S. and Canada. There is even one area in South America where they alternate their growing cycles on adjacent fields, giving rise to a very obvious checkerboard pattern. Of course, national boundaries do not stand out by themselves as on a map, but some national boundaries can be seen where there are different land usage policies in effect on each side of a border, giving rise to different surface texture or color. In this way the southern border of Israel can be made out, as can part of the division between the U.S. and Canada. The stories

about the Great Wall of China being visible from space may be true, but I have yet to see it.

One of the most readily visible signs of human presence is the occurrence of contrails from aircraft in the upper atmosphere. These are crystals of ice formed from water, a byproduct of the combustion process in the aircraft engines [that] is collected into the wake vortices of the aircraft. [Contrails] are very long-lasting, and can be seen over virtually all parts of the world as white streaks across the sky. They can be striking around cities that are major air traffic hubs, and can oftentimes be seen radiating out from these cities, like spokes in a wheel.

The view of the Earth at night is equally spectacular, and cities can be made out very clearly with all their streetlights. Some areas stand out very noticeably such as Japan, where the high population density is given away by the abundance of nightlights. In fact there are so many lights you can delineate the shape of the Japanese island chain with ease. The presence of myriad small points of light offshore, probably fishing boats, betrays Japan's heavy reliance on seafood.

There is a host of natural phenomena that is spectacular at night. In the temperate zones, we can see vast thunderstorm fronts stretching for miles and being lit up by huge flashes of lightning. Occasionally, I have seen lightning start at one point on a storm front and trigger a cascade of lightning flashes propagating along the storm front, like a falling row of dominoes.

Of course, stars are visible at night, but without any atmospheric attenuation, so they can be seen clearly. They look much as they do when viewed from an isolated desert region away from city lights, but of course they do not twinkle. Perhaps one of the most sublime of all the cosmic sights I have seen to date is the Aurora Australis over the southern polar regions. Only visible at night, it is an eerie curtain of pale green phosphorescence that waves and twists above the Earth, stretching for hundreds of miles.

Meteors are visible from space, too. However, we have the unique vantage of being able to look down on the Earth and see meteors streaking into the atmosphere way below us. Having that perspective is a compelling reminder that we are indeed flying in space.

Unfortunately, this orbital vantage also gives us a unique view of the deleterious effects of human habitation. As I write this, there are huge areas in Central America that are burning. A giant pall of smoke is blanketing the entire southwestern peninsula of the North American continent and is being carried in the winds over much of the United States and as far north as Canada. Indeed, at the northern extreme of one of our orbits, while crossing the Great Lakes, I could see the smoke haze coming up from the distant south and blanketing the land below us. This kind of perspective from space allows us to appreciate that all lands are connected into a common biosphere and that the environmental policies in one country have far-reaching effects in other countries.

Andy

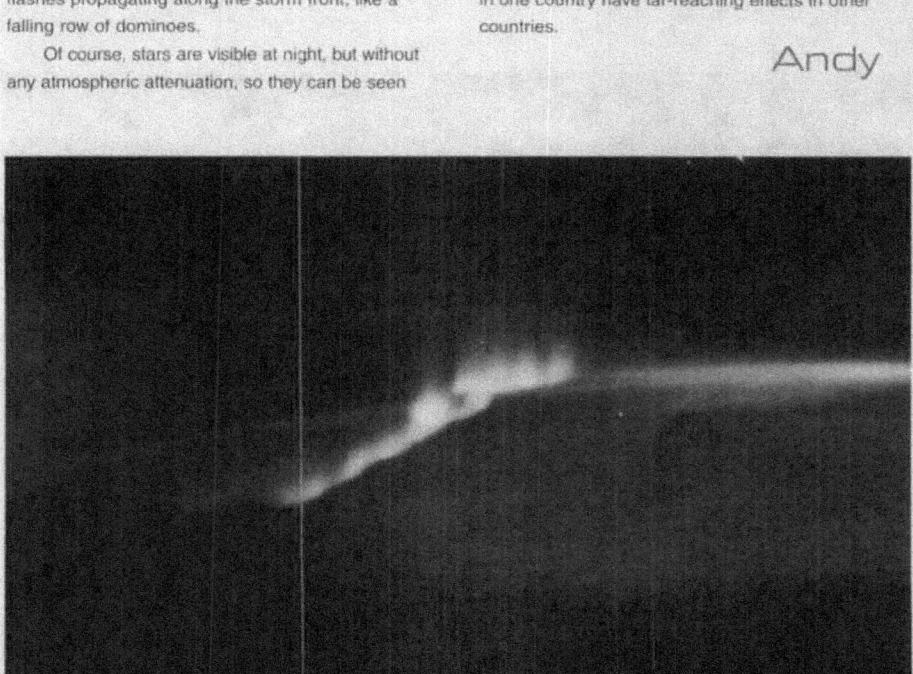

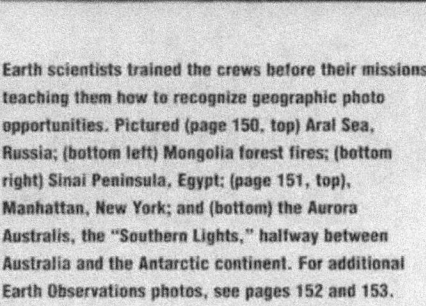

Earth scientists trained the crews before their missions, teaching them how to recognize geographic photo opportunities. Pictured (page 150, top) Aral Sea, Russia; (bottom left) Mongolia forest fires; (bottom right) Sinai Peninsula, Egypt; (page 151, top), Manhattan, New York; and (bottom) the Aurora Australis, the "Southern Lights," halfway between Australia and the Antarctic continent. For additional Earth Observations photos, see pages 152 and 153.

151

Earth Observations
More than mere beauty

Abu Dhabi Port Facility, United Arab Emirate, Persian Gulf

Richat wind erosion structure, Mauritania

Nile River, Lake Nasser, and the Red Sea, Egypt

The view of Earth from space is more than spectacular. It is valuable to science and the world's economy. The human eye discerns subtle changes on the Earth's surface. It notes patterns that would go undetected by robotic cameras, and it follows changes in both natural climate and human land use over the continuum of time.

During Shuttle-Mir, the Mir crews photographed the Earth as it changed below them. Winter to summer. Storm to calm. At different times and in different places, the land was dusty ... smoggy ... erupting ... on fire. Lakes dried up. Plankton bloomed. The crews took some 22,000 photographs, using 35mm and 70mm cameras to capture both natural phenomena and changes caused by people.

Earth scientists trained the crews before their missions, teaching them what to watch for and giving them lists of photo targets. The Mir-21 crew of Shannon Lucid, Yury Usachev, and Yuri Onufriyenko captured seasonal changes in the Northern Hemisphere, and they filmed wildfires in Mongolia and the Kalahari Desert. The Mir-22 crew of John Blaha, Valeri Korzun, and Alexander Kaleri documented flooding of the lower Nile, a drought in southern Africa, and the spring thaw in the southern Andes. Both of these crews helped record baseline conditions before the famed 1997 El Niño climate event. Mir-23's Jerry Linenger, Vasily Tsibliev, and Aleksandr Lazutkin witnessed the retreat of winter ice from North America, and Linenger photographed large dust storms over the Tibetan Plateau. After recovering from the June 1997 Progress collision, the Mir-24 crew of Mike Foale, Anatoly Solovyev, and Pavel Vinogradov was able to follow the El Niño event, and to provide new images of aerosol concentrations off southern Africa. Foale also took the first videos from space of glowing—or *noctilucent*—high-altitude clouds.

After David Wolf joined the Mir-24 crew, he continued tracking El Niño. Other images from Wolf's Mir increment included remarkable photographs of the lush Somali coast after record rains and severe smog conditions over Italy. Mir-25's Andy Thomas, Talgat Musabayev, and Nikolai Budarin completed the El Niño study, and documented the massive wildfires in Mexico and Central America. Australian-born Thomas also did an extensive photo survey of his native land.

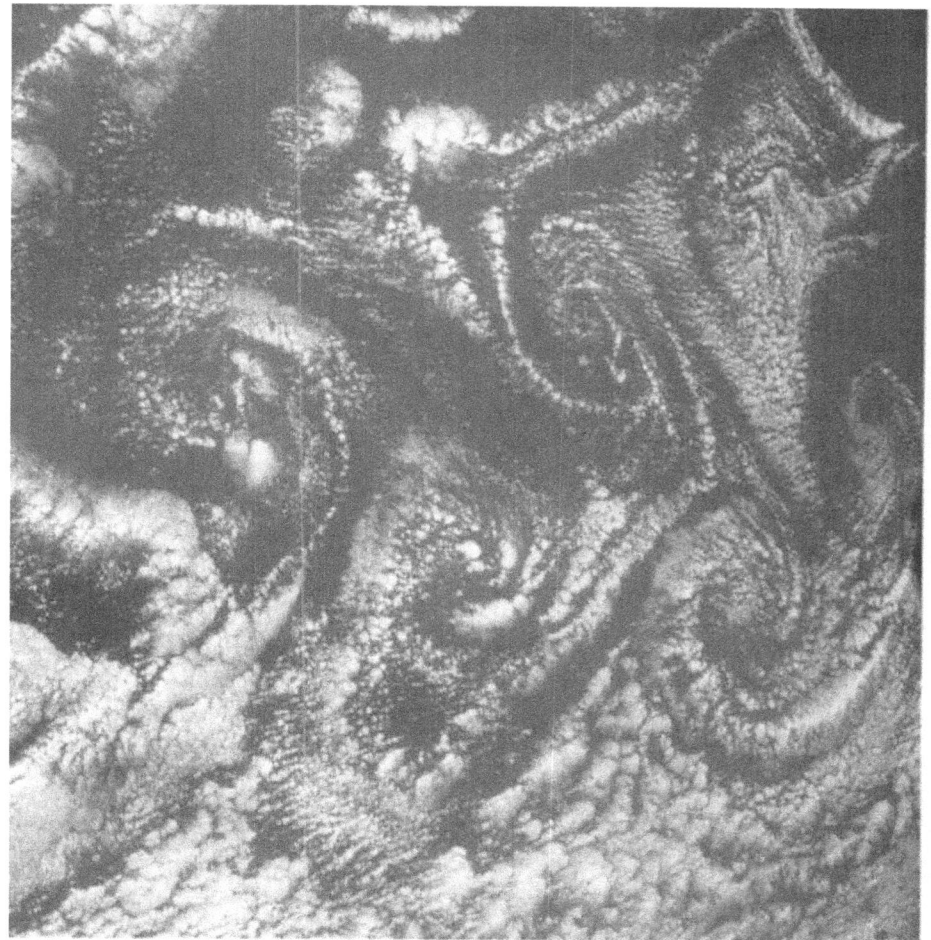

Cloud vortexes

Moonrise over Asia

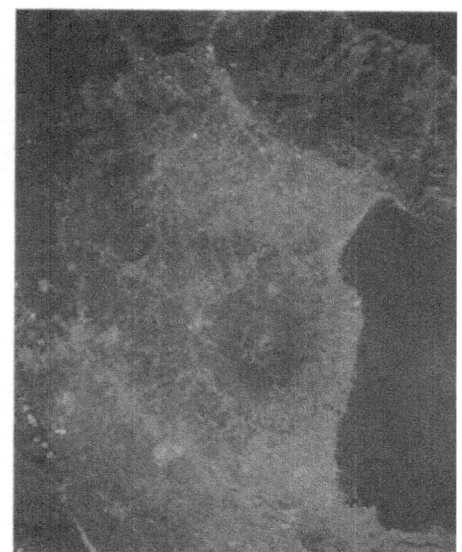

Vesuvius volcano, Italy

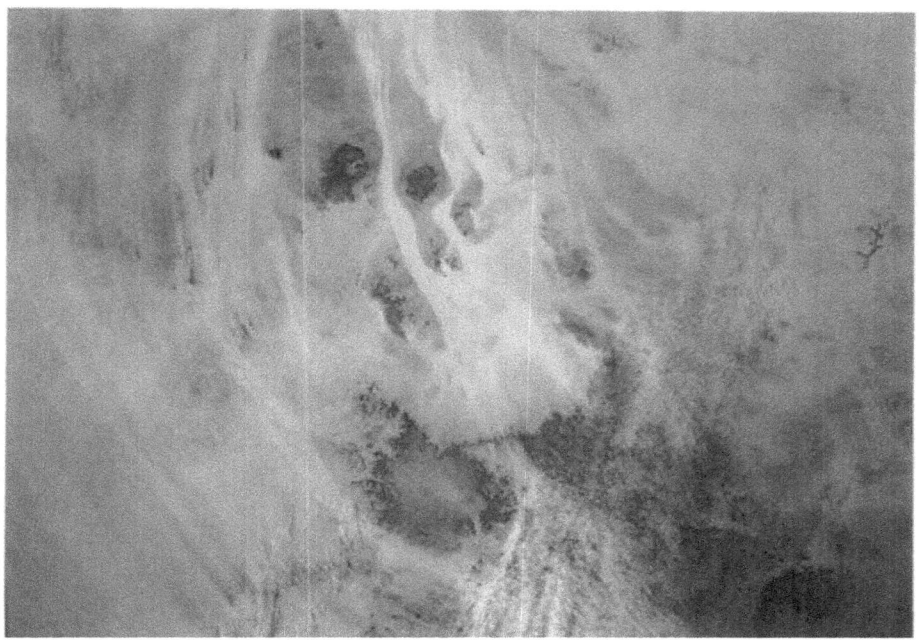

Wind-eroded volcanoes and rock outcrops,
Sahara Desert, Libya

153

STS-91
Closing Out Shuttle-Mir
June 2 – 12, 1998

Space Shuttle *Discovery*

Launched:	June 2, 1998, 6:06 p.m. EDT Kennedy Space Center, Pad 39-A
Orbit:	173 nautical miles
Inclination:	51.6 degrees
Landed:	June 12, 1998, 2:00 p.m. EDT Kennedy Space Center
Mission:	9 days, 19 hours, 54 minutes

STS-91 CREW

Commander Charles J. Precourt, Jr.
Fourth Shuttle flight and third visit to Mir

Pilot Dominic L. Gorie
First Shuttle flight

Payload Commander Franklin Chang-Díaz, Ph.D.
Sixth Shuttle flight

Mission Specialist Wendy B. Lawrence
Third Shuttle flight and second visit to Mir

Mission Specialist Janet L. Kavandi, Ph.D.
First Shuttle flight

Cosmonaut Valery V. Ryumin
Russian Space Agency
First Shuttle flight and fourth space mission

Mission Specialist Andrew S. W. Thomas, Ph.D.
Third Shuttle flight; returning from Mir

PAYLOADS

Orbiter Docking System
Space Habitation (Double) Module
Alpha Magnetic Spectrometer Investigation
Space Experiment Module
"Get Away Specials"
Protein Crystal Growth
Solid Surface Combustion
Spektr Gas Release

When Commander Charlie Precourt and Pilot Dom Gorie landed *Discovery* at precisely 2 p.m. EDT on June 12, 1998, they ended the operational phase of the Shuttle-Mir Program and delivered Andy Thomas back to Earth after his 4½ months onboard Mir. The landing culminated 975 days spent in orbit by the seven U.S. Mir astronauts, including 907 days spent as Mir crewmembers and an 812-day continuous U.S. presence in space. NASA now had firsthand experience of staging and running a multiyear space station program, as well as a new pool of long-duration spaceflight veterans to turn to for guidance and advice.

The landing also marked the end of the beginning of the International Space Station Program. For the next several years, most of NASA's Shuttle flights would support the International Space Station. Russia's venerable and aging Mir space station would see its last regular crew leave the outpost on August 28, 1999, while discussion continued about when—and whether—to deorbit the space station Mir.

One STS-91 crewmember, who understood both the significance and the nostalgia of this last Phase 1 mission, was 58-year-old Mission Specialist Valery Ryumin, the Russian Director of Shuttle-Mir. In 1957, Ryumin was a trainee at the Soviet's Bureau of Experimental Machine Building when its top-secret project, Sputnik, became the world's first artificial satellite. He went on to help develop all Soviet orbital stations, beginning with Salyut-1. A veteran of three spaceflights, Ryumin passed his one-year cumulative time in orbit mark during STS-91, which was his first space mission in 18 years. He was able to learn Shuttle operations and to examine the conditions onboard Mir.

After launch and before rendezvous, the STS-91 crew opened *Discovery's* payload doors, activated the SPACEHAB module and the Alpha Magnetic Spectrometer payload, filled water bags for transfer to Mir, and examined the docking system. The crew also watched for the Russian space station.

A happy Thomas first saw *Discovery* "as a point of light out on the horizon, like a bright star. Then you just make out that it was not a star, that it actually was the Shuttle. Then, of course, it came closer and you could see it clearly. It just got closer and closer, and then at one point you feel the whole station shudder, and you know that they've made contact and have latched on. It was a great moment." Hatch openings and a ceremony followed with the now-traditional shared meal. With all the preparations for docking, neither crew had eaten a meal in over 12 hours.

Once onboard, Ryumin was shocked by Mir's onboard clutter. He told Commander Precourt, "Charlie, this place is in bad shape." Mir was "awful ... worse than I imagined ... unbelievable ... unsafe." But as Precourt recounted, after a few days the former Soviet Army tank commander had reacclimatized to what Mike Foale had called Mir's "frat house" conditions. Ryumin said, "Well, you know, I think I could get used to staying up here.... Maybe I'll just stay." As others had, Ryumin warmed to Mir's peculiar charm.

On the other hand, Precourt had been to Mir three times and had seen it grow with the additions of the Priroda, Spektr, and docking modules. He said that, other than the fact that Spektr was sealed off, "the condition of the inside of the space station was better than it was on my previous two visits. The air was cleaner, it was a better controlled

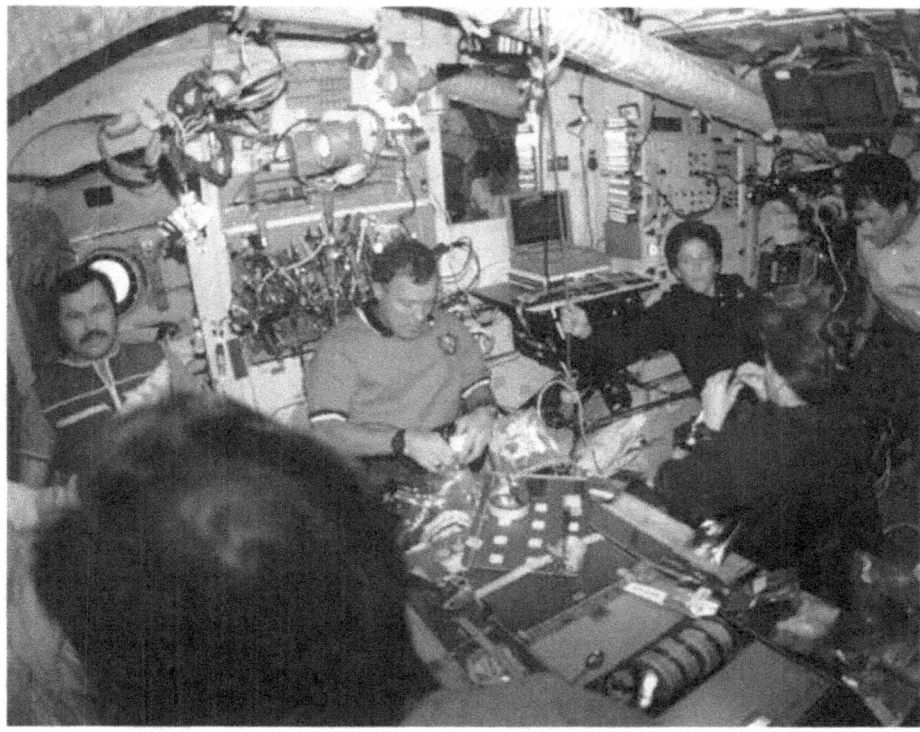

Commander Charlie Precourt (below, left) and Mir-25 Commander Talgat Musabayev greet one another at the docking module hatch. (Above) STS-91 and Mir-25 crewmembers share a meal in the Mir space station Base Block.

temperature, it was drier, [and] the walls of the surfaces of the structure everywhere were nice and dry."

The Shuttle crew delivered more than 1,100 pounds of water and almost 4,700 pounds of cargo, experiments, and supplies to Mir. They also conducted science investigations, including the Alpha Magnetic Spectrometer, whose Principal Investigator was Nobel Laureate Samuel Ting. This 3½-ton particle detector was designed to look for antimatter left over after the creation of the universe, according to the Big Bang theory, and signs of "dark matter" theorized to constitute most of the universe.

On Flight Day 5, the crew released a fluorescent tracer gas into the depressurized Spektr module, hoping to locate the breach in Spektr's hull that resulted from the collision with a Progress resupply ship in 1997. If lighting conditions were right, the gas would appear as a dull green cloud. The crew could not detect any leaks at this time or when the experiment was repeated after undocking. However, the procedure may prove useful in the future on the International Space Station.

Before the undocking on Flight Day 7, Mir Commander Talgat Musabayev presented Precourt with a two-foot wrench. "Charlie, take this wrench," Musabayev said. "It's sort of a relay stick from an old lady, station Mir, to the International Space Station." Precourt responded, "We're going to need this for all the work that we have ahead of us."

Discovery landed at Kennedy Space Center on June 12, 1998, bringing the operational aspect of the Shuttle-Mir Program to an end. On the whole, the Shuttle-Mir rendezvous and dockings had been so flawless that their difficulty and significance went unnoticed by many.

The landing culminated 975 days spent in orbit by the seven U.S. Mir astronauts, including 907 days spent as Mir crewmembers and an 812-day continuous U.S. presence in space.

155

Seven U.S. astronauts resided on the Mir space station as participants in the Shuttle-Mir partnership with Russia. They are (left to right, bottom row) Norman E. Thagard, John E. Blaha, Jerry M. Linenger, and David A. Wolf; (top row) Andrew S. W. Thomas, Shannon W. Lucid, and C. Michael Foale.

Looking Back, Looking Forward:
Considering Shuttle-Mir

The benefit of the space program is the unexpected. I can guarantee you that as long as we continue pushing the boundaries, pushing the frontiers, we will benefit and we'll be surprised by how we'll benefit.

— Frank L. Culbertson, Jr., Shuttle-Mir Program Manager

With the return of NASA-7 Astronaut Andy Thomas to Earth, the highest stage had been successfully shared by the United States and Russia. The operational curtain of Shuttle-Mir fell on August 25, 1998, when the Mir-25 cosmonauts brought back the final science results. However, the next high-tech, high-stakes performance by these two former adversaries was already under way. The U.S.-financed, Russian-built Zarya module of the International Space Station would launch on a Proton rocket from Baikonur on November 20, 1998.

In retrospect, how should the Shuttle-Mir Program be remembered? How does this partnership fit into the relatively short story of human spaceflight and the larger story of human aspiration?

Historically, Shuttle-Mir shares similarities with all of NASA's earlier projects and programs. As did Project Mercury of 1958-63, Shuttle-Mir focused U.S. public attention on individual astronauts. For several years before Shuttle-Mir, Space Shuttle crews had been including up to seven members, and few Americans could recite their names. However, like the Mercury pioneers, each of the Mir astronauts metaphorically "carried the mail" for all of NASA. Although each Mir astronaut was supported by hundreds of NASA employees and each served under a Russian commander, Americans could again identify with a single individual and experience vicariously the astronaut's aloneness and dangers.

As with the Gemini Project of 1962-66, Shuttle-Mir served as a critical stepping-stone to a next, higher goal. Gemini was started after the Apollo Program's quest for the Moon had begun; Shuttle-Mir was devised to prepare for the much more ambitious International Space Station Program. Shuttle-Mir's rapid pace was comparable to Gemini's (NASA launched ten manned missions in less than 20 months). During Shuttle-Mir, six of the seven NASA Mir residencies followed immediately, one after another. In both programs, tight schedules and rendezvous requirements forced ground operations to be honed extra sharp. The Gemini XI launch "window" lasted only two seconds. Every Shuttle-Mir Shuttle launch occurred during its first launch window.

Rendezvous and spacewalking techniques were practically invented for the Gemini flights; they were further perfected through Shuttle-Mir. Gemini and Shuttle-Mir both sought to extend astronauts' stays in space. Gemini VII required astronauts Frank Borman and James A. Lovell, Jr., to spend 14 days in their cramped, two-person capsule. NASA-2 Astronaut Shannon Lucid spent 188 days in orbit. Furthermore, spacecraft attitude control presented some very tense moments during both Gemini and Shuttle-Mir. During Gemini VIII's docking with an Agena booster, a thruster became stuck, causing the joined spacecraft to begin tumbling. The crew of Neil Armstrong and David Scott hurriedly undocked their capsule, but it then rolled even faster at a dizzying rate of one revolution per second. Forced to use their reentry control thrusters to stop the motion, the crew had to execute an emergency return to Earth a mere ten hours after their launch. A similar experience occurred during Shuttle-Mir Astronaut Mike Foale's residence on Mir, when a collision with a Progress resupply vehicle initiated a period of tumbling and power losses. Foale had to use his knowledge of physics and the stellar constellations to help devise a way to stop Mir's tumbling.

Apollo and Shuttle-Mir were somewhat similar, especially in the area of public awareness. Public interest in Apollo waned after the first two lunar landings, but the near loss of Apollo 13 brought back the attention of television networks and the public, whose interest again abated during the next, several successful Moon missions. Similarly, Shuttle-Mir's mishaps—notably a fire and a collision—riveted the public's attention. But, the smoother sailing of the last two American Mir residencies received much less media ink. For both Apollo and Shuttle-Mir, the continuing good news was often presented as no news at all.

The Skylab Project of 1973-74 and the Apollo-Soyuz Test Project orbital link-up in 1975 together constituted a kind of proto-Shuttle-Mir. Skylab was America's first and only long-duration experiment before Shuttle-Mir. Like the modules of the Russian Mir Space Station, Skylab was delivered to orbit by a heavy launcher. During Skylab, three 3-person crews spent, respectively, 28, 59, and 84 days in space. Both programs provided NASA an opportunity to perform extended experiments in the medical and physical sciences. Besides teaching NASA how to live for a long duration in orbit, Skylab helped NASA gain other knowledge that was used during Shuttle-Mir, such as performing spacewalks to repair vehicles in orbit. Another similarity of the two space habitation programs was the extreme amount of work required of the crews—not just in orbit but on the ground where

24-hour monitoring and support continued until the crewmembers returned safely to Earth.

Apollo-Soyuz, the goodwill gesture between the U.S. and the Soviet Union, resulted in the first time spacecraft from two nations rendezvoused and docked in orbit. Important hardware was developed, such as a common docking system, and American and Russian space experts had their first good looks at each other's programs. In a forerunner to Shuttle-Mir, Soviet cosmonauts and their backups trained at the NASA Johnson Space Center; and the American crew and their backups trained in Star City. Flight controllers from both nations conducted joint simulations. Perhaps most importantly to the future Shuttle-Mir Program, individuals from both space programs carried forward a tremendous working relationship that provided a solid foundation of respect.

The last American space program before and during Shuttle-Mir involved NASA's Space Shuttle. The Orbiter's abilities were absolutely necessary for Shuttle-Mir, but in ways more differences than similarities existed between the Shuttles and Mir. Space Shuttles are basically a kinetic means of access to orbit, while space stations are passive orbital platforms. Generally, a space shuttle performs the "getting there" while a space station provides the "being there."

While obviously the Shuttle-Mir Program left its mark on the history of human spaceflight, how did this space partnership fit into the story of humankind? A major historical distinction may be that it was the second grand exploration project sponsored for the purpose of better international relations.

The ancient Greeks and Romans sailed for glory and to expand their trading empire. The European nations from the 13th century onward likewise explored for power and influence. Even the Lewis and Clark expedition, 200 years ago, was intended by President Thomas Jefferson to survey U.S. claims to interior North America as well as to study its geography and biology. The one possible precedent for Shuttle-Mir as an expeditionary means to establish and improve international relations was Admiral Cheng Ho's Grand Treasure Fleets expeditions from China in 1405-33. Cheng Ho's seventh command included as many as 317 ships and 7,000 crewmembers. Its mission: to spread Chinese influence through the giving of gifts to every nation the command could find.

Six hundred years later, Shuttle-Mir gave its

Shuttle-Mir Accomplishments

The Shuttle-Mir Program showed that space exploration is no longer a competition between nations. NASA officials pointed out that the Shuttle-Mir Program also:

- Developed flexibility by operating in space with several launch vehicles and a space station.
- Conducted long-term operations with multiple control centers and U.S. and Russian teams at each other's facilities.
- Conducted spacewalks outside both Russian and American vehicles with astronauts and cosmonauts testing each other's spacesuits, a preparation for joint walks to assemble the International Space Station.
- Trained astronauts, cosmonauts, and other team members in each other's language, methods, and tools to facilitate operations in orbit and make mission training more efficient.
- Created a joint U.S. and Russian process for analysis, safety assessment, and certification of flight readiness.
- Led to refinements in software, hardware, and procedures that will be used in operations onboard the International Space Station.
- Established that noncritical systems may fail and be replaced through routine maintenance, without compromising safety or mission success.
- Showed that multiple oxygen-generation systems are essential for safe, uninterrupted operations.
- Mated U.S. and Russian hardware in orbit and verified complex robotics operations during the delivery and assembly of the Russian-built docking module.
- Collected data on the effects of long-duration exposure of hardware to the space environment.
- Learned how to conduct long-term research and maintenance on a space station through flexible scheduling of crew time on orbit.
- Developed a process for mission planning and psychological support for astronauts on orbit during extended periods.

own lasting gift to the world—the example of the great Cold War adversaries working together.

Also, the Shuttle-Mir Program fits nicely into the age-old forms of myth, legend, and tale. Shuttle-Mir presented so many twists and turns contained within a grand narrative arc that, upon reflection, it resembled a classical drama. For its prologue, Shuttle-Mir had the U.S.-Soviet Cold War and Space Race, and a situation similar to Shakespeare's Romeo and Juliet with its "two households, both alike in dignity" yet sharing an "ancient grudge." For Act One, astronaut Norm Thagard ventured forth like a modern Marco Polo, traveling alone in a foreign culture as an ambassador between East and West.

Shuttle-Mir's Act Two began deceptively blissful with Shannon Lucid's record-setting flight. But, soon after came the fire and the collision.

The plot thickened with the second-guessing by the program's opponents and supporters. The final act rewove many narrative threads, and the Shuttle-Mir story ended with a note of triumph. Even NASA's critic, Congressman James Sensenbrenner, offered a back-handed compliment when he said, "The Shuttle-Mir Program has been very useful in giving our astronauts good training in crisis management."

While the program was moving toward completion with its goals accomplished, "crisis management" was again what NASA managers were engaged in. International Space Station budget overruns and Russian delays in constructing station components were causing some in the U.S. Congress to question the whole International Space Station Program. NASA Administrator Dan Goldin was forced to consider the benefits and costs of moving ahead—with and without the Russians. On top of that, a former chief of astronaut safety claimed to have been ignored when he had raised concerns for the safety of the astronauts onboard Mir. Finally, seeming to symbolize NASA's situation, some of Andy Thomas' last Mir photos were of the smoke from Mexican wildfires, which was casting a pall over the Johnson Space Center in Houston.

Yet with all the problems of getting Phase 2 started, the end of Phase 1 saw the Shuttle-Mir teams working well and "clicking." An example came just before STS-91 *Discovery's* launch, when Mir's attitude control computer failed again. According to NASA Training Manager Tommy Capps, "Suddenly, we brought a team together. That could not have happened a few years before.... We had [Russian flight director] Victor Blagov [at Kennedy Space Center] to work it. We had a full team sitting together—NASA and Russia—constantly communicating with each other ... and [pulling] together as a team."

"It was a very emotional experience," said Capps. "Really, it was a sobering experience for me—because I really got to thinking about how far we had come." Earlier in the program, "We would have had a lot of sidestepping, and sashaying back and forth." But by STS-91, "We were there. We were clicking. And, we were a program."

Unemotional statistics reflect Shuttle-Mir's operational accomplishments. Seven U.S. astronauts spent a total of 907 days onboard Mir. Forty-three different American astronauts flew to the space station. (They comprised nearly half of all visitors to Mir.) In all the Shuttle flights to Mir, no crew had to "suit up" for launch more than once, in spite of the restrictive launch windows. The three Space Shuttle Orbiters carried to Mir 28,000 pounds of supplies (67,000 pounds, including water and the docking module), and they brought back 17,500 pounds of equipment and materials. Operationally, rendezvous and docking techniques were developed and perfected.

As Shuttle-Mir was beginning, "We had no idea how much we would learn," according to NASA's Shuttle-Mir Program Manager Frank Culbertson. "I think the most valuable lesson is that you're going to have things happen that are going to require problem solving continuously. The best-laid plans, the best-designed systems—

you're still going to have difficulties. Long-duration spaceflight is hard, and the most valuable lesson we learned from that is to expect it to be difficult. Plan for that. Train for that. And then, be prepared to handle the unexpected."

Concerning in-orbit operations, Charlie Precourt, Shuttle commander, echoed Culbertson's sentiments. He said, "What I hope the American public can glean from the Shuttle-Mir Program is that all hardware breaks down. We have to learn

> The first time I went up and flew next to Mir, I was struck by a couple of things. Number one, how brilliant and white and perfect it looked. But more than that I thought to myself, "Well, here's a country that was in a race with us, they didn't get to the Moon ... they have [many] more budget problems than we have, and yet they took the next step."
>
> – Astronaut James Wetherbee

to take our hardware to space and not bring it home in a hurry—like an airliner that might be flying home that has a problem." According to Precourt, spacecraft need to be able to be repaired in space. "If we don't learn to do it out there, we won't ever be able to stay there very long. Because hardware does fail. Our ultimate goal is to be able to go to the Moon and Mars, and put bases there for scientific research and for exploration purposes, and stay there and survive.

"The fact that the Mir went through ups and downs and we were able to live through that ... is a great testament to what we were able to do together, and it should make people think twice when they try to criticize the Russians and their system for what it is or is not capable of doing," added Precourt.

NASA-2 Astronaut Shannon Lucid said that it's not only the hardware, "[It's] the people you fly with. And, if the people are compatible and you get along, then you'll have a great flight."

Hardware, software, processes, planning, management, operations, and human nature—all of these presented challenges during Shuttle-Mir, and all offered lessons to learn.

At the end of Shuttle-Mir, Culbertson looked back at its four goals. He said that the first goal was originally "to learn to work with the Russians," but that was quickly changed to "to learn to work with each other"; and this certainly did not mean merely teaching each other. It meant "observing each other and learning from each other," and realizing that both sides had a lot to offer. At different times, the Russians and the Americans surprised, dismayed, and perplexed one another. Still, Culbertson could later tell a story of how neither side was "stranger" than the other. Once, Culbertson's wife asked him why many Americans made comments about the Russian engineers. They said the Russians seemed to bring their own foods to Houston. They didn't get out much. They occasionally walked to grocery stores. They ate mostly in their rooms. And, they seemed to dress, well, funny. Then, she asked him what the Americans did in Moscow. Culbertson told her, "Most of us take our own food, eat mostly in our rooms, walk everywhere except to work, and keep mostly to ourselves. And, we certainly dress funny." In orbit and on the ground, the getting used to each other—and each other's ways—was a major challenge and great accomplishment of Shuttle-Mir.

The second Phase 1 goal—the mitigation of risk for the International Space Station—was sometimes achieved in unforeseen ways. Events such as the fire and collision injected unexpected danger into the Program, and showed both NASA and Russian managers where they were lacking. In Culbertson's words, "We got more than we bargained for." Fortunately, they didn't pay too high a price.

Thinking about the emergencies onboard Mir caused Culbertson to look forward to the International Space Station. He said, "You could probably take most of the things that happened during the course of the Phase 1 Program and predict that each of them will happen, in one form or another, during the 15 or so years of [International Space Station] life." That would include a future argument about when the International Space Station Program should end, and about how safe and useful the aging station will be after ten years in orbit. "All of that's going to happen," said Culbertson. He hoped enough of the people who had experienced Shuttle-Mir would "still be around," to deal with it appropriately.

The third Shuttle-Mir goal was to conduct long-duration spaceflight studies for the United States. This was NASA's first opportunity since Skylab in 1973, and Norm Thagard broke Skylab's record on the very first NASA Mir increment. NASA collected much data on human physiology in microgravity, but also learned about the psychological aspects of people living a long time in space. "That aspect has been eye-opening," Culbertson said.

So were the psychological effects of the support team members on the ground, who were also dealing with a different language and culture and with long hours of hard work, separation, and deprivation. Culbertson said, "That is an unexpected lesson that we're going to have to work on very hard, to ensure that we don't burn out our people—both in orbit and on the ground—and that we don't neglect them. And, that we provide them with sufficient support so that they'll want to keep doing this job over and over and over, because it's harder than we thought it would be."

Shuttle-Mir's fourth goal was to conduct a science research program. Although much good science was returned from Shuttle-Mir from a science program that spanned many disciplines, Culbertson said he occasionally "took a lot of heat" from the scientists. The scientists would protest, "We're last priority," and Culbertson would respond, "No, you're fourth priority"—after other matters that at the moment were more critical to overall success. "You have to keep it in perspective," Culbertson said. "If it goes well, it gets a good bit of attention. If it goes poorly, it gets a heck of a lot of attention. And, people feel like they're failing if one of the research experiments doesn't work." Thanks to Shuttle-Mir, Culbertson said, the whole concept was changing on how to conduct research in space.

"Mir operations and life were hard," Culbertson said. "A lot of people have said it's the hardest thing they've ever done. And, my prediction is that the [International Space Station] is going to be even harder for a lot of reasons that people don't understand yet.... So, this next ten years is going to be really, really fascinating.

"And, the story is going to be even more interesting than the one we're telling here."

Comparing the Increments

NASA-3 Astronaut John Blaha said, "I'm trying to tell people ... that they hear a different message from different people who flew on the long flights.... Each one of the seven long Mir flights really was different."

Blaha recommended, "Instead of looking at the seven people who flew these missions as different people, realize that the missions were different. I think the people are actually more similar."

He had a good point. The seven NASA Mir astronauts enjoyed—and endured—many weeks onboard the space station Mir. How to compare their long orbital sojourns? The similarities and differences put together a picture.

All seven Mir astronauts were products of NASA's intensive selection and rigorous training. Three came out of the military; two came from academia. All had highly technical backgrounds in physical or biological sciences. Three were medical doctors. Three were jet pilots. All were focused, determined, disciplined individuals, accustomed to success. All seven were articulate—and unafraid to speak their minds. They were also very alike—in courage and grit.

Overachievers? They were achievers, certainly. But, "over-achievement" means achievement of one goal at the cost of something else. All handled their challenges successfully, and all met with continued success after Mir.

Three—Thagard, Blaha, and Linenger—left NASA shortly after their Mir odysseys. Thagard and Blaha had flown many Shuttle missions and so were nearing the ends of their space careers anyway. After Mir, both became involved in the Challenger Center for Space Education. Blaha entered the business world in San Antonio, Texas. Thagard became a professor at Florida State University. Linenger entered private life in Michigan.

Four—Lucid, Foale, Wolf, and Thomas—remained NASA astronauts to work in the International Space Station Program and to continue training for spaceflight.

The Mir missions of these seven astronauts were more like each other's than they were like any other missions in NASA's history. All had to deal with breakdowns in equipment and communications. All seven endured months of microgravity, isolation, separation, confinement, and constant danger, all the while in an unfamiliar social and linguistic environment.

But on closer inspection, each increment did have its idiosyncrasies.

Thagard's was the first—the groundbreaker—separated from everything that had gone before it and separated, too, from the six contiguous missions.

Lucid's flight was blessed with crew harmony but was challenged by an unexpected six-week extension.

Blaha served under a "new" commander—new to Blaha and new to spaceflight.

Linenger faced a dangerous fire early in his flight, witnessed a near-miss with another spacecraft, and endured poor communications with managers on the ground.

Foale survived a frightening collision and the danger that followed.

Wolf launched to Mir during a very public controversy over whether he should go at all.

Thomas' increment has been called "smoother sailing"; but it, too, had its challenges. Besides, "smoother" is used to compare it to the other six increments, which were hardly smooth at all.

In his remarks on the seven increments, Blaha went on to comment on the Russian element. He said, "The Russian commander each of us flew with was a different human being. We know from our Apollo and Shuttle missions that the commander has a lot to do with [the] tone of a mission." Further, Blaha said, "Anything that occurs on a short Shuttle flight is magnified. [If it occurs on a long-duration mission], it is magnified again if it's in a person's second language." Misunderstandings can be more difficult to overcome.

In sum, the seven NASA increments on Mir differed in important ways. They were influenced by the order of the increments, the personalities onboard, and the events that transpired during each term. The astronauts themselves were indeed similarly trained, but each astronaut was a unique individual with a strong personality. Each succeeded in his or her own way.

One could paraphrase the great Russian writer Leo Tolstoy and say, "Easy successes are all alike, but any hard-won victory is like no other."

International Space Station—Phase 2

Four months after astronaut Andy Thomas ended the final Shuttle-Mir increment, on-orbit construction of the International Space Station (ISS) began with the Proton rocket launch November 20, 1998, of the U.S.-owned, Russian-built Zarya control module from Baikonur Cosmodrome.

Two weeks later, *Endeavour* (STS-88) launched from Kennedy Space Center to deliver and attach the U.S.-built Unity connecting module. STS-96 *Discovery* delivered tools and cranes to the ISS in June 1999. STS-101 *Atlantis* launched in May 2000 to perform station

maintenance and deliver more supplies. The Russian Zvezda service module launched and docked with the ISS in July 2000 to become the third major component. In September 2000, STS-106 *Atlantis* visited the ISS to outfit Zvezda.

Five other STS missions traveled to the ISS in the next seven months to continue station assembly. Also during that period, the Expedition One crew launched on October 31, 2000, on a Russian Soyuz. The second Expedition crew arrived March 10, 2001, aboard *Discovery* to continue the newest phase of human spaceflight exploration.

The ISS Program represents a 16-nation partnership. When complete the one-million pound space station will include more than 100 components and six laboratories, and will provide more space for research than any spacecraft built. It will support crews up to seven for missions from three to six months. More than 40 space flights over five years and at least three space vehicles—the Space Shuttle, the Soyuz, and the Russian Proton—will deliver the various components to Earth orbit.

International Space Station as of December 2000.

Mir space station, 1997

Mir Space Station

The space station Mir became a legend in its own time, reflecting Russia's past space glories and her future as a leader in space.

The Russian Space Station Mir endured 15 years in orbit, three times its planned lifetime. It outlasted the Soviet Union that launched it into space. It hosted scores of crewmembers and international visitors. It raised the first crop of wheat to be grown from seed to seed in outer space. It was the scene of joyous reunions, feats of courage, moments of panic, and months of grim determination. It suffered dangerous fires, a nearly catastrophic collision, and darkened periods of out-of-control tumbling.

Mir soared as a symbol of Russia's past space glories and her potential future as a leader in space. And, it served as the stage—history's highest stage—for the first large-scale, technical partnership between Russia and the United States after a half-century of mutual antagonism.

Mir did all of that and, like most legends, was controversial and paradoxical. At different times and by different people, Mir was called "venerable" and "derelict." It was also "robust," "accident-prone," and "a marvel," as well as "a lemon."

For Russians, the very name "Mir" held meaning, feeling, and history. Mir translates into English as "world," "peace," and "village"; but a single-word translation misses its full significance. Historically, after the Edict of Emancipation in 1861, the word "mir" referred to a Russian peasant community that owned its own land. A system of state-owned collective farms replaced the mir after the Russian revolution of 1917. [Refer to the CD-ROM for the essay by Frank Culbertson, Shuttle-Mir Program Manager, on the meanings of "Mir: What's in a Name?"]

As with most legends, Mir was literally beyond the reach of most men and women; but it could be seen by many as a bright light arcing across the night sky. Mir undoubtedly provoked many thoughts around the globe about who we—as a human race—are and where we are going.

The cosmonauts and astronauts who were fortunate enough to travel to Mir were always impressed by its appearance. Regardless, Mir remained difficult to describe. Someone once called Mir a 100-ton Tinker Toy*, a term that recalled Mir's construction. Adding modules over the years, and then sometimes rearranging them, the Russians had built the strangest, biggest structure ever seen in outer space. Traveling at an average speed of 17,885 mph, the space station orbited about 250 miles above the Earth. Mir was both great and graceful—and incongruous and awkward—all at the same time.

In outward appearance, Mir has also been compared to a dragonfly with its wings outstretched, and to a hedgehog whose spines could pierce a spacewalker's suit. NASA-4 Astronaut Jerry Linenger compared Mir to "six school buses all hooked together. It was as if four of the buses were driven into a four-way intersection at the same time. They collided and became attached. All at right angles to each other, these four buses made up the four Mir science modules.... Priroda and Spektr were relatively new additions ... and looked it—each sporting shiny gold foil, bleached-white solar blankets, and unmarred thruster pods. Kvant-2 and Kristall ... showed their age. Solar blankets were yellowed ... and looked as drab as a Moscow winter and were pockmarked with raggedy holes, the result of losing battles with micrometeorite and debris strikes over the years."

On the inside, Mir often surprised people, too, even when they thought they were ready for the view. By the time Americans arrived on Mir—nearly a decade into its life—the station had become cluttered with used-up and broken equipment and floating bags of trash. During Mir's lifetime, no adequate remedy was ever developed to deal with the stowage situation. Mir looked like a metal rabbit warren, or, as Mike Foale put it, "a bit like a frat house, but more organized and better looked after."

Still, Mir was home and shelter to its crews, and how it looked to them depended on their perspectives and situations. The ivory-like controls of the Base Block reminded David Wolf of classic science-fiction stories, such as *The Time Machine* by H.G. Wells. After a fly-around in the cramped Soyuz capsule, Linenger wrote: "Looking into the station I could see a lone ray of light shining through the port widow and outlining the dining table. We had left some food out for dinner. It was the only time during my stay in space that Mir looked warm, inviting, and spacious. It reminded me of opening the door to a summer cottage that had been boarded up for the winter, looking inside, and seeing familiar surroundings."

Mir set every record in long-duration spaceflight. Physician Valeri Polyakov lived aboard Mir for a single, continuous-orbit stay of 437 days, 17 hours, and 38 minutes. He completed his stay in 1995 as American Norm Thagard began his Mir residency. Polyakov's experiences contributed greatly to the biomedical studies of long-term human spaceflight conducted by the Institute of Biomedical Problems, where he served as Deputy Director. Combined with an earlier Mir expedition flight, the Russian cosmonaut spent a total 678 days, 16 hours, and 33 minutes on the Russian space station. However, his achievement for total time in space was surpassed in 1999 by Sergei Avdeyev, who endured a total 747 days, 14 hours, and 12 minutes during three space missions. During Shuttle-Mir, Shannon Lucid set the space endurance record for women in 1996 when she spent 188 days, 4 hours, and 00 minutes in orbit.

Just as "mir"—the word—had many meanings for Russians, Mir—the place—provoked many different feelings. In February 1995, Russian Cosmonaut Vladimir Titov flew aboard the "near Mir" flight, STS-63, when the Shuttle rendezvoused with Mir. Six years earlier, Titov had spent a year aboard Mir as an expedition member, when Mir consisted of only the Base Block, the two Kvant modules, a Soyuz, and a Progress spacecraft. About seeing Mir again,

Titov said, "It was very wonderful, a wonderful view." STS-63 did not dock, but Titov visited Mir again as a crewmember of STS-86.

Alas, the sturdy Mir was built on a sinking foundation. Without repeated boostings, all things in low Earth orbit must eventually come down. With the new International Space Station requiring much of the Russian space program's attention and financing, the Mir space station was doomed to be deorbited. A strong effort rallied in Russia to keep Mir aloft; and, at one point, Russian State Duma representatives were calling for the firing of Yuri Koptev from his post as the head of Russia's aerospace agency. However, on December 30, 2000, Russian Prime Minister Mikhail Kasyanov signed a resolution calling for Mir to be sunk into the ocean early in 2001.

Concerns circled the globe about Mir crashing into populated areas. Mir's path crossed over nearly every city on Earth. Its orbits tracked over everything between 51 degrees North and South latitude, roughly within the limits of the Aleutian Islands to the north and the southern Andes Mountains to the south. Pieces of previous large spacecraft had landed in Canada, Australia, and southern South America, albeit fortunately without any damages or casualties.

For Mir, Russia acquired insurance in the event that its deorbit caused some physical damage. Japan kept a close watch because the final orbit would bring the Mir over the island nation. The U.S. government provided Russia with tracking and trajectory data, atmospheric conditions, and even solar activity, which can cause the Earth's atmosphere to expand farther into space. Although there was considerable certainty that debris could be limited to falling in the ocean, Yuri Semenov, RSC-Energia President, was quoted as saying, "We don't have a 100-percent safety guarantee."

After more than 86,000 total orbits, Mir reentered Earth's atmosphere on Friday, March 23, 2001, at 9 a.m. Moscow time. The 134-ton space structure broke up over the southern Pacific Ocean. Some of its larger pieces blazed harmlessly into the sea about 1,800 miles east of New Zealand. Observers in Fiji reported spectacular gold-and-white streaming lights. An amazing saga and a highly successful program finally had come to a watery end.

Anatoly Solovyev had lived a total of 651 days on Mir and served as Mir-24 commander for Americans Foale and Wolf. He was quoted in Star City as saying, "I am especially sad these days. An entire era of our Soviet space program is ending, into which we invested not only our money but, what is more important, our intellectual potential."

The Russians' investment began when a Soviet Proton launcher boosted Mir's Base Block (core module) into orbit on February 20, 1986. This module resembled the existing Salyut-7 space station, but Mir's design called for expansion through the addition of future modules. Mir's first crew arrived in mid-March 1986, and the inaugural crew of Leonid Kizim and Vladimir Solovyev stayed aboard until May 5, 1986. This Solovyev would later become the Russian cochair of the Flight Operations Working Group for the Shuttle-Mir Program. And, it was he who took charge of the Moscow Mission Control Center immediately after the Progress resupply vehicle collided with Mir during NASA-5 Astronaut Foale's residency.

In 1987, the Soviets added Mir's first expansion module, Kvant-1, and had the world's first modular space station. They still needed a more versatile way of transporting crews and equipment to and from Mir—something like the American Space Shuttle. In 1988, the Soviets launched the *Buran*, a winged, reusable space vehicle and a close copy of the U.S. Shuttle. Its first flight was near perfect. However, at this point in history the Soviet Union was crumbling. No further *Buran* flights were attempted; four planned orbiters remained unfinished.

The Soyuz-TM vehicle and Progress-M resupply (cargo) vehicle became the transports of crews and supplies to the Mir. The Kvant-1 featured a docking port to accommodate the arrival of these spacecraft. The system worked well as the Russian space station was unoccupied on only five brief occasions until its deorbit on March 23, 2001. During its existence, the station had remained almost continuously occupied for nine years.

Mir continued to expand during the next years with the additions of modules for research and residence. Kvant-2 arrived in November 1989 with an airlock that allowed crewmembers access to the outside of the complex for extravehicular activities. Kristall, launched at the end of May 1990, housed Earth observation instruments and was used for semiconductor and biological experiments. Five years later, Spektr, a remote-sensing module for geophysical sciences, was added to the Mir.

On June 29, 1995, U.S. Space Shuttles began docking with the Russian space station. Before the first docking, the Mir-19 crew used the Lyappa manipulator arm to relocate the Kristall, thus allowing ample clearance for *Atlantis*. In November 1995, a new docking module arrived via STS-74 and was attached to the Kristall to provide means for future dockings without interference. The next year, on April 23, 1996, the final module, the Priroda, was added to the Mir.

The complex retained a docked Soyuz-TM vehicle at all times as this spacecraft served as the crew's "lifeboat." The vehicle carried a maximum of three persons, took two to three days to reach its destination, and could remain docked with the Mir for approximately 200 days before its orbital lifetime limit expired.

The resident Soyuz was used for an occasional, scheduled "fly-around" of the T-shaped Mir, but crews primarily ventured outside for extravehicular activities (EVAs). During Mir's lifetime, crewmembers spent more than 325 hours as part of 75 planned spacewalks to conduct research and repairs on the exterior of the structure. Additional hours were spent during three intravehicular walks inside the unpressurized Spektr module. Participants in the Mir EVAs included 29 Russian cosmonauts, three U.S. astronauts, two French astronauts, and one European Space Agency astronaut who was a citizen of Germany. Cosmonaut Anatoly Solovyev donned the Russian Orlan spacesuit for 16 spacewalks for a total time of 77 hours, 46 minutes—more EVA time than any other spacewalker in the world.

After the Russian space station moved into its second decade, the Mir became notorious as an accident-prone spacecraft, even as it remained unparalleled in continuous service. A 15-minute fire in an oxygen-generating device imperiled the station in February 1997. Failures of the Elektron electrolysis oxygen-generating units and problems with attitude and environmental controls often seemed to alternate with computer malfunctions and power outages. The June 1997 collision with the Progress supply vehicle breached the integrity of the Spektr's hull and rendered that module uninhabitable.

But, Mir remained; and its space explorers endured. Over its lifetime, the space station hosted 125 cosmonauts and astronauts from 12 different nations. It supported 17 space expeditions, including 28 long-term crews. Its residents arrived via the 31 spacecraft that docked with Mir; nine of the dockings involved the Space Shuttle. Additionally, 64 uncrewed cargo vessels ferried supplies and equipment periodically to Mir. And, Mir served as a floating laboratory for 23,000 scientific and medical experiments.

Although Mir was gone by early 2001 and the International Space Station (ISS) was growing rapidly in orbit, the U.S. and Russia were still using spacecraft as statecraft. On March 23—the same day as Mir's deorbit—Russia expelled four U.S. diplomats and said it would expel 46 more in retaliation for the American expulsion of 50 Russian diplomats for espionage-like activities. It wasn't the Cold War all over again, but international tensions were certainly continuing, and the need remained for a worthy program for U.S. and Russian cooperation.

One could still apply to the ISS the same hopes that Shuttle Commander Charlie Precourt had held for it during Shuttle-Mir. Precourt had predicted that the ISS would "provide the psychological impetus for politicians to force themselves to find an agreement to disputes that otherwise they wouldn't—because they'll all look up there and say, 'Well, we have an investment in that, too. We have to keep this relationship going in a proper direction.'" Although the U.S.-Russian relationship was still going in the "proper direction," toward continued cooperation in space, the proper use and the funding of ISS were still in question. NASA cost overruns for the ISS clouded the program's future, and Russia's foreign department was threatening to reduce its participation in the ISS. To make the situation even more complicated, the ISS partner nations were discussing whether Russia should launch a wealthy American "space tourist" to the space station. The launch occurred, and the "space tourist" spent eight days in space.

Notwithstanding all the diplomatic wrangling, Mir's demise also coincided neatly with the successful finish to the first U.S.-Russian expedition to the ISS. On March 22, 2001, Expedition One crewmembers Sergei Krikalev, Yuri Gidzenko, and Commander William Shepherd returned to the Johnson Space Center in Houston. They received a ride from STS-102 Shuttle Commander James Wetherbee and a crew that included former Mir resident Andy Thomas. STS-102 had ferried the Expedition Two crew of James Voss, Susan Helms,

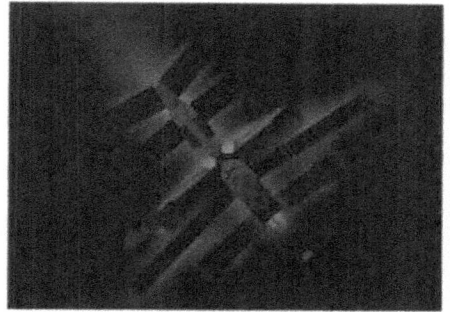
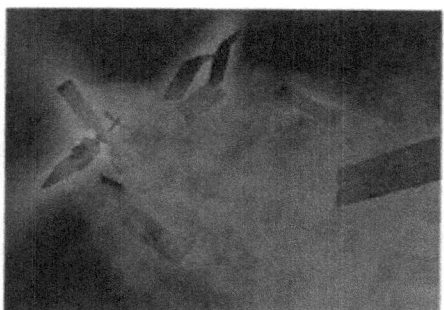

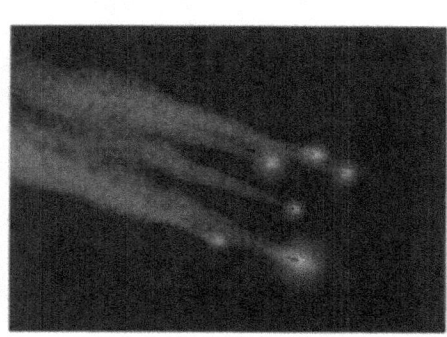

Mir Reentry

The journey of the 15-year-old Russian space station ended March 23, 2001, as Mir reentered the Earth's atmosphere near Nadi, Fiji, and fell into the South Pacific. Its downfall—planned and controlled—began around 8 a.m. Moscow time. Engines of a cargo ship docked to Mir were fired, causing the station's orbit to brake and starting the Mir's descent. The computer-generated images (above) illustrate the breakup of the 143-ton station as it descended to Earth (top left). At approximately 100 km, Mir entered the atmosphere and friction began to heat the outer surfaces (top right). The initial breakup began at about 95 km, when aerodynamic forces tore off the solar panels (bottom left). At around 85 km, all peripheral pieces were torn away, and the main modules began to buckle (bottom right). The surviving fragments fell into the South Pacific east of New Zealand. Witnesses to the fiery downfall attributed sonic booms to the estimated 20 to 25 tons of remnants moving quickly toward the Earth's surface.

and Commander Yury Usachev to the station.

Just as the ISS itself grew out of the lessons of Mir, many of the principal people in the ISS Program drew from their experiences during the Shuttle-Mir Program. And in Houston and in Moscow, American and Russian managers, engineers, and technicians who had worked in the Shuttle-Mir Program were working to make the ISS a success.

The International Space Station was growing, but the memories of Mir refused to fade. Indeed, people had anticipated its demise for long enough that, even before it fell, it had entered nostalgia. In a 1998 interview, Vladimir

Semyachkin reflected on Mir. He had developed the motion control systems and navigation systems for all vehicles and stations that were produced and launched into space by RSC-Energia. Semyachkin, as much as anyone, had wrestled with Mir's problems. He said: "It's a shame.... Our child, who we gave birth to so many years ago, ... we're going to have to put it to sleep. But, on the other hand, we understand that sometimes there's nothing to be done.... One cannot sit, as it were, on two chairs at the same time. Nevertheless, despite this sorrow with ... regard to Mir, we nonetheless do look forward to the future with a great deal of hope."

Mir Base Block

The Mir Base Block (core module) evolved from the earlier Soviet Salyut to serve as the heart of the space station. Launched in February 1986, the 13.1-meter-long, 20.4-metric ton core (pictured below) contained the primary living and working area, and life support and power as well as the main computer, communications, and control equipment—all in 90 cubic meters of habitable volume. Mir's environment was generally maintained at temperatures of 64° to 82° and humidity of 20 to 70 percent. The core had four main compartments.

The Working Compartment was actually two cylinders connected by a conical section. It provided operations and living areas. Operations included monitoring, command, and scientific activities. The living area provided the necessities for long-duration missions, including a galley with a table, cooking elements, trash storage; a bicycle exerciser and treadmill with medical monitoring equipment; video equipment; and individual crew areas, each with a porthole, hinged chair, and sleeping bag. The personal hygiene area, with toilet and sink, was located in one end of the Working Compartment. Mir had several portholes, with shutters outside to protect them from orbital debris impacts. Two television screens permitted face-to-face communications with the ground. Four more television screens monitored the other Mir modules.

The Transfer Compartment was a spherical structure at the front end of Mir, providing one end-docking port for visiting spacecraft plus four radial berthing ports, set in a 90-degree arrangement for access to the station's added modules. An approaching module used the Kurs (course) automatic docking system to dock with the forward port. Crews could then use the module's manipulator arm system to move it to a radial port, thus freeing the forward port for future use. The Transfer Compartment had no simulated "up and down" indicators; it was an area of Mir where astronauts reported sensations of disorientation.

The nonpressurized Assembly Compartment, on the other end of the Base Block, contained the station's main engine and fuel tanks; it supported antennas, lights, and optical sensors. The pressurized Intermediate Compartment tunneled through the Assembly Compartment to connect the Working Compartment to the aft docking port, where the Kvant-1 module was permanently docked.

Kvant-1

Kvant means "quantum." When Kvant-1 (pictured right) was docked permanently to Mir's aft docking port in April 1987, it increased Mir's usable volume and expanded its scientific capabilities. Kvant-1 supported research in the physics of galaxies, quasars, and neutron stars by measuring electromagnetic emissions. The module also supported biotechnology experiments and had some station control and life support functions. The 11-metric ton Kvant-1 measured 4.4 meters by 6.3 meters long, with 40 cubic meters of pressurized volume. The module was equipped with six gyrodynes that provided accurate pointing of the station and significantly reduced the amount of fuel used for attitude control. Its aft docking port was available for Soyuz and Progress vehicles.

Kvant-2

Kvant-2 was a scientific and airlock module, providing biological research, Earth observations, and extravehicular activity capabilities. The Kvant-2 (pictured right) enhanced Mir with drinking water and oxygen provisions, motion control systems, and power distribution as well as shower and washing facilities. Its airlock contained a self-sustained cosmonaut maneuvering unit that increased the range and complexity of extravehicular activity tasks. The 19.6-metric ton Kvant-2 measured 4.4 meters by 13.7 meters long with 61.3 cubic meters of volume and 27.4 meters of solar arrays. It was the first module equipped with the Lyappa manipulator arm, used to move the modules after they docked with Mir. Kvant-2 docked with Mir in November 1989.

Kristall

Kristall means "crystal." This module (pictured left) supported biological and materials production technologies in the microgravity environment. These included semiconductors, cellular substances, and medicines. Kristall also supported astrophysical and technical experiments. It had a radial docking port, originally designed as a means of docking the Russian Shuttle-type orbiter *Buran*, and was used for the first STS-71 docking in 1995. Added in June 1990, the 19.6-metric ton Kristall measured 4.4 meters by 13 meters long, with 60.8 cubic meters of volume and 36 meters of solar arrays.

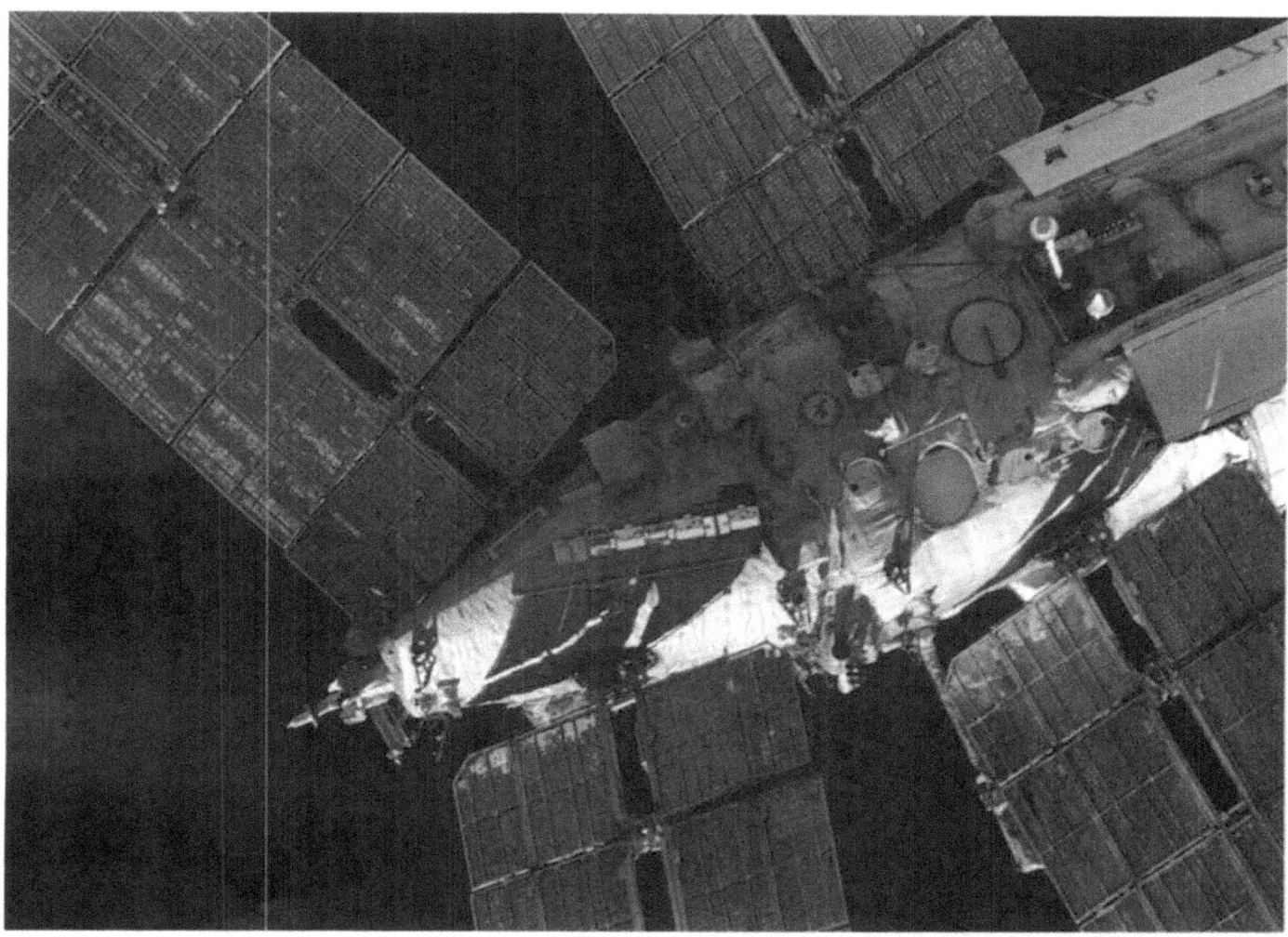

Spektr

Spektr means "spectrum," and this module allowed for better investigations and monitoring of Earth's natural resources and atmosphere. Spektr (pictured above) also supported research into biotechnology, life sciences, materials science, and space technologies. American astronauts sometimes used Spektr as their living quarters. Launched in May 1995 during Norm Thagard's mission to Mir, Spektr carried more than 1,600 pounds of U.S. equipment, mainly for biomedical research. Included with its arrival were two pairs of solar arrays to boost power to the station and a Lyappa manipulator arm to assist in moving the modules on Mir. The 19.3-metric ton Spektr module measured 14.4 meters by 4.4 meters, with a pressurized volume of 62 cubic meters, and had four solar arrays. On June 25, 1997, an uncrewed Progress resupply vehicle collided with the Spektr module, causing solar array and hull damage, and depressurization. The Mir crew closed the hatch to the leaking Spektr, preventing further pressure loss onboard Mir.

Priroda

Priroda means "nature," and this module's main purpose was Earth remote-sensing including the weather; the ocean-atmosphere system; land, mineral, and crop conditions; and humankind's impacts and opportunities in the environment. Priroda (pictured below) also collected information from remote-sensor buoys in nuclear power, seismic activity, and other areas to create an integrated monitoring and warning system. Launched in April 1996, Priroda, the last of the Mir modules, arrived during Shannon Lucid's stay on Mir. The 19.7-metric ton Priroda measured 4.4 meters by 12 meters long and had a pressurized volume of 66 cubic meters.

Progress

Resupply Vehicle

The unmanned, automated Progress was derived from the Soyuz crew transfer vehicle. It was designed to resupply and refuel the Russian Salyut and Mir space stations. A Progress (pictured left) typically approached and docked automatically at Mir's aft docking port using the Kurs (course) system. During Shuttle-Mir, Progress vehicles supplied Mir about once every two months. They are also used for the International Space Station.

Like the Soyuz, the Progress freighter measured about seven meters long by 2.7 meters wide. It had 6.6 cubic meters of volume and carried about 2,400 kg of cargo. The refueling compartment of the Progress was replaced with a nonpressurized cargo compartment to enable the transport of materials to be used on the outside of Mir. Progress was also designed to carry small satellites, which could be released by the Mir crews. After being loaded with trash, waste, and other unnecessary items, a Progress undocked and deorbited, to burn up upon reentry over the Pacific Ocean.

Early in the Soviet space station program, crews always got into their Soyuz descent vehicles as a manually guided Progress approached in case of an accident. The practice was discontinued after successes increased confidence.

Docking
Module

The Russian-built Docking Module (DM) (pictured right) was delivered by STS-74 on November 14, 1995. Attached to the Kristall, the DM provided clearance for the Shuttle to dock easily with Mir without interference from the station's solar panels.

The DM featured a pressure-sealed body and an androgynous peripheral docking system (APDS), compatible with the Kristall and Shuttle Orbiter docking systems (ODS). The Shuttle ODS, an external airlock extension, was fitted to the forward payload bay bulkhead and was accessible by the crew via the mid-deck airlock. When docked, the APDS provided locking, structural stiffness, and an airtight seal between the two structures.

The DM was 4.7 meters long from tip to tip of the identical APDSs on either end; its diameter was 2.2 meters; and it weighed approximately 4.1 metric tons. The module carried to the Mir two solar arrays: one Russian and one jointly developed by the U.S. and Russia to augment Mir's power supply. The DM carried the arrays retracted and stowed to be later deployed by cosmonauts.

Solar
Arrays

Electrical power in spacecraft can be produced in many ways; for example, from batteries or by nuclear and chemical fuel cells, or by solar cell arrays. Solar panels (pictured right) provided most of the power on Mir, and the station sometimes appeared as if it were "more panel than module." Nevertheless, as one Russian engineer put it, "we never managed to get rid of our energy hunger." Whenever the station lost attitude control—due to computer or gyrodyne failures—restoring solar power generation was usually the main concern, because almost all the other systems ultimately depended on it.

Interior

Mir's interior reflected an environment for work and residency; but more than anything, it contained essential systems, equipment, and materials for survival in space. On a daily basis, crewmembers completed tasks necessary for the station to maintain its existence as well as to provide the conditions vital for the crew. Over the years, Mir grew more and more crowded and experienced problems with stowage and inventory. The cluttered interior surrounds Aleksandr Lazutkin, Mir-23 Flight Engineer, as he exercises on the ergometer located in the Base Block (below, top right). Shuttle Commander Charlie Precourt floats through the crowded pathway to the Kristall that was also used to transfer supplies from the Orbiter to the Mir (bottom right). Mission Specialist Chris Hadfield demonstrates the limited space in the Docking Module passageway (bottom left). NASA-4 Astronaut Jerry Linenger (top left), has room to work in the Priroda, the last module to arrive to the complex.

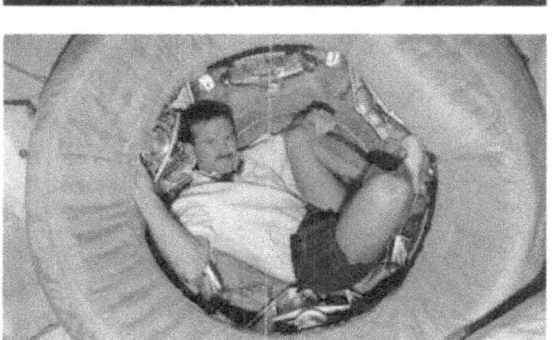

Soyuz-TM

Soyuz means "union," which refers both to the former Soviet Union and to the uniting of spacecraft. The Soyuz-TM spacecraft (pictured left) typically ferried three crewmembers to and from Mir. It remained docked with Mir to be available as an escape vehicle in case of emergency, and was sometimes used to make "fly-around" inspections of the station.

The Soyuz has launched in greater numbers than any other spacecraft in history, providing access to space for more than 30 years. Soyuz-TM has a mass of 7.2 metric tons, a length of seven meters, a maximum diameter of 2.7 meters, and a pressurized volume of 11 cubic meters. The spacecraft consists of three main sections: the orbital module that contains life support, rendezvous, and docking systems, and serves as a crew habitat during nondynamic flight phases; the instrument assembly module, a cylindrical shell that has the orbital flight systems; and the descent module, containing the Soyuz main systems control station and is the area where the crew stays during launch and conducts orbital maneuvers, rendezvous and docking, and descent. Two solar arrays (10.6-meter span) provide the vehicle's electrical power and could

Soyuz in Правда

Secrecy usually veiled the Soviet space program, but the Soviets sometimes used their space successes to promote an image of technological robustness to both their own citizens and the outside world. At times, detailed descriptions of space exploits and equipment were published, including this one of an early Soyuz model, excerpted from a November 17, 1968, issue of the Soviet newspaper Pravda.

The Soyuz consists of the following main modules: the orbital module ... a descent [module], intended for putting crews into orbit and returning them to Earth; and the service module, which houses the ... engines.

The orbital module is in the fore part of the ship and is connected with the descent capsule. The service module is placed behind the descent capsule. When the ship is being placed into orbit, it is protected against aerodynamic and thermal overloads by a nose faring, which is jettisoned after the passage through the dense layers of the atmosphere.

The cosmonaut's cabin [descent module] ... is covered on the outside by a ... heat-resistant covering to protect it from intensive aerodynamic heating during descent to Earth. After the vehicle has been slowed down by the atmosphere in its descent from orbit, the braking parachute opens ... then the main parachute, which is used for landing, opens. Directly before landing, at a height of about one meter above the Earth, the solid-fuel braking engines of the soft-landing system are switched on.

[In the] service module ... a hermetically sealed ... container carries the equipment for the thermo-regulation system, the system of unified electric power supply, the equipment for long-range radio communications and radio telemetry, and instruments for the system of orientation and control. The non-pressurized part of the service module contains the liquid-fuel propulsion installation [system] that is used for maneuvering in orbit and ... for ... descent back to Earth. The installation has two engines (the main one and the spare one). The ship has a system of low-thrust engines for orientation.

The [sensors] for the orientation system are located outside the service module. Mounted on ... the service module are the solar [arrays]. To ensure that the solar [arrays] are constantly illuminated, they are oriented towards the Sun by rotating the ship.

The ... spaceship is equipped with an automatic docking system. The onboard systems of the ship may be controlled either by the cosmonaut from the control panel or automatically. The ship's equipment allows for the craft to be piloted ... quite independently of ground control.

be interconnected with Mir's electrical system to furnish an additional 1.3 kW.

Typically, the journey from the Baikonur launch site to Mir took 50 hours. Transporting crews and cargo to and from the Russian space station, the Soyuz docked on the axial port on the transfer compartment of the core module. Probe-and-drogue devices were used to mate the Soyuz with Mir; the probing rod located on the Soyuz entered the receiving cone located on the space station. After "capture," an electric drive retracted the probe and pulled the two parts of the docking mechanism together. Hooks secured the two assemblies around a docking interface seal. When docking was completed, hatches were opened and the crews began a handover period to exchange information and tasks. At the end of this duration, the departing crew left in the resident Soyuz. The replacement crew moved the Soyuz to another port to allow docking access for the next spacecraft. Because the Soyuz has a limited life in space of five to six months, the rotation of vehicles guaranteed that Mir crewmembers had transportation back to Earth. On its return trip, the Soyuz capsule deployed parachutes after reentering the atmosphere, then fired braking rockets when it was just above the ground in Kazakhstan.

Soyuz Recovery

Following landing of the tiny Soyuz descent module, crewmembers traditionally sign the outside of the capsule in gratitude for a safe landing and return to Earth. (Top photo)

As the Soyuz lands on its single parachute, braking rockets fire about one meter above ground to soften the landing impact. Normally four of the six rocket clusters fire, with the other two clusters held in reserve should the main parachute fails and the smaller back-up chute deploy. The two reserve clusters must be disabled prior to transporting the capsule. Here technicians carefully remove one of the engine clusters with its multiple cartridges. The cartridges are then carried a safe distance from the capsule and detonated. (Middle photo)

After being transported from the landing site to the nearby airport in Dzhezkazgan by an Mi-8 helicopter, the Soyuz is loaded onto a truck and taken to a temporary storage facility for further processing. It will then be packaged and flown back to Moscow for thorough post-flight analysis by design and construction engineers. (Bottom photo)

Mir Expeditions

While NASA's Space Shuttle typically performs missions of up to two weeks in duration, Russia's Mir expeditions generally lasted about half a year. During Mir Principal Expedition16, cosmonauts unloaded the first U.S. science equipment, which had been launched on Progress 224, and conducted the first U.S. research onboard Mir with that hardware. During Mir-17, Cosmonaut Valeri Polyakov set an individual record for time in space of well over a year. The visits to the Russian space station as part of the Shuttle-Mir Program began with Mir Principal Expedition 17 and ended with Mir-25.

Mir Principal Expedition 17
Launched and landed in Soyuz-TM 20
October 3, 1994–March 22, 1995

Mir-17 hosted the first European Space Agency mission aboard Mir, Euromir 94, which ended November 4. Another international milestone was the rendezvous of the Space Shuttle *Atlantis* with Mir during the STS-63 mission. Cosmonaut Researcher Valeri Polyakov (seen above in the window), established a new human spaceflight duration record with his 438-day stay on the station. Also onboard Mir-17 was Commander Aleksandr Viktorenko and Flight Engineer Elena Kondakova.

Mir Principal Expedition 18

Launched in Soyuz-TM 21
March 14, 1995

Landed in Space Shuttle *Atlantis*
July 7, 1995

The major objectives of the Mir-18 mission were to conduct joint U.S.-Russian medical research and weightlessness effects investigations; to reconfigure the station for the arrival of the Spektr science module; and to welcome the Space Shuttle *Atlantis*. The mission saw the first American to be part of a Mir crew, NASA-1 Astronaut Norman Thagard (left), the addition of the first new module (Spektr) since Kristall arrived in 1990, and the first docking of a U.S. Space Shuttle (STS-71) with the Mir complex. Pictured with Thagard are Commander Vladimir Dezhurov (center) and Flight Engineer Gennady Strekalov (right).

Mir Principal Expedition 19

Launched in *Atlantis* (STS-71)
June 27, 1995

Landed in Soyuz-TM 21
September 11, 1995

The only complete Mir mission of 1995 with an all-Russian crew, Mir-19 had many international elements. As the first Mir crew launched on a Space Shuttle (STS-71, *Atlantis*), Commander Anatoly Solovyev (right) and Flight Engineer Nikolai Budarin (left) began their work in conjunction with the visiting U.S. crew and departing Mir-18 international crew. Two of their extravehicular activities involved deploying and retrieving international experiments. They ended their stay by welcoming the incoming international Euromir 95 crew.

Mir Principal Expedition 20
Launched and landed in Soyuz-TM 22
September 3, 1995–February 29, 1996

Mir-20 was the second Mir mission with a Euromir designation and the second with a European Space Agency astronaut as part of the crew. Cosmonaut Researcher Thomas Reiter (center) was the first non-Russian Mir crewmember with the added designation of Flight Engineer. Mir-20 was also the second Mir mission to include a U.S. Space Shuttle docking (STS-74, *Atlantis*). During that phase of the mission, the station complex housed crewmembers from four countries. Representing Russia were Commander Yuri Gidzenko, (left) and Flight Engineer Sergei Avdeyev (right). Reiter represented Germany as a member of the European Space Agency. Chris Hadfield, representing the Canadian Space Agency, arrived with the STS-74 crew—U.S. astronauts Kenneth Cameron, James Halsell, Jr., Jerry Ross, and William McArthur, Jr.

Mir Principal Expedition 21
Launched and landed in Soyuz-TM 23
February 21–September 2, 1996

Commander Yuri Onufriyenko (right) and Flight Engineer Yury Usachev (center) began their mission without the third crewmember, NASA-2 Astronaut Shannon Lucid (left), who would join them on March 23 during STS-76, the third *Atlantis*-Mir docking mission. On one of the seven extravehicular activities during Mir-21, two STS-76 Astronauts, Linda Godwin and Rich Clifford, walked outside Mir; the first U.S. extravehicular activity (EVA) outside the two mated spacecraft and the first U.S. spacewalk outside a space station since Skylab—22 years earlier. On two other EVAs, Onufriyenko and Usachev installed a new solar array on the Kvant module. In May 1996, the last permanent module—Priroda, with its large complement of Earth science experiments—was added to the complex.

Mir Principal Expedition 22
Launched and landed in Soyuz-TM 24
August 17, 1996–March 2, 1997

Commander Valeri Korzun (top right), Flight Engineer Alexander Kaleri (top left), and French Researcher Claudie Andre-Deshays (bottom center) joined NASA-2 Astronaut Shannon Lucid (top center) and the Mir-21 crew, Commander Yuri Onufriyenko (bottom right) and Flight Engineer Yury Usachev (bottom left), on Mir. Andre-Deshays returned to Earth with the Mir-21 crew after about two weeks onboard the station. Lucid's stay with the Mir-22 expedition crew was lengthened about six weeks due to a Space Shuttle launch postponement. STS-79 launched on September 16, 1996, delivering NASA-3 Astronaut John Blaha and returning home with Lucid onboard on September 26. STS-81 launched on January 12, 1997, with the next U.S. resident, Jerry Linenger, the first American to perform a spacewalk in a Russian Orlan spacesuit.

Mir Principal Expedition 23
Launched and landed Soyuz-TM 25
February 10, 1997–August 14, 1997

Mir-23 had more troubles than any other Mir expedition, including an onboard fire that occurred early in the mission with six crewmembers onboard—Mir-22 crew Valeri Korzun, Alexander Kaleri, and German Researcher Reinhold Ewald; Mir-23 crew Commander Vasily Tsibliev (left) and Flight Engineer Aleksandr Lazutkin (right); and NASA-4 Astronaut Jerry Linenger (center). Challenges continued for the Mir-23 crew after NASA-5 Mir Astronaut Mike Foale arrived on STS-84 when a Progress resupply vehicle collided with Mir's Spektr module, causing a loss of station control and depressurization.

Mir Principal Expedition 24
Launched and landed Soyuz-TM 26
August 5, 1997–February 19, 1998

Mir-24 hosted three American astronauts, NASA-5 Astronaut Mike Foale, NASA-6 Astronaut David Wolf (center), and NASA-7 Astronaut Andy Thomas. Mir continued to have systems problems, but overall conditions improved. Wolf performed an extravehicular activity with Commander Anatoly Solovyev (right). Also onboard Mir-24 was Flight Engineer Pavel Vinogradov (left).

Mir Principal Expedition 25
Launched and landed Soyuz-TM 27
January 29, 1998–August 25, 1998

Mir-25 was the last Shuttle-Mir expedition. French Astronaut Leopold Eyharts flew to Mir with the Mir-25 crew and returned to Earth with the Mir-24 crew. Russian Phase 1 Director Valery Ryumin visited Mir with the crew of STS-91, which brought NASA-7 Astronaut Andy Thomas back to Earth. Mir-25 crewmembers were Commander Talgat Musabayev and Flight Engineer Nikolai Budarin. Pictured are (top row, left to right) Ryumin, STS-91 crewmember Wendy Lawrence, Thomas, Janet Kavandi, Budarin; (bottom row, left to right) STS-91 Commander Charlie Precourt, Musabayev, and STS-91 crewmember Dominic Gorie. Not pictured is STS-91 crewmember Franklin Chang-Díaz.

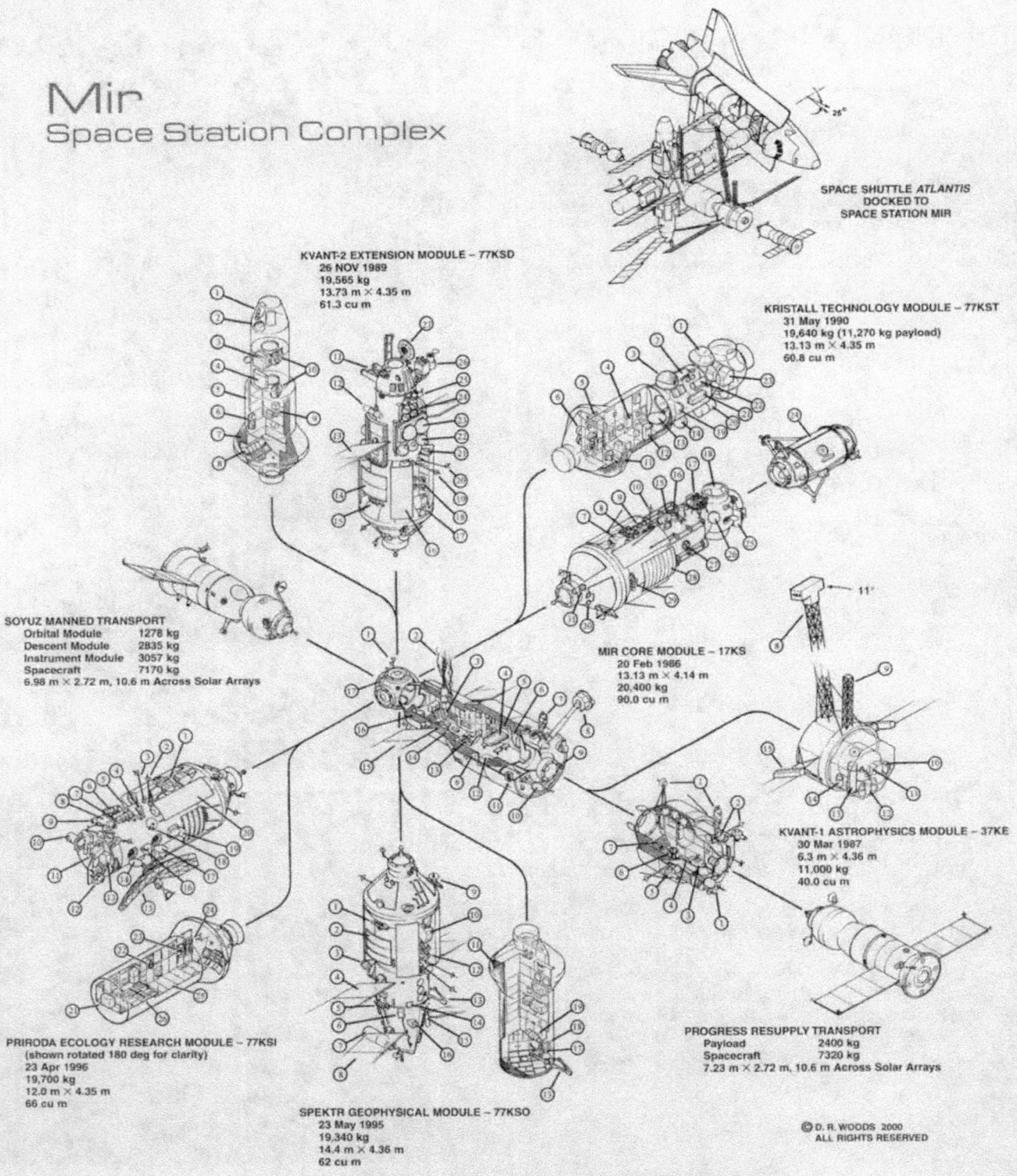

Mir
Space Station Complex

SPACE SHUTTLE *ATLANTIS*
DOCKED TO
SPACE STATION MIR

KVANT-2 EXTENSION MODULE – 77KSD
26 NOV 1989
19,565 kg
13.73 m × 4.35 m
61.3 cu m

KRISTALL TECHNOLOGY MODULE – 77KST
31 May 1990
19,640 kg (11,270 kg payload)
13.13 m × 4.35 m
60.8 cu m

SOYUZ MANNED TRANSPORT
Orbital Module   1278 kg
Descent Module   2835 kg
Instrument Module   3057 kg
Spacecraft   7170 kg
6.98 m × 2.72 m, 10.6 m Across Solar Arrays

MIR CORE MODULE – 17KS
20 Feb 1986
13.13 m × 4.14 m
20,400 kg
90.0 cu m

KVANT-1 ASTROPHYSICS MODULE – 37KE
30 Mar 1987
6.3 m × 4.36 m
11,000 kg
40.0 cu m

PRIRODA ECOLOGY RESEARCH MODULE – 77KSI
(shown rotated 180 deg for clarity)
23 Apr 1996
19,700 kg
12.0 m × 4.35 m
66 cu m

SPEKTR GEOPHYSICAL MODULE – 77KSO
23 May 1995
19,340 kg
14.4 m × 4.36 m
62 cu m

PROGRESS RESUPPLY TRANSPORT
Payload   2400 kg
Spacecraft   7320 kg
7.23 m × 2.72 m, 10.6 m Across Solar Arrays

MIR CORE MODULE [Base Block] – 17KS

1. Kurs Rendezvous Antenna
2. Auxiliary Solar Array (340 kg) 10.6 m
3. Module Control Consoles
4. Meal and Work Table
5. Cooking Elements
6. Personal Sleeping Compartment (2)

7. Igla Rendezvous Antenna
8. SDRN Luch Satellite Antenna (11/14 GHz)
9. Maneuvering Thruster (2) 300 kgf/ea
10. Igla Rendezvous Antenna
11. Attitude Control Thrusters – 6 Clusters
    32 Thrusters (total) 14 kgf/ea

12. Exercise Treadmill
13. Veloergometer Bicycle Exerciser
14. 50-Liter Refrigerator
15. Nine-Panel Solar Array (2) 29.73-m Span
16. Salyut-5B/Argon-16B Control Computers
17. Lyappa Module Rotation Socket

KVANT-1 ASTROPHYSICS MODULE – 37KE

1. Igla Rendezvous Antenna
2. Kurs Rendezvous Antenna
3. Sluice Airlock Chamber on Glazar UV Telescope
4. MKF-6 Multi-Spectral Camera
5. Attitude Control Star Sensor (2) 80 kg/ea
6. Visual Measurement and Photographic Device
7. Module Control Consoles

8. Sofora Truss (14.5 5 1.5 5 1.5 m) with
   VDU Roll Control Thruster Complex (700 kg)
9. Rapana Truss (26 kg) 5 m
10. Phoswich X-Ray Spectrometer 15-200 kev
11. GSPC – Gas Scintillation Proportional
    Counter Spectrometer 2-100 kev

12. Sirene-2 High-Pressure Gas Scintillation
    Proportional Spectrometer 2-100 kev
13. TTM Coded Mask Telescope 2-30 kev 60 kg
14. Pulsar X-Ray Spectrometer
    Telescope (5) 20-800 kev
15. MSB-2 and MSB-4 Extendable Solar
    Arrays (2) from Kristall (250 kg/ea) 7.5-m Span

KVANT-2 EXTENSION MODULE – 77KSD

1. Orlan-DMA EVA Spacesuit with 50-m Umbilical
2. KAP-350 Topography Photo Camera
3. MKF-6MA Multi-Spectral Camera (6 Bands)
4. Control Panel
5. NiCd Electrical Storage Batteries
6. Volna-2 Capillary Action Fuel Development Unit
7. Liquid and Waste Management System
8. Module Control Consoles
9. Incubator-2 Biotechnical Complex
10. Section Separation Hatch
11. Ryabina-2 Celestial Radiation Source Detector
12. Phasa AFM-2 Near Earth Atmosphere Telescope

13. VEP-3 and VEP-4 Solar Arrays, 24.13-m Span
14. Thermal Radiator (2)
15. Attitude Control Thrusters (4 Clusters)
    – 5 Approach Thrusters (40 kgf/ea)
    – 4 Stabilization Thrusters (1.5 kgf/ea)
16. Thermal Shield Over Propellant Tank
17. Course Correction and Rendezvous
    – Thruster (2) 415 kgf/ea
18. Attitude Control Star Sensors (2)
19. Kub Kontur Command Antenna
20. Kurs Rendezvous Antenna
21. Propellant Tank (4) 600 kg

22. Gyrodyne Control Moment Units (3) 490 kg
23. Rodnik Water Supply (2)
    300 liter / 420 kg ea
24. EVA Repressurization Air Supply (4) 28 kg
25. Attitude Control Star Sensors (3)
26. ASPG-M Gimballed Platform (110 kg)
    – MKS-M2 Optical Spectrometer
    – ITS-7D IR Spectrometer
    – ARIZ X-Ray Sensor
    – Gamma-2 Video Spectropolarimeter
27. EVA Hatch (1-m diam)

KRISTALL TECHNOLOGY MODULE – 77KST

1. Marina-2 Thermal Cover
2. TSB Thermostat
3. Glazar-2 Housing
4. Krater-B Electric Furnace - Gallium Arsenide
5. Optizon-1 Electric Furnace - Silicon
6. Electric Crucibleless Furnaces - Metal Melt
   – Zona-02
   – Zona-03
7. Course Correction and Rendezvous
   Thruster (2) 415 kgf/ea
8. Earth Horizon Attitude Control IR Sensors (2)
9. Helium Pressurization Tanks (6)
10. Kurs Rendezvous Antenna

11. Marina-2 Spectrometer
12. Svet Botanical Research Complex
13. Section Separation Hatch
14. Rodnik Water Supply (2)
15. Glazar-2 UV Telescope
16. Ksenia
17. Marina Spectrometer
18. APAS-89 Docking Port
19. TUB Thermostat
20. B16M (2)
21. Ainur Electrophoretic Complex
22. ChSK-1 Crystalizer

23. Priroda-5 High-Resolution Cameras (2)
    – SA-20M-I
    – SA-20-11
24. Space Shuttle Docking Module – 316GK
    4085 kg 4.75 x 2.22 m
25. Thruster Plume Shield
26. Granat Spectrometer
27. Solar Array Mount
    (Arrays Transfered to Kvant-1)
28. Thermal Radiators (2)
29. Attitude Control Thrusters (30)
30. Solar Attitude Sensor (4 Sets)
31. Lyappa Module Rotation Arm

SPEKTR GEOPHYSICAL MODULE – 77KSO

1. Attitude Control Thrusters
2. Thermal Radiators (2)
3. Astra-2 Atmospheric Trace Element Sensor
4. Seven-Segment Primary Solar Array (2)
5. Attitude Control Thrusters
6. Ryabina-4P Cosmic Radiation Sensor
7. Miras Atmospheric Spectrometer

8. Eight-Segment Auxiliary Solar Array (2)
9. Kurs Rendezvous Antenna
10. Course Correction and Rendezvous
    Thruster (2) 415 kgf/ea
11. Taurus / Grif-1 X-Ray / Gamma Ray
    Induced Radiation Sensor
12. Attitude Control Solar Sensors (4 Sets)

13. Pelican Manipulator Arm
14. Komza Interstellar Gas Sensor
15. Phasa Telespectrometer
16. Phoenix IR Spectrometer
17. Sluice Airlock Chamber
18. Astra View Port
19. Priroda Cameras (2)

PRIRODA INTERNATIONAL ECOLOGY RESEARCH MODULE – 77KSI

1. MOM-02P Earth Imager
2. KUB Kontur Command Antenna
3. Kurs Rendezvous Antenna
4. IKAR-N RP-600 Microwave Radiometer (Fixed)
5. Canopus Equipment
6. Marina-2 Radiometer
7. IKAR-D: R-30, R-80, R-135, R-225P
   Microwave Radiometer (Scanning)
8. Ozone-M Spectrometer
9. DK-35 Photometer
10. DOPI

11. ISTOK-1 Microwave Radiometer
12. MSU-5K Multi-Spectral Scanner
13. Greben Ocean Radar Altimeter
14. MOZ-Obzor Multi-Zonal Spectrometer
15. IKAR-D: R-400 Microwave
    Radiometer (Scanning)
16. Travers Synthetic Aperture
    Mapping Radar
17. Alisa Aerosol Lidar Ocean Altimeter
18. IKAR-D: RP-225 Microwave
    Radiometer (Scanning)

19. Delta-ZP Multi-Channel
    Scanning Radiometer
20. Meteor/Thermal Shield Over Propellant Tank (4)
21. SIGB
22. Fire Extinguisher
23. Microgravity Glovebox Facility
24. Gas Analysis System/Metabolic
    Analysis Physiology – GASMAP
25. Storage Lockers
26. Biotechnical Experiment System Modules

Acknowledgments

We have the story of Shuttle-Mir thanks to the hundreds of individuals in the U.S. and Russian space programs who successfully worked through countless technical, cultural, and philosophical challenges. Their efforts resulted in placing this episode in its rightful place in spaceflight history.

To preserve the history of Shuttle-Mir, Program Manager Frank Culbertson commissioned the production of the illustrated history, its companion CD-ROM, and related Web site. NASA Johnson Space Center Director George W. S. Abbey authorized the resources and actively supported the effort. William A. Larsen coordinated the project for JSC, and the JSC Oral History Project provided materials and resources vital for the human element focus of the project. The contributions of the NASA Public Affairs Office were likewise incalculable. PAO news releases, weekly status reports, and Shuttle-Mir team interviews provided continuity for the chronological narrative.

Rebecca Wright organized and directed this Shuttle-Mir History Project from its beginning and contributed to its success in many ways, including interviewing, writing, editing, design, management, and production. Sandra L. Johnson designed and created the companion CD-ROM. Paula J. Vargas provided artistic direction for the project, and researched and selected its photos, videos, and graphics. Summer Chick Bergen was invaluable in tracking down and validating sources, and compiling the chapter notes.

Sid Jones computer-crafted every page of the publication, Sharon I. Hecht edited the pages, and Sue McDonald produced the index. Gary Seloff provided guidance and contract support, as did S. Michael Smith and Donna Baumer. Carol L. Butler and Kevin M. Rusnak assisted in many areas of research and validation. David R. Woods created the labeled illustration of Mir, and D.S.F. Portree researched and gathered background materials. NASA's Chief Historian Roger D. Launius was always available with advice and support.

Besides making this history possible by living its story, many Shuttle-Mir Program team members gave hours of their time for oral histories and text reviews. The seven Mir astronauts, of course, were essential in providing the remembrances that constitute the bulk of the text. Also from NASA JSC, Michael R. Barratt, John J. Uri, Jessie Gilmore, Connie Van Praet-Cremins, Joseph P. Loftus, Jr., and Kamlesh P. Lulla provided important texts and support.

Other NASA JSC individuals and organizations assisting with the Shuttle-Mir History Project include Mary Wilkerson, Allen Bond, Irene Jenkins, Mike Gentry, William Stafford, Dexter C. Herbert, Pat Ryan, Brad Sayles, Jason Fennelly, and the Digital Imagery Lab.

Also, Don Bourque, L.D. Stevenson, Howard D. Ross, Jacque Havelka and Cherise Moore, NASA Kennedy Space Center Photo Repository, Tom Wiseman, Analytical Graphics, Inc., VideoCosmos, Russia, and Cable News Network (CNN).

Bibliographic Essay

The history of Shuttle-Mir comes first from those who experienced it. This book presents the human side through a detailed chronology and background information. Much of the material was provided by the NASA Johnson Space Center Oral History Project for which dozens of Shuttle-Mir participants (see list next page) offered their words, their stories, and their memories.

Historian Stephen E. Ambrose wrote in the introduction to his book, *Citizen Soldiers*, "Long ago my mentors … taught me to let my characters speak for themselves by quoting them liberally. They were there. I wasn't. They saw with their own eyes; they put their lives on the line. I didn't. They speak with an authenticity no one else can match. Their phrases, their word choices, their slang are unique—naturally enough, as their experiences were unique."[1]

Shuttle-Mir was likewise unique. And, its oral histories will continue through the years to illustrate the humanity and illuminate the importance of the Program.

Also, this book reflects the changing times. The Internet came of age during the Shuttle-Mir Program, and many of the book's sources reflect the Internet's capabilities. For historical background, NASA history offices maintain an ever-growing library of electronic texts. NASA's various Centers maintain Internet Web sites pertinent to their missions, such as the Shuttle launch records at Kennedy Space Center and human spaceflight information at the Johnson Space Center (JSC).

During and after the Program, JSC hosted a Shuttle-Mir Web site that included weekly updates and interviews. Russia's space agency also hosted a Web site; and newspapers, such as *The New York Times* and *Florida Today*, maintain news story archives available on the Web. Of the many Internet sources, two of the most interesting and helpful were Chris van den Berg's MirNews, which reported on Mir-to-ground radio transmissions, and Mark Wade's Encyclopedia Astronautica. Many memos and daily reports from NASA's internal electronic mail system proved useful. The author's file contains many of these documents, as well as other source materials such as personal correspondence and interviews. The author's file is archived in the history collection at NASA JSC.

Other Shuttle-Mir resources were—and are—available electronically. They include the *Phase 1 Program Joint Report* and Sue McDonald's *Mir Mission Chronicle*. These are contained on another sign of the changing times, this book's companion CD-ROM (S-M CD-ROM). Other bibliographical materials accessible on the companion disk include crew interviews, mission status reports, and weekly updates; reports from the Inspector General and Program Manager; letters from the Russian space station; and documentation focusing on the science program, as well as imagery, videos, and related information.

Additionally, two memoirs published during the writing of this book provided valuable insight and background. These were *Off the Planet* by NASA-4 Astronaut Jerry Linenger, and *Michael, Mir, and Me* by Colin Foale, father of NASA-5 Astronaut Mike Foale.

Finally, much use was made of JSC's Shuttle-Mir Program file archives and materials collected by Phase 1 Program Manager Frank Culbertson. And, nearly everyone involved in Shuttle-Mir was happy to contribute and clarify information for the production of this book.

---

[1] Stephen E. Ambrose, *Citizen Soldiers: The U.S. Army from the Normandy Beaches to the Bulge to the Surrender of Germany, June 7, 1944 to May 7, 1945* (New York: Simon & Schuster Inc., 1998), 13.

Shuttle-Mir Oral History Project Participants

All Shuttle-Mir oral histories are archived on CD and transcript in the NASA Lyndon B. Johnson Space Center History Collection as part of the Shuttle-Mir Oral History Project. Unless otherwise noted, oral history interviews were conducted by Rebecca Wright.

Albrecht, Mark J., National Space Council, 20 April 1999.

Aleksandrov, Aleksandr Pavlovich, Co-Chair Crew Training and Exchange Working Group, Extravehicular Activity (EVA) Working Group, 25 March 1998, interview by Mark Davison.

Barratt, Michael R., Flight Surgeon, 14 April 1998.

Billica, Roger D., Leader, Medical Operations Working Group, 17 June 1998.

Blagov, Victor Dmitrievich, Russian, Director of Flight Leadership, 28 May 1998.

Blaha, John E., Astronaut, NASA-3 Mir Resident, 24 August 1998.

Bogomolov, Valeri Vasilevich, Institute of Biomedical Problems, 30 May 1998.

Brice, Travis R., Russian Projects Office, 7 May 1998.

Brown, William C. "Charlie," Chair, Crew Exchange Working Group, 6 May 1998.

Capps, Tommy E., Training Manager, 18 August 1998.

Cardenas, Jeffery A., Co-Chair/Chair Mir Operations and Integration Working Group, 24 April 1998 and 21 May 1998.

Castle, Robert E., Jr., Flight Director, 24 June 1998.

Charles, John B., Shuttle-Mir Mission Scientist, 28 August 1998.

Chiodo, Christine A., Operations Lead, Russia, 4 August 1998.

Cremins, Tom E., Deputy, Assistant to the Director, Russia, 13 February 1998, interview by Michelle Kelly.

Culbertson, Frank L., Jr., Phase 1 International Space Station Program Manager, 24 March 1998, interview by Mark Davison.

Dailey, Brian, National Space Council, 20 April 1999.

Davis, Sally P., Russian Interface Officer (RIO), 14 August 1998.

Dunbar, Bonnie J., Astronaut, 16 June 1998.

Dye, Paul F., Flight Director, 27 May 1998.

Engelauf, Philip L., Flight Director, 24 June 1998, interview by Mark Davison.

Engle, Joe H., Stafford-Utkin Task Force, 15 July 1998.

Flynn, Christopher F., Flight Surgeon, 8 September 1998.

Foale, C. Michael, Astronaut, NASA-5 Mir Resident, 16 June 1998, 7 July 1998, and 31 July 1998.

Fullerton, Richard K., Co-Chair, Extravehicular Activity Working Group, 21 May 1998, interview by Mark Davison.

Gahring, Scott D., NASA-7 Operations Lead, 31 July 1998.

Gerstenmaier, William H., NASA-2 Operations Lead, 22 September 1998.

Gilmore, Jessie M., Executive Secretary/Assistant, Phase 1 ISS Program Office, 1 September 1998.

Glazkov, Yuri Nikolayevich, Gagarin Cosmonaut Training Center, 31 May 1998, interview by Mark Davison.

Holland, Albert W., Psychologist, 13 August 1998.

Jett, Brent W., Astronaut, Director of Operations-Russia, 17 June 1998, interview by Mark Davison.

Johnson, Gary W., Deputy Director for Russian Projects SR&QA, 17 February 1998, interview by Michelle Kelly.

Kargopolov, Yuri Petrovich, Gagarin Crew Training Center, 29 May 1998, interview by Mark Davison.

Kitmacher, Gary H., Lead, Priroda Module, 29 June 1998.

Lawrence, Wendy B., Astronaut, 21 July 1998,

Lomanov, Anatoli V., Deputy Director-Russian Program Coordination & Organization, 31 May 1998.

Lopez-Alegria, Michael E., Astronaut, Director of Operations-Russia, 7 July 1998.

Lucid, Shannon W., Astronaut, NASA-2 Mir Resident, 17 June 1998, interview by Mark Davison.

Lutomski, Michael G., 12 March 1998, presentation.

Marshburn, Thomas H., Flight Surgeon, 4 September 1998.

Moore, Isaac W. "Caasi," NASA-3 Operations Lead, 19 June 1998.

Moore, Patricia "Patti," NASA-6 Operations Lead, 30 July 1998.

Morgun, Valeri V., Gagarin Crew Training Center, 17 July 1998.

Mott, Michael, (former) Associate NASA Administrator, 23 April 1999.

Nise, James R., Contracts Manager, 12 February 1998, interview by Michelle Kelly.

Noah, Donald S., Manager of Space Shuttle Integration Engineering Office, 4 August 1998.

Nygren, Richard W., Chair, Mir Operations and Integration Working Group, 23 July 1998.

Pool, Sam L., Assistant Director, Space Medicine, Space and Life Sciences Directorate, 3 August 1998.

Precourt, Charles J., Astronaut, 12 July 1998.

Rahn, Debra, Public Affairs Officer for International Relations, NASA Headquarters, 8 July 1998, interview by Summer Chick Bergen.

Readdy, William F., Astronaut, Director of Operations-Russia, 8 June 1998, interview by Mark Davison.

Reed, Lisa M., NASA Johnson Space Center Training Lead, 19 June 1998.

Reeves, William D., Flight Director, 22 June 1998, interview by Mark Davison.

Ryumin, Valery Viktorovich, Russian Program Director, Shuttle-Mir Program, 27 April 1998, interview by Paul Rollins.

Sandars, George W., Chair, Flight Operations and Systems Integration Working Group, 12 June 1998.

Sang, Anthony C., NASA-4 Operations Lead, 18 March 1998, interview by Mark Davison.

Sega, Ronald M., Astronaut, Director of Operations-Russia, 6 July 1998 and 9 September 1998, interview by Carol L. Butler.

Semyachkin, Vladimir, Korolev Rocket and Space Corporation (RSC)-Energia, Rendezvous and Docking, 17 July 1998.

Sharipov, Salihzan Shakirovich, Cosmonaut, 8 March 1998, interview by Mark Davison.

Stegemoeller, Charles, Spektr Module, 6 August 1998.

Thagard, Norman E., Astronaut, NASA-1 Mir Resident, 16 September 1998.

Thomas, Andrew S. W., Astronaut, NASA-7 Mir Resident, 22 July 1998.

Titov, Vladimir Georgyevich, Cosmonaut, 21 July 1998.

Tsygankov, Oleg, Extravehicular Activity & Maintenance, Korolev Rocket and Space Corporation (RSC)-Energia, 25 March 1998, interview by Mark Davison.

Uri, John J., NASA Phase 1 Mission Scientist, 15 May 1998.

Van Laak, James E., Deputy Director, Phase 1 Program Office, 12 March 1998.

Vorobiev, Pavel Mikhailovich, Flight and Cargo Schedules, 30 May 1998.

Watson, J. Kevin, NASA Liaison, 9 April 1998, interview by Mark Davison.

Wetherbee, James D., Astronaut, 6 August 1998.

Wolf, David A., Astronaut, NASA-6 Mir Resident, 23 June 1998.

Zimmerman, Keith E., NASA-5 Operations Lead, 23 July 1998.

Chapter Notes

The Chapter Notes provide a detailed listing of the materials used for this publication. Listed under each title are the general sources for that section. Also listed under the title are the sources for the direct quotes within that section. Many of these source materials are available on the companion CD-ROM.

History's Highest Stage

Albrecht Shuttle-Mir oral history.

Dailey Shuttle-Mir oral history.

Edward Clinton Ezell and Linda Neuman Ezell, *The Partnership: A History of the Apollo-Soyuz Test Project*, The NASA History Series (Washington, DC: NASA, 1978).

Roger D. Launius, *Frontiers of Space Exploration* (Westport, CT: Greenwood Press, 1998).

Roger D. Launius and Bertram Ulrich, *NASA & the Exploration of Space*, Foreword by John Glenn (New York: Stewart, Tabori & Chang, 1998).

Sue McDonald, *Mir Mission Chronicle: November 1994-August 1996*, NASA/TP-98-207890 (Hanover, MD: NASA Center for Information, 1998), 9-21. (Also available on S-M CD-ROM).

David S. F. Portree, *Mir Hardware Heritage*, RP-1357/JSC-26770 (Houston: NASA Johnson Space Center, 1995).

Don Puddy to Rebecca Wright, 25 August 1998, facsimile, Morgan author file, Shuttle-Mir Collection, History Collection, Scientific and Technical Information Center, Lyndon B. Johnson Space Center, Houston, TX.

Judy A. Rutherman, comp., *U.S. Human Spaceflight, A Record of Achievement 1961-1998*, Monograph in Aerospace History, No. 9 (Washington, DC: NASA, 1998).

John Noble Wolford, "Apollo and Soyuz Blast into Orbit for a Rendezvous," New York Times, 16 July 1975, 1.

Direct Quotes
Page 1
"In the Open West ... the sky seemed almost alien.":
Roger D. Launius, "Sputniks and the Origins of the Space Age," NASA History Homepage, Online, <http://history.nasa.gov/sputnik/sputorig.html> (Last Updated 16 October 2000; Accessed December 2000).

"The first man-made satellite ... his name will be Ivan.":
Ezell and Ezell, *The Partnership*, 27-28.

"[If] we are to win the battle ... which road they should take.":
"Special Message to the Congress on Urgent National Needs, President John F. Kennedy, Delivered in person before a joint session of Congress, May 25, 1961," John Fitzgerald Kennedy Library and Museum Homepage, Online, <http://www.jfklibrary.org/j052561.htm> (Last Updated 27 March 2001; Accessed 3 April 2001).

Page 2
"commit itself ... to put him there.":
"Special Message to the Congress on Urgent National Needs," John Fitzgerald Kennedy Library and Museum Homepage.

"Let both sides seek ... let us explore the stars.":
"Inaugural Address, President John F. Kennedy, Washington, D.C., January 20, 1961," John Fitzgerald Kennedy Library and Museum Homepage, Online, <http://www.jfklibrary.org/j012061.htm> (Last Updated 27 March 2001; Accessed 3 April 2001).

"to join with us ... secrets of the universe.":
Ezell and Ezell, *The Partnership*, 29.

"go to the Moon together.":
Ezell and Ezell, *The Partnership*, 29.

"joint expedition to the Moon.":
"Address Before the 18th General Assembly of the United Nations, President John F. Kennedy, New York, September 20, 1963," John Fitzgerald Kennedy Library and Museum Homepage, Online, <http://www.jfklibrary.org/j092063.htm> (Last Updated 27 March 2001; Accessed 3 April 2001).

"No part ... without the consent of the Congress.":
Ezell and Ezell, *The Partnership*, 56.

"The exploration and use of outer space ... shall be the province of all mankind.":
The Treaty on Principles Governing the Activities of States in the Exploration and Use of Outer Space, including the Moon and Other Celestial Bodies, S-M CD-ROM.

Page 3
"I believe ... together in this effort.":
Homer Edward Newell, "Beyond the Atmosphere: Early Years of Space Science, APPENDIX J, [441] Statement by President Nixon on the Space Program, 7 March 1970. Released from the Office of the White House, Press Secretary, Key Biscayne, Florida," NASA Headquarters History Homepage, Online, <http://www.hq.nasa.gov/office/pao/History/SP-4211/appen-j.htm> (Last Updated 25 April 2001; 27 April 2001).

"apart from our formal negotiations ... the Space Shuttle would become available.":
George Low to President Richard M. Nixon, 25 January 1971, letter.

"The best of wishes ... My heart is with you.":
"Apollo-Soyuz Test Project Mission Commentary Part 4 (MC54/1)," NASA History Office Apollo-Soyuz Test Project Homepage, Online, <http://www.history.nasa.gov/astp/astp4.pdf> (Last Updated 11 July 2000; Accessed 3 April 2000).

"You're making history ... with the Soviet Union.":
"Apollo-Soyuz Test Project Mission Commentary Part 4 (MC54/1)," NASA History Office Apollo-Soyuz Test Project Homepage.

"We came together at a slow rate ... then it quickly stopped.":
"Former Astronaut Recalls Historic Apollo-Soyuz Mission," Release: 00-53, NASA Dryden Flight Research Center Homepage, Online, <http://www.dfrc.nasa.gov/PAO/PressReleases/2000/00-53.html> (Release Dated 21 July 2000; Accessed 8 September 2000).

"Glad to see you.":
Steve Garber, ed., "The Flight of Apollo-Soyuz," NASA Headquarters Homepage, Online, <http://www.hq.nasa.gov/office/pao/History/apollo/apsoyhist.html> (Last Updated 14 October 1998; Accessed 3 April 2000).

"Very, very happy to see you.":
Garber, ed., "The Flight of Apollo-Soyuz," NASA Headquarters Homepage.

Page 4
"It was a positive experience ... Apollo-Soyuz was the first for this kind of cooperation.":
"Former Astronaut Recalls Historic Apollo-Soyuz Mission," Release: 00-53, NASA Dryden Flight Research Center Homepage.

"Russia was such a closed society ... vertically separated.":
Thomas P. Stafford, interview by William Vantine, 15 October 1997, CD, transcript, Johnson Space Center Oral History Project, Scientific and Technical Information Center, Lyndon B. Johnson Space Center, Houston, TX.

"a joint simulated space rescue mission":
"President, Senators, Proposing Space Rescue Operations Plans," *Aviation Week & Space Technology*, Vol. 121, No. 2, 9 July 1984: 16-17.

Page 5
"If we did not do something ... the use of those technologies.":
Dailey Shuttle-Mir oral history.

Page 6
"all JSC activities ... Russian Federation programs.":
"Key Personnel Assignments," Johnson Space Center Announcement, No. 92-107, 20 August 1992, Morgan author file.

"We had from November ... after we communicated.":
  Capps Shuttle-Mir oral history.

"We started trying to do things ... they had no problem whatsoever driving.":
  Capps Shuttle-Mir oral history.

Page 7
"The Russians don't delegate ... it's kind of difficult in that regard.":
  Nise Shuttle-Mir oral history.

The Goals of Shuttle-Mir

Ezell and Ezell, *The Partnership*.

Launius, *Frontiers of Space Exploration*.

John M. Logsdon, "Together in Orbit: The Origins of International Cooperation in the Space Station," NASA Monographs in Aerospace History, No. 11, (Washington, DC: NASA, 1998).

Howard E. McCurdy, *The Space Station Decision: Incremental Politics and Technological Choice* (Baltimore, MD: Johns Hopkins University Press, 1990).

"National Air and Space Museum," Smithsonian National Air and Space Museum Homepage, Online, <http://www.nasm.si.edu/> (Last Updated 8 February 2001; Accessed 3 March 2001).

David S. F. Portree, "NASA's Origins and the Dawn of the Space Age," NASA Monographs in Aerospace History, No. 10 (Washington, DC: NASA, 1998).

All direct quotes taken from the Culbertson Shuttle-Mir oral history.

Profile: Frank Culbertson

"Biographical Data: Frank L. Culbertson, Jr.," NASA Johnson Space Center Homepage, Online, <http://www.jsc.nasa.gov/Bios/htmlbios/culberts.html> (Last Updated February 2000; Accessed April 2000).

All direct quotes taken from the Flynn Shuttle-Mir oral history.

Profile: Valery Ryumin

"Biographical Data: Valery Victorovitch Ryumin," NASA Johnson Space Center Homepage, Online, <http://www.jsc.nasa.gov/Bios/htmlbios/ryumin.html> (Last Updated February 1998; Accessed 2 March 2000).

Marcia Dunn, "Ex-cosmonaut Returns to Orbit," ABCNEWS.com Homepage, Online, <http://abcnews.go.com/sections/science/DailyNews/mir_ryumin980601.html> (Article Dated 1 June 1998; Accessed 4 April 2000).

Precourt Shuttle-Mir oral history.

STS-60: *A Cosmonaut Flies on the Shuttle*

"STS-60 (60)," Kennedy Space Center Homepage, Online, <http://science.ksc.nasa.gov/shuttle/missions/sts-60/mission-sts-60.html> (Last Updated 31 August 2000; Accessed September 2000).

"STS-60 Press Kit-February 1993, Wake Shield Facility SPACEHAB-2," Kennedy Space Center Homepage, Online, <http://science.ksc.nasa.gov/shuttle/missions/sts-60/sts-60-press-kit.txt> (Last Updated February 1993; Accessed 12 January 2000).

Direct Quotes
Page 8
"not exactly correct ... without an accent.":
"STS-60 Transcript of Russian Premier Crew Call," NASA Spacelink Homepage, Online, <http://spacelink.nasa.gov/NASA.Projects/Human.Exploration.and.Development.of.Space/Human.Space.Flight/Shuttle/Shuttle.Missions/Flight.060.STS-60/Russian.Premier.Calls> (Last Updated 9 February 1994; Accessed September 2000).

Profile: Sergei K. Krikalev

"Biographical Data: Sergei Konstantinovich Krikalev," NASA Johnson Space Center Homepage, Online, <http://www.jsc.nasa.gov/Bios/htmlbios/krikalev.html> (Last Updated November 2000; Accessed 3 March 2001).

"STS-60 (60)," Kennedy Space Center Homepage.

"STS-60 Transcript of Russian Premier crew call," NASA Spacelink Homepage.

All direct quotes taken from the Brice Shuttle-Mir oral history.

Meanwhile on Earth, February 1994 - March 1995

"Cameron to Manage NASA Activities at Star City, Russia," Release: J94-27, NASA Spacelink Homepage, Online, <http://spacelink.nasa.gov/NASA.News/NASA.News.Releases/Previous.News.Releases/94.News.Releases/94-02.News.Releases/94-02-23> (Release Dated 23 February 1994; Accessed 19 May 2000).

Cremins Shuttle-Mir oral history.

McDonald, *Mir Mission Chronicle*, 9-21.

"Mir-22 Mission Status Reports: August 18, 1996 - March 2, 1997," Morgan author file.

Portree, *Mir Hardware Heritage*.

Readdy Shuttle-Mir oral history.

Rumerman, comp., *U.S. Human Spaceflight, A Record of Achievement 1961-1998*.

Direct Quotes
Unless otherwise noted, direct quotes taken from the Cremins Shuttle-Mir oral history.

Page 11
"Then the State Department transferred [the money] ... It typically had a two-month lag in it.":
  Brice Shuttle-Mir oral history.

"little support group ... we'd have tea with her colleagues there.":
  Dunbar Shuttle-Mir oral history.

"absolutely spectacular":
  Readdy Shuttle-Mir oral history.

"Everybody here kind of worked together as a family ... and got it done.":
  Gilmore Shuttle-Mir oral history.

Profile: Tommy Holloway

Culbertson Shuttle-Mir oral history.

"Holloway Named Space Station Manager, Dittemore to Head Shuttle; Brinkley leaves NASA for private sector," Release: J99-9, NASA Johnson Space Center Homepage, Online, <http://www.jsc.nasa.gov/pao/media/rel/1999/J99-9.html> (Release Dated 6 April 1999; Accessed 25 February 2000).

"Space Flight Banquet to Honor NASA's Tommy W. Holloway," *Citizen* (Houston), 5 August 1999, 13.

All direct quotes taken from the Culbertson Shuttle-Mir oral history.

Space Shuttles

Jim Dumoulin, "Orbiter Vehicles," NASA Kennedy Space Center Science, Technology and Engineering Homepage, Online, <http://science.ksc.nasa.gov/shuttle/resources/orbiters/orbiters.html> (Last Updated 29 April 1995; Accessed 3 March 2001).

Space Shuttle Life

"Living on the Space Shuttle," FS-1995-08-001JSC, NASA Johnson Space Center Homepage, Online, <http://www.jsc.nasa.gov/pao/factsheet/factsheets/9508001.pdf> (Last Updated 10 March 2000; Accessed 3 March 2001).

STS-63: *First Rendezvous*

"STS-63 (67)," Kennedy Space Center Homepage, Online, <http://science.ksc.nasa.gov/shuttle/missions/sts-63/mission-sts-63.html> (Last Updated 31 August 2000; Accessed September 2000).

"STS-63 Press Kit," Kennedy Space Center Homepage, Online, <http://science.ksc.nasa.gov/shuttle/missions/sts-63/sts-63-press-kit.txt> (Last Updated February 1995; Accessed 17 February 2000).

Direct Quotes
Page 14
"in a room that's on fire.":
    "Female Frontiers-An Interview with Eileen Collins," NASA Quest Homepage, Online, <http://quest.nasa.gov/space/frontiers/interview/index.html> (Accessed 27 April 2001).

Page 15
"We launched ... they didn't want the Shuttle coming too close to the Mir.":
    Reeves Shuttle-Mir oral history.

"like a snowstorm for five miles up into space,":
    Wetherbee Shuttle-Mir oral history.

"very sharp and astute ... and asked all the right questions.":
    Reeves Shuttle-Mir oral history.

"[I]t wasn't until the morning ... make the close approach.":
    Engelauf Shuttle-Mir oral history.

"like a big ocean liner coming in ... gave the Russians a lot of confidence on [STS-] 63.":
    Wetherbee Shuttle-Mir oral history.

"Oh, beautiful, beautiful, beautiful!":
    NASA Select TV broadcast of STS-63 rendezvous with Mir, 6 February 1995, Morgan author file.

"just blown away by the sight ... Mir looked so brilliant and white and bright.":
    Wetherbee Shuttle-Mir oral history.

The Snowball in the Fax

All direct quotes taken from the Reeves Shuttle-Mir oral history.

NASA-1 Norm Thagard: *An End and a Beginning*

McDonald, *Mir Mission Chronicle*, 9-21.

"Mir-18 Morning Status Summaries," March 14, 1995-July 7, 1995, Morgan author file.

"Mir-22 Mission Status Reports," Morgan author file. "Shuttle-Mir Science Program Mir 18 Flight Anomaly Log," *Mir 18 Management Report*, Morgan author file.

"Spaceviews Sites of the Week: June 1997," SEDS Homepage, Online, <http://www.seds.org/spaceviews/award/9706.html> (Last Updated 2000; Accessed December 2000).

Thagard Shuttle-Mir oral history.

Mark Wade, ed., "Chronology-1960-Quarter 4," Encyclopedia Astronautica Homepage, Online, <http://www.friends-partners.org/~mwade/chrono/19604.htm#5836> (Last Updated 22 December 2000; Accessed December 2000).

Mark Wade, ed., "Chronology-1983-Quarter 3," Encyclopedia Astronautica Homepage, Online, <http://www.friends-partners.org/~mwade/chrono/19833.htm> (Last Updated 22 December 2000; Accessed December 2000).

Dennis Webb, "MEAT Operations in Moscow, Weekly Summary for March 18 through 24, 1995," Morgan author file.

All direct quotes taken from the Thagard Shuttle-Mir oral history.

Baikonur Cosmodrome

"Baikonur Cosmodrome Launch Facilities," NASA Human Spaceflight Homepage, Online, <http://spaceflight.nasa.gov/station/assembly/elements/fgb/baikonur.html> (Last Updated 22 October 2000; Accessed 7 January 2001).

Engle Shuttle-Mir oral history.

"Launch Facilities," Marshall Space Flight Center Liftoff to Space Exploration Homepage, <http://liftoff.msfc.nasa.gov/rsa/pads.html> (Last Updated 8 November 1996; Accessed 3 March 2001).

"Russian space.com," Russian space.com Homepage, Online, <http://russianspace.com/baikonur.html> (Last Updated 30 September 2000; Accessed 1 November 2000).

Stegemoeller Shuttle-Mir oral history.

Direct Quotes
Page 20
"an enormous, vast, desolate ... I guess it's a sign of the times.":
    Engle Shuttle-Mir oral history.

"Never have so many done so much ... putting it in the module.":
    Stegemoeller Shuttle-Mir oral history.

Profile: Norman E. Thagard

"Biographical Data: Norman E. Thagard (M.D.)," NASA Johnson Space Center Homepage, Online, <http://www.jsc.nasa.gov/Bios/htmlbios/thagard.html> (Last Updated August 1995; Accessed 30 November 1998).

All direct quotes taken from the Thagard Shuttle-Mir oral history.

Meanwhile on Earth, NASA-1

Barrett Shuttle-Mir oral history.

"Future Mir Astronauts," News Release 95-39, NASA Spacelink Homepage, Online, <http://spacelink.nasa.gov/NASA.News/NASA.News.Releases/Previous.News.Releases/95.News.Releases/95-03.News.Releases/95-03-30.Future.Mir.Astronauts> (Release Dated 30 March 1995; Accessed 19 May 2000).

Shannon Lucid to Clay Morgan, 18 September 2000, email, Morgan author file.

Direct Quotes
    Unless otherwise noted, direct quotes taken from the Barrett Shuttle-Mir oral history.

Page 30
"There really wasn't any specialization in the staff ... sliding into an operational or semi-operationally focused job.":
    Van Laak Shuttle-Mir oral history.

Quick Work on the Ground

Fullerton Shuttle-Mir oral history.

"MIR-19: Weekly Management Summary," July 10 - 14, 1995, Morgan author file.

George C. Nield and Pavel Mikhailovich Vorobiev, ed., *Phase 1 Joint Report*, NASA/SP-1999-6108 (Hanover, MD: NASA Center for Information, 1998). (Also available on S-M CD-ROM.)

Charles A. (Chip) Shearrow, (JSC-GA) to Gary A. Seloff (JSC-GS), 16 August 2000, email, Morgan author file.

STS-71: *First Docking*

Precourt Shuttle-Mir oral history.

Reed Shuttle-Mir oral history.

"STS-71 (69)," Kennedy Space Center Homepage, Online, <http://science.ksc.nasa.gov/shuttle/missions/sts-71/mission-sts-71.html> (Last Updated 31 August 2000; Accessed September 2000).

"STS-71 Fact Sheet: *Atlantis*/First Shuttle-Mir Docking Mission," KSC Release No. 84-95, KSC Fact Sheets Homepage, Online, <http://www-pao.ksc.nasa.gov/kscpao/release/1995/48-95.htm> (Release Dated May 1995; Accessed December 2000).

"STS-71 Press Kit," Kennedy Space Center Homepage, Online, <http://science.ksc.nasa.gov/shuttle/missions/sts-71/sts-71-press-kit.txt> (Last Updated 31 June 1995; Accessed 21 February 2000).

Direct Quotes
Page 32
"science fiction movies ... that thing is pretty phenomenal.":
    Precourt Shuttle-Mir oral history.

Page 33
"cosmic ballet,":
    McDonald, *Mir Mission Chronicle*, 22.

Contact and Capture: Emotions of the Moment

All direct quotes taken from the Reed Shuttle-Mir oral history.

Rendezvous & Docking

"STS-71 Press Kit," Kennedy Space Center Homepage.

Wetherbee Shuttle-Mir oral history.

All direct quotes taken from the Precourt Shuttle-Mir oral history.

STS-74: *A New Docking Module*

"Mir EVA activities," SeeSat-L Archives Homepage, Online, <http://www2.satellite.eu.org/sat/seesat/Jul-1995/0073.html> (Last Updated 20 July 1995; Accessed 24 April 2001).

"STS-74 (73)," Kennedy Space Center Homepage, Online, <http://science.ksc.nasa.gov/shuttle/missions/sts-74/mission-sts-74.html> (Last Updated 31 August 2000; Accessed 15 March 2000).

"STS-74 Fact Sheet: STS-74/*Atlantis*, Shuttle/Mir Mission-2, Docking Module, Solar Arrays-October 1995," KSC Fact Sheets Homepage, Online, <http://www-pao.ksc.nasa.gov/kscpao/nasafact/74facts.htm> (Last Updated 27 November 2000; Accessed December 2000).

"STS-74 Pre-Flight Interview with James Halsell," NASA Human Spaceflight Homepage, Online, <http://spaceflight.nasa.gov/shuttle/archives/sts-74/crew/halsell_int.html> (Last Updated 29 September 1995; Accessed 15 March 2000).

"STS-74 Pre-Flight Interview with Kenneth Cameron," NASA Human Spaceflight Homepage, Online, <http://spaceflight.nasa.gov/shuttle/archives/sts-74/crew/cameron_int.html> (Last Updated 22 September 1995; Accessed 15 March 2000).

"STS-74 Pre-Flight Interview with William McArthur," NASA Human Spaceflight Homepage, Online, <http://spaceflight.nasa.gov/shuttle/archives/sts-74/crew/mcarthur_int.html> (Last Updated 19 September 1995; Accessed 15 March 2000).

"STS-74 Press Kit," Kennedy Space Center Homepage, Online, <http://science.ksc.nasa.gov/shuttle/missions/sts-74/sts-74-press-kit.txt> (Last Updated 31 August 2000; Accessed 15 March 2000).

Direct Quotes
Page 37
"like looking at the top of a building ... accurately judge position or orientation.":
"STS-74 Pre-Flight Interview with Kenneth Cameron," NASA Human Spaceflight Homepage.

Flying Through the Furlough

"STS-74 Fact Sheet," KSC Fact Sheets Homepage.

"STS-74 Pre-Flight Interview with Kenneth Cameron," NASA Human Spaceflight Homepage.

"STS-74 Press Kit," Kennedy Space Center Homepage.

"Utterly Unofficial STS-74 Web Site," Reston Communications Homepage, Online, <http://www.reston.com/sts74/sts74page.html> (Accessed 1 November 2000)

Direct Quotes
Page 37
"NOTE: A number of NASA Web servers were shut down ... It certainly wasn't their fault.":
"Utterly Unofficial STS-74 Web Site," Reston Communications Homepage.

"I've never talked to an unemployed American before.":
Stegemoeller Shuttle-Mir oral history.

Training and Operations

Training

Blaha Shuttle-Mir oral history.

Brown Shuttle-Mir oral history.

Culbertson Shuttle-Mir oral history.

Dunbar Shuttle-Mir oral history.

Foale Shuttle-Mir oral history, 16 June 1998.

Foale Shuttle-Mir oral history, 31 July 1998.

Glazkov Shuttle-Mir oral history.

Johnson Shuttle-Mir oral history.

Lucid Shuttle-Mir oral history

Nield and Vorobiev, ed., *Phase 1 Joint Report*.

Direct Quotes
Page 39
"In Russia, they do it the old-fashioned way ... administer an oral exam to the student.":
Blaha Shuttle-Mir oral history.

"Their advantage ... go home and study it.":
Brown Shuttle-Mir oral history.

"The impression ... It was like a totally new language.":
Culbertson Shuttle-Mir oral history.

Page 40
"Flights are hard; ... training is harder.":
Phase 1 Review News Conference, 15 August 1998, Morgan author file.

Training in Orbit

"Mir-22 Weekly Update-October 18, 1997: Interview with Jeff Cardenas," S-M CD-ROM.

Nise Shuttle-Mir oral history.

Direct quote from Uri Shuttle-Mir oral history.

Gagarin Cosmonaut Training Center

Mark Wade, ed., "astronautix.com Encyclopedia Astronautica," Encyclopedia Astronautica Homepage, Online, <http://www.friends-partners.org/~mwade/spaceflt.htm> (Last Updated 22 December 2000; Accessed December 2000).

Survival Training

Mike Foale, "Mir-21 Weekly Update-June 14, 1996: Black Sea Training," on NASA Human Spaceflight Homepage, Online, <http://spaceflight.nasa.gov/history/shuttle-mir/mir21/status/week11/foale.html> (Last Updated 14 June 1996; Accessed 3 March 2001).

Dunbar Shuttle-Mir oral history.

Direct Quotes
Page 41
"This is all done with the hatch closed ... the training suits leak worse than real ones.":
Foale, "Black Sea Training," on NASA Human Spaceflight Homepage.

"the most comfortable thing ... taking care of the fire actually was kind of fun.":
Dunbar Shuttle-Mir oral history.

Americans and 'Warm Fuzzies'
Training Cosmonauts for the Space Shuttle

All direct quotes taken from the Reed oral history.

Life in Star City, Russia and "Space City," Texas

Yuri Karash, "Star City: Berkeley for Cosmonauts," Space.com Homepage, Online, <http://www.space.com/news/spacehistory/star_city_000411.html> (Article Dated 12 April 2000; Accessed 12 April 2000).

Direct Quotes
Page 42
"Star City really is a little tiny city ... grandkids and grandparents.":
Readdy Shuttle-Mir oral history.

"It's a very historical place ... peaceful, quiet, forests, trees, fresh air.":
Barratt Shuttle-Mir oral history.

Page 43
"Things got steadily better ... Moscow is unrecognizable compared to the way it was":
Foale Shuttle-Mir oral history, 17 June 1998.

"Brenda, these are really great books ... we'd eat dinner together and talk.":
Blaha Shuttle-Mir oral history.

"Moscow was not very comfortable. ... everything was for us like a little bit new.":
    Titov Shuttle-Mir oral history.

"We can do that? ... Keep your car full of gas and just go.":
    Brice Shuttle-Mir oral history.

Bilingual Blues

Michael R. Barratt, M.D., *Mir 18 Supplemental Medical Kit Checklist*, NASA Life Sciences Directorate, October 1994.

Blaha Shuttle-Mir oral history.

Brice Shuttle-Mir oral history.

Dunbar Shuttle-Mir oral history.

Foale Shuttle-Mir oral history, 16 June 1998.

Foale Shuttle-Mir oral history, 31 July 1998.

Jett Shuttle-Mir oral history.

Steven Jones interview, 5 August 1999, Morgan author file.

Lucid Shuttle-Mir oral history.

Marshburn Shuttle-Mir oral history.

The New Encyclopaedia Britannica, 15th ed., s.v. "Russian Language."

Thagard Shuttle-Mir oral history.

Direct Quotes
Page 44
"because you had to learn all of these things ... all of a sudden.":
    Dunbar Shuttle-Mir oral history.

"biggest mistake":
    Blaha Shuttle-Mir oral history.

"fairly disdainful ... those things you like doing in your free time.":
    Foale Shuttle-Mir oral history, 31 July 1998.

Page 45
"We made a lot of jokes about it ... We can't stand to listen to your Russian anymore!":
    Lucid Shuttle-Mir oral history.

"It's on and off.":
    Marshburn Shuttle-Mir oral history.

Operations: A Tale of Two Systems

Nield and Vorobiev, ed., *Phase 1 Joint Report*.

"Shuttle/Mir Operation," NASA Human Spaceflight Homepage, Online, <http://spaceflight.nasa.gov/history/shuttle-mir/ops/> (Last Updated 7 August 1998; Accessed 3 March 2001).

Direct Quotes
Page 46
"that you get up in the morning ... you have your weekends to yourself.":
    Engelauf Shuttle-Mir oral history.

"[We] might tell the crew member ... the astronaut might go back to the first experiment":
    Van Laak Shuttle-Mir oral history.

"Things are a little slower paced ... whereas in the Shuttle Program ten days is an entire mission.":
    Castle Shuttle-Mir oral history.

Operations Leads and Russian Interface Officers

Cardenas Shuttle-Mir oral history, 21 May 1998.

Culbertson Shuttle-Mir oral history.

Davis Shuttle-Mir oral history.

Engelauf Shuttle-Mir oral history.

Direct Quotes
Page 47
"a good solid team ... to execute it together.":
    Culbertson Shuttle-Mir oral history.

"trying to give him a thumbnail sketch of what's coming up on the day.":
    Cardenas Shuttle-Mir oral history, 21 May 1998.

"As far as marching down the road ... Keep the control teams in line.":
    Davis Shuttle-Mir oral history.

DOR: Director of Operations-Russia

Brown Shuttle-Mir oral history.

"Cameron to Manage NASA Activities at Star City, Russia," Release: J94-27, NASA Spacelink Homepage.

All direct quotes taken from the Readdy Shuttle-Mir oral history.

Mission Control Center-Houston

"Flight Control of STS-74," Release: J95-069, NASA Johnson Space Center Homepage, Online, <http://www.jsc.nasa.gov/pao/media/rel/Past-News-Releases-93-94-95/95-069-DOC.html> (Release Dated 6 November 1995; Accessed 3 March 2001).

Mission Control Center-Moscow

Cardenas Shuttle-Mir oral history, 24 April 1998.

Cardenas Shuttle-Mir oral history, 21 May 1998.

Culbertson Shuttle-Mir oral history.

Davis Shuttle-Mir oral history.

Engelauf Shuttle-Mir oral history.

Nield and Vorobiev, ed., *Phase 1 Joint Report*.

All direct quotes taken from the Davis Shuttle-Mir oral history.

Long Duration Psychology: The Real Final Frontier?

Glaskov Shuttle-Mir oral history.

Holland Shuttle-Mir oral history.

Jerry Linenger, "Letters to My Son," S-M CD-ROM.

Nield and Vorobiev, ed., *Phase 1 Joint Report*.

Andy Thomas, "Letters from the Outpost," S-M CD-ROM.

Direct Quotes
Page 50
"It was frightening for one or two seconds ... 'Oh, this is a surprisingly robust station.'":
    "Astronaut thought he was going to die instantly," BBC News Homepage, Online, <http://news1.thdo.bbc.co.uk/low/english/sci/tech/newsid_18000/18134.stm> (Article Dated 30 October 1997; Accessed 3 March 2001).

"the one thing that makes spaceflight ... can really make your work day difficult.":
    Andy Thomas, "Letters from the Outpost-April 1998: The First Days on Mir," S-M CD-ROM.

"life in space is never monotonous.":
    Jerry Linenger, "Letters to My Son-March 4, 1997: Life in space is never monotonous," S-M CD-ROM.

"The situation of work underload is one of the worst situations ... to subject himself to.":
    Holland Shuttle-Mir oral history.

"I loved the greenhouse experiment ... no power for about two months.":
    "Mir-24 Postflight Press Conference with Mike Foale," NASA Human Spaceflight Homepage, Online, <http://spaceflight.nasa.gov/history/shuttle-mir/mir24/status/week12/foale.html> (Last Updated 31 October 1997; Accessed 3 March 2001).

Page 51
"A simple walk would be fine ... the sound of wind through the trees overhead.":
    Jerry Linenger, "Letters to My Son-April 27, 1997: Cabin Fever," S-M CD-ROM.

"We put our crew members into … an isolation chamber … you have to learn to adapt to each other.":
Glaskow Shuttle-Mir oral history.

"It's really probably good … we learned a lot more.":
Holland Shuttle-Mir oral history.

"One of the things that was astounding … sent them over there.":
Holland Shuttle-Mir oral history.

STS-76: *Starting a Continuous U.S. Presence*

Lucid Shuttle-Mir oral history.

"STS-60 Press Kit-February 1993, Wake Shield Facility SPACEHAB-2," NASA Spacelink Homepage. Online, <http://spacelink.msfc.nasa.gov/NASA.Projects/Human.Exploration.and.Development.of.Space/Human.Space.Flight/Shuttle/Shuttle.Missions/Flight.060.STS-60/Press.Kit> (Last Updated February 1993; Accessed 12 January 2000).

"STS-76 (76)," Kennedy Space Center Homepage. Online, <http://science.ksc.nasa.gov/shuttle/missions/sts-76/mission-sts-76.html> (Last Updated 31 August 2000; Accessed September 2000).

"STS-76 Fact Sheet: STS-76/*Atlantis*, Shuttle/Mir Mission-3," KSC Release No. 19-96, KSC Fact Sheets Homepage. Online, <http://www-pao.ksc.nasa.gov/kscpao/nasafact/76facts.htm> (Release Dated February 1996; Accessed December 2000).

"STS-76 Press Kit," Kennedy Space Center Homepage. Online, <http://science.ksc.nasa.gov/shuttle/missions/sts-76/sts-76-press-kit.txt> (Last Updated March 1996; Accessed 10 December 1999).

"STS-91: The Final Mission to Mir," SPACEHAB Homepage. Online, <http://www.spacehab.com/mission/previous_missions/91/missions_91.htm> (Copyright 2000; Accessed April 2000).

Direct Quotes
Page 53
"After hearing of the experiences … things that were tacked along the sides.":
Sega Shuttle-Mir oral history, 9 September 1998.

"The whole station is chameleon-like … beautiful hues of white, tan, gold, and blue.":
Scarfoss interview, 28 April 2001, Morgan author file.

Mir Environmental Effects Payload

"STS-76 Press Kit: Mir Environmental Effects Payload (MEEP)," NASA Human Spaceflight Homepage. Online, <http://spaceflight.nasa.gov/shuttle/archives/sts-76/glance/presskit/meep.html> (Last Updated 13 March 1996; Accessed 3 March 2001).

NASA-2 Shannon Lucid: *Enduring Qualities*

Michael Cabbage, "Lucid flies to Mir this week: communication key to preventing isolation," *Florida Today* Homepage. Online, <http://www.flatoday.com> (Article Dated 18 March 1996; Accessed 4 April 2000).

"Cosmonauts take space walk and stage Pepsi commercial," *Florida Today* Homepage. Online, <http://www.flatoday.com> (Article Dated 13 June 1996; Accessed 6 April 2000).

Todd Halvorson, "Cosmonaut Lucid craving M&M candies," *Florida Today* Homepage. Online, <http://www.flatoday.com> (Article Dated 11 April 1996; Accessed 4 April 2000).

Todd Halvorson, "Experts weighing impact of Russian elections on space partnership," *Florida Today* Homepage. Online, <http://www.flatoday.com> (Article Dated 9 June 1996; Accessed 5 April 2000).

Todd Halvorson, "Lucid adapting to Mir; Clifford, Godwin await spacewalk," *Florida Today* Homepage. Online, <http://www.flatoday.com> (Article Dated 26 March 1996; Accessed 4 April 2000).

Todd Halvorson, "Lucid takes a break from Mir duties," *Florida Today* Homepage. Online, <http://www.flatoday.com> (Article Dated 9 May 1996; Accessed 6 April 2000).

Todd Halvorson, "Lucid's Great Adventure Begins Aboard Mir," *Florida Today* Homepage. Online, <http://www.flatoday.com> (Article Dated 24 March 1996; Accessed 4 April 2000).

Lucid Shuttle-Mir oral history.

"Lucid's stay will be more appealing," *Florida Today* Homepage. Online, <http://www.flatoday.com> (Article Dated 20 March 1996; Accessed 4 April 2000).

McDonald, *Mir Mission Chronicle*, 9-21.

"Preflight Interview: Shannon Lucid, Mission Specialist," NASA Human Spaceflight Homepage. Online, <http://www.spaceflight.nasa.gov/shuttle/archives/sts-76/crew/intlucid.html> (Last Updated 12 March 1996; Accessed 16 September 1999).

"NASA-2/Mir-21 and Mir-22 Mission Status Reports: April 5, 1996-September 27, 1996," S-M CD-ROM.

Nise Shuttle-Mir oral history.

"Russian Cosmonauts to stay extra 40 days in space," *Florida Today* Homepage. Online, <http://www.flatoday.com> (Article Dated 21 June 1996; Accessed 6 April 2000).

Direct Quotes
Unless otherwise noted, direct quotes taken from the Lucid Shuttle-Mir oral history.

Page 56
"They're both very, very nice people … so that works out real well.":
"Preflight Interview: Shannon Lucid, Mission Specialist," NASA Human Spaceflight Homepage.

"because we know that women … keep the place pretty tidy … cultural level":
Halvorson, "Lucid adapting to Mir," *Florida Today* Homepage.

Page 58
"mainly Russian … Shannon won't forget her English.":
Halvorson, "Cosmonaut Lucid craving M&M candies," *Florida Today* Homepage.

"One of the more mundane things … brought up on future Shuttle flights.":
"NASA-2/Mir-21 Mission Status Report Weekly Update-April 26, 1996: Interview with Tom Sullivan," S-M CD-ROM.

Page 59
"After a lot of work … earned ourselves a holiday.":
Shannon Lucid, "Mir-21 Weekly Update-June 14, 1996: Pink Socks and Jello, Shannon Lucid writes home," on NASA Shuttle Homepage. Online, <http://www.shuttle.nasa.gov/history/shuttle-mir/mir21/status/week11/pinksox.html> (Last Updated 14 June 1996; Accessed 11 February 1999).

Page 62
"I couldn't ask for anything more … we understand each other.":
Todd Halvorson, "Lucid on Mir mission: 'You can't beat it!," *Florida Today* Homepage. Online, <http://www.flatoday.com> (Article Dated 28 May 1996; Accessed 4 April 2000).

"There I was on Mir … and the world seemed to shrink down to that.":
"Shuttle/Mir: Shannon Lucid Post Flight News Conference," NASA Shuttle Homepage. Online, <http://www.shuttle.nasa.gov/history/shuttle-mir/ops/crew/lucid/shannon1.html> (Last Updated 25 October 1996; Accessed 11 February 1999).

Page 63
"the worst case … You should be aware of the situation.":
Frank Culbertson to Valery Ryumin, 3 July 1996, memorandum, Morgan author file.

"The two things I had planned … when I do get home.":
Todd Halvorson, "Lucid breaks U.S. spaceflight duration record," *Florida Today* Homepage. Online, <http://www.flatoday.com> (Article Dated 15 July 1996; Accessed 4 April 2000).

"like going to the bookstore … I'll continue to have an enjoyable time.":
Todd Halvorson, "Marathon stay on Mir doesn't faze Lucid; however, she misses chips, sun," *Florida Today* Homepage. Online, <http://www.flatoday.com> (Article Dated 11 July 1996; Accessed 4 April 2000).

Page 64
"The Russians are used to taking care of the crew … we passed those on to Moscow.":
"NASA-2/Mir-21 Mission Status Report Weekly Update-July 26, 1996: Interview: Charlie Stegemoeller," S-M CD-ROM.

"My family would be surprised … I hope I can bring some of that back with me.":
Robyn Suriano, "Lucid: I've developed patience," *Florida Today* Homepage. Online, <http://www.flatoday.com> (Article Dated 27 August 1996; Accessed 6 April 2000).

"I don't think you've taken the record ... Everybody loves her.":
"Russians heap praise on Lucid's efforts," *Florida Today* Homepage. Online. <http://www.flatoday.com> (Article Dated 15 September 1996; Accessed 6 April 2000).

Profile: Shannon W. Lucid

"Biographical Data: Shannon W. Lucid (Ph.D.)," NASA Johnson Space Center Homepage. Online. <http://www.jsc.nasa.gov/Bios/htmlbios/lucid.html> (Last Updated June 2000; Accessed December 2000).

Lucid Shuttle-Mir oral history.

Books Onboard Mir

Lucid Shuttle-Mir oral history.

Thomas Shuttle-Mir oral history.

Direct Quotes
Page 61
"Here I was, reading *David Copperfield* ... transcending the centuries—transcending culture":
Lucid Shuttle-Mir oral history.

"I'd always wanted to read *Huckleberry Finn*, since it's a landmark book ... and very controversial.":
Thomas Shuttle-Mir oral history.

"It was in Russian," she said, "a little New Testament.":
Lucid Shuttle-Mir oral history.

Pink Socks and Jell-O

Lucid. "Mir-21 Weekly Update June 14, 1996: Pink Socks and Jello," on NASA Shuttle Homepage.

Women on Mir; Women in Space

Karash, "Star City: Berkeley for Cosmonauts," Space.com Homepage.

"Mir – The Guest Book," Space.com Homepage. Online. <http://www.space.com> (Last Updated 27 August 1999; Accessed 6 March 2000).

Wade, ed., "astronautix.com Encyclopedia Astronautica," Encyclopedia Astronautica Homepage.

Meanwhile on Earth, NASA-2

Gerstenmaier Shuttle-Mir oral history.

Julie Meredith, "Hearing on the Mir Safety, September 18, 1997, before the Committee on Science, House of Representatives," NASA Headquarters Homepage. Online. <http://www.hq.nasa.gov/congress/culbert.html> (Last Updated 18 September 1997; Accessed 18 October 2000).

All direct quotes taken from the Gerstenmaier Shuttle-Mir oral history.

STS-79: *First American Handover*

Readdy Shuttle-Mir oral history.

"Space Shuttle Mission Chronology-STS-79," Kennedy Space Center Homepage. Online. <http://www-pao.ksc.nasa.gov/kscpao/chron/sts-79.htm> (Last Updated 20 September 2000; Accessed September 2000).

"STS-79 (79)," Kennedy Space Center Homepage. Online. <http://science.ksc.nasa.gov/shuttle/missions/sts-79/mission-sts-79.html> (Last Updated 31 August 2000; Accessed September 2000).

"STS-79 Fact Sheet: STS-79/*Atlantis*, Shuttle/Mir Mission-4," KSC Release No.85-96, KSC Fact Sheets Homepage. Online. <http://www-pao.ksc.nasa.gov/kscpao/release/1996/85-96.htm> (Release Date August 1996; Accessed October 2000).

"STS-79 Press Kit," Kennedy Space Center Homepage. Online. <http://science.ksc.nasa.gov/shuttle/missions/sts-79/sts-79-press-kit.txt> (Last Updated September 1996; Accessed 29 March 2000).

Direct Quotes
Page 68
"a lot bigger than my JSC office.":
Akers interview, Morgan author file.

Page 69
"Depending on whether you want to look at the glass ... allowed her to set the world record for time in space.":
Readdy Shuttle-Mir oral history.

NASA-3 John Blaha: *Pulling it Together*

Blaha Shuttle-Mir oral history.

"NASA-3/Mir-22 Mission Status Reports: September 20, 1996-January 10, 1997," S-M CD-ROM.

"Relief for Russian-American Space Crew Delayed," *Florida Today* Homepage. Online. <http://www.flatoday.com> (Article Dated 8 November 1996; Accessed 18 April 2000).

Direct Quotes
Unless otherwise noted, direct quotes taken from the Blaha Shuttle-Mir oral history.

Page 71
"Every Shuttle flight I've flown, ... I will define what 'ever' is.":
"Mir 22 Weekly Update-September 13, 1996: Blaha Prepares for Mir Mission," NASA Human Spaceflight Homepage. Online. <http://spaceflight.nasa.gov/history/shuttle-mir/mir22/status/week2/blaha.html> (Last Updated 7 October 1996; Accessed 16 May 2000).

Page 72
"I remember ... I'm going to have to fly on the one that's really up there.":
"Preflight Interview: John Blaha, Mission Specialist," NASA Human Spaceflight Homepage. Online. <http://www.spaceflight.nasa.gov/shuttle/archives/sts-79/crew/intrblaha.html> (Last Updated 13 September 1996; Accessed 15 March 2000).

Page 73
"It ain't Apollo 13 ... saved that experiment.":
"Mir-22 Weekly Update-Quick Connect: International and internal cooperation pave the way for fast equipment repair, saving an important Mir experiment," NASA Humans Spaceflight Homepage. Online. <http://spaceflight.nasa.gov/history/shuttle-mir/mir22/status/week6/bts.html> (Last Updated 16 October 1996; Accessed 15 March 2000).

"I will always remember ... what an incredible spaceship America built.":
John Blaha, "Mir 22 Space Mission," 4 December 1999, Morgan author file.

"Actually, I was surprised ... cleaned up on the ground ":
"NASA-3/Mir-22 Mission Status Report Weekly Update-October 11, 1996: Interview with John Blaha," S-M CD-ROM.

Page 74
"Valery and Sasha ... working on things to accomplish all of the work.":
"NASA-3/Mir-22 Mission Status Reports-October 25, 1996: Mir-22 Crew News Conference," S-M CD-ROM.

"to settle down in the evening ... They're like medicine to me.":
"Preflight Interview: John Blaha, Mission Specialist," NASA Human Spaceflight Homepage.

December 20, 1996 news conference excerpt:
"NASA-3/Mir-22 Mission Status Reports-December 20, 1996: Mir-22 Crew News Conference," S-M CD-ROM.

Profile: John E. Blaha

Blaha Shuttle-Mir oral history.

"Biographical Data: John E. Blaha (Colonel, USAF, Ret.)," NASA Johnson Space Center Homepage. Online. <http://www.jsc.nasa.gov/Bios/htmlbios/blaha.html> (Last Updated November 1999; Accessed 5 December 1999).

Ham on Mir: Amateur radio

"ARISS Objectives on the International Space Station," Online. <http://ariss.gsfc.nasa.gov/objectives.html> (Last Updated 24 April 1997, 3 March 2001).

Philip Chien, ed., "The "Unofficial" MIR FAQ," The American Radio Relay League, Inc. Homepage, Online, <http://www.arrl.org/sarex/mir-faq.html> (Last Updated 22 November 2000; Accessed 24 January 2001).

Life on Mir

Blaha Shuttle-Mir oral history.

Foale Shuttle-Mir oral history, 16 June 1998.

Foale Shuttle-Mir oral history, 7 July 1998.

Foale Shuttle-Mir oral history, 31 July 1998.

Lucid Shuttle-Mir oral history.

Precourt Shuttle-Mir oral history.

Wolf Shuttle-Mir oral history.

Direct Quotes
Page 76
"I was surprised ... Two of the modules are very new inside.":
Blaha Shuttle-Mir oral history.

"still food supplies there ... that nobody's ever going to eat.":
Precourt Shuttle-Mir oral history.

"very easy to lose each other ... often just for a ten-minute tea break.":
Foale Shuttle-Mir oral history, 16 June 1998.

"ate all our meals together and spent a lot of time talking to each other over mealtimes.":
Lucid Shuttle-Mir oral history.

"I didn't realize ... every single day except a few.":
Wolf Shuttle-Mir oral history.

"I put all my personal things in Spektr ... every day for 188 days.":
Lucid Shuttle-Mir oral history.

"has been an amazing experience ... I think Andy can expect the same.":
"Endeavour arrives at Mir to pick up homesick astronaut," Florida Today Homepage, Online, <http://www.flatoday.com> (Article Dated 24 January 1998; Accessed 25 February 2000).

John Blaha Mails a Letter Home

John Blaha, "Mir-22 Weekly Update-December 13, 1996: John Blaha Pens a Letter Home," S-M CD-ROM.

Life in Microgravity

Linenger, "Letters to My Son," S-M CD-ROM.

Jerry M. Linenger, Off The Planet: Surviving Five Perilous Months Aboard the Space Station Mir (New York: McGraw-Hill, 2000).

Pool Shuttle-Mir oral history.

Thagard Shuttle-Mir oral history.

Thomas, "Letters from the Outpost," S-M CD-ROM.

Thomas Shuttle-Mir oral history.

Dave Wolf, "Letters Home," S-M CD-ROM.

Direct Quotes
Page 80
"flying in your dreams ... But you don't go anywhere.":
Personal conversation with Barbara Morgan, Morgan author file.

"And yet, ... really never got beyond stomach awareness":
Thagard Shuttle-Mir oral history.

"It was obvious that they were all relieved ... they could remain still and feel miserable alone.":
Linenger, Off The Planet, 97.

"One of the things that surprised me ... just to remind you of where you really were.":
Thomas Shuttle-Mir oral history.

"Everything floats up here ... who would have thought of it.":
Jerry Linenger, "Letters to My Son-January 25, 1997: Propulsion and suction," S-M CD-ROM.

Page 81
"They give you one wet towel ... cleaning up after exercise.":
Thagard Shuttle-Mir oral history.

"Sasha [Lazutkin] is running ... about 1 hertz ... resonate at a dangerous level.":
Jerry Linenger, "Letters to My Son-February 4, 1997: Like being in a rowboat," S-M CD-ROM.

"The most frustrating thing is that you are forever losing things ... we are accustomed to seeing things.":
Thomas, "Letters from the Outpost-April 1998: First Days on Mir," S-M CD-ROM.

"We just don't expect the pliers ... in front of our face.":
Dave Wolf, "Letters Home-December 22, 1997: The Great Blackcurrant Jelly Juice Spill," S-M CD-ROM.

"Flying in space ... now that we've begun to fly with them.":
Pool Shuttle-Mir oral history.

Meanwhile on Earth, NASA-3

Travis Brice to Clay Morgan, 18 September 2000, email, Morgan author file.

Chiodo Shuttle-Mir oral history.

Caasi Moore Shuttle-Mir oral history.

Mark Wade, ed., "Astronaut and Rocketplane Pilot Index," Encyclopedia Astronautica Homepage, Online, <http://www.friends-partners.org/mwade/astros/astindex.htm> (Last Updated 22 December 2000; Accessed December 2000).

Direct Quotes
Unless otherwise noted, direct quotes taken from the Caasi Moore Shuttle-Mir oral history.

Page 82
"We had the room decorated in lights ... It was either going to be Frank Culbertson or Jim Van Laak.":
Chiodo Shuttle-Mir oral history.

The Flight Docs

Barratt Shuttle-Mir oral history.

Billica Shuttle-Mir oral history.

Flynn Shuttle-Mir oral history.

Marshburn Shuttle-Mir oral history.

Pool Shuttle-Mir oral history.

Wolf Shuttle-Mir oral history.

Direct Quotes
Page 82
"The flight surgeons take on a new importance ... on these long-duration missions.":
Wolf Shuttle-Mir oral history.

"[I]t made it so that the crew member knew this doctor ... whatever the situation was.":
Billica Shuttle-Mir oral history.

"the American system ... So that is a fairly big difference in the two systems.":
Pool Shuttle-Mir oral history.

Page 83
"We often had disagreements ... declare a person healthy or certified for training or spaceflight.":
Barratt Shuttle-Mir oral history.

"A lot of it is an attitude and an approach ... it may slip through the cracks.":
Marshburn Shuttle-Mir oral history.

"[He] had a lot of medical experiments ... to really do your job well.":
Flynn Shuttle-Mir oral history.

STS-81: *Bringing Back the Harvest*

Jeff Shuttle-Mir oral history.

"STS-81 (81)," Kennedy Space Center Homepage, Online. <http://science.ksc.nasa.gov/shuttle/missions/sts-81/mission-sts-81.html> (Last Updated 31 August 2000; Accessed September 2000).

"STS-81 Fact Sheet: STS-81/*Atlantis*, Shuttle/Mir Mission-5/SPACEHAB-DM," KSC Release No. 145-96, KSC Fact Sheets Homepage, Online, <http://www-pao.ksc.nasa.gov/kscpao/release/1996/145-96.htm> (Release Dated December 1996; Accessed December 2000).

"STS-81 Press Information and Time Line," Boeing Homepage, Online, <http://www1.boeing.com/defense-space/space/liftoff/sts81/> (Last Updated 18 July 2000; Accessed 29 March 2000).

"STS-81 Press Kit," Kennedy Space Center Homepage, Online, <http://science.ksc.nasa.gov/shuttle/missions/sts-81/sts-81-press-kit.txt> (Last Updated January 1997; Accessed 29 March 2000).

"STS-81 Status Reports," NASA Spacelink Homepage, Online, <http://spacelink.msfc.nasa.gov/NASA.Projects/Human.Exploration.and.Development.of.Space/Human.Space.Flight/Shuttle/Shuttle.Missions/Flight.081.STS-81/STS-81.Status.Reports/.index.html> (Last Updated January 1997; Accessed March 2000).

Direct Quotes
Page 85
"exploring a cave ":
        Grunsfeld interview, Morgan author file.

"kind of grabbed me ... what you would expect for a station.":
        Grunsfeld interview, Morgan author file.

"seemed real comfortable over in the Mir ... a very critical situation like he had with the fire.":
        Jeff Shuttle-Mir oral history.

SPACEHAB

"STS-60 Press Kit-February 1993, Wake Shield Facility SPACEHAB-2," NASA Spacelink Homepage.

"STS-81 Fact Sheet" KSC Fact Sheets Homepage.

"STS-91: The Final Mission to Mir," SPACEHAB Homepage.

NASA-4 Jerry Linenger: *Fire and Controversy*

Linenger, "Letters to My Son," S-M CD-ROM.

Linenger, *Off The Planet*.

"NASA-4/Mir-22 and Mir-23 Mission Status Reports: January 24, 1997-May 16, 1997," S-M CD-ROM.

Direct Quotes
Page 87
"what got me here":
        Jerry Linenger, "Letters to My Son-January 23, 1997: I am still an earthling," S-M CD-ROM.

Page 88
"The suit is a tight fitting pressure bladder ... we made our way to the Soyuz.":
        "The Grunsfeld Report Mission Update: Dispatch #5, Onboard Mir 17 January 1997," NASA Human Spaceflight Homepage.

"Space is a frontier ... What a privilege!":
        Linenger, "Letters to My Son-January 23, 1997: I am still an earthling," S-M CD-ROM.

Page 89
"I am a physician ... to carry out the experiment.":
        "STS-81 Jerry M. Linenger Preflight Interview," NASA Human Spaceflight Homepage, Online, <http://www.spaceflight.nasa.gov/shuttle/archives/sts-81/crew/intlinenger.html> (Last Updated 7 January 1997; Access 26 March 2001).

"I try my best ... basically a one-person show.":

Jerry Linenger, "Letters to My Son-January 28, 1998: We all depend on each other," S-M CD-ROM.

"Let me tell you about my house ... Radio, ham radio, and telemetry.":
        Jerry Linenger, "Letters to My Son-January 30, 1997: You would enjoy playing hide and seek up here," S-M CD-ROM.

"I felt and heard the springs ... Then the thrusters.":
        Jerry Linenger, "Letters to My Son-February 6, 1997: An Unforgiving Place," S-M CD-ROM.

"Smooth, yet firm, push-off ... Glad to open the door again.":
        Jerry Linenger, "Letters to My Son-February 7, 1997: An afternoon spin in a space-ship," S-M CD-ROM.

Page 90
"a distinct, burnt-dry smell ... seeing familiar surroundings.":
        Linenger, *Off the Planet*, 95.

"There is no free time ... I am not sure five months is going to be long enough up here.":
        "NASA-4/Mir-22 Mission Status Report Weekly Update-February 14, 1997," S-M CD-ROM.

Page 91
"Today I saw huge dust storms ... I feel real good.":
        Jerry Linenger, "Letters to My Son-February 10, 1997: Human being questions," S-M CD-ROM.

"a stout, winged insect.":
        Linenger, *Off the Planet*, 95.

"Sasha ... , Valery, and I have never collided ... new guys adapt and learn the terrain.":
        Jerry Linenger, "Letters to My Son-February 15, 1997: Hurricane, a good bilingual nickname," S-M CD-ROM.

"There was also caviar ... sitting around the table.":
        "Mir Mortals," BBC Homepage, Online. <http://www.bbc.co.uk/science/horizon/mir-mortals.shtml> (Last Updated 23 April 1998; Accessed 28 March 2000).

Page 92
"We immediately started fighting that fire ... how fast and rapid the smoke spread throughout the complex.":
        "NASA-4/Mir-23 Mission Status Report Weekly Update-March 21, 1997: Interview-Fighting Fire Aboard Mir, Jerry Linenger describes the experience," S-M CD-ROM.

"flying across and splattering ... It had everything it needed.":
        "Mir Mortals," BBC Homepage.

"When I saw the ship was full of smoke ... You can't just open a window to ventilate the room.":
        "Mir Mortals," BBC Homepage.

"I did not inhale anything ... they protected us from inhalation injury.":
        "Weekly Update-March 21, 1997: Linenger Interview-Fighting Fire Aboard Mir," S-M CD-ROM.

Page 93
"When I started spraying foam ... 'Valery, how do you feel?'":
        "Mir Mortals," BBC Homepage.

"in the distant modules at the very end of the cones.":
        "Weekly Update-March 21, 1997: Linenger Interview-Fighting Fire Aboard Mir," S-M CD-ROM.

"We even thought someone had switched the lights out in Kvant. That's how black it was.":
        "Cosmonaut recounts fire, collision aboard Mir," Reuter Information Service, Reuters Homepage, Online, <http://www.reuters.com/> (Article Dated 16 September 1997; Accessed March 2000).

"Being a physician ... get into the oxygen masks quickly.":
        "Weekly Update-March 21, 1997: Linenger Interview-Fighting Fire Aboard Mir," S-M CD-ROM.

Page 94
"We've got some great experiments ... I am very preoccupied with my work.":
        "NASA-4/Mir-23 Mission Status Reports-March 21,1997," S-M CD-ROM.

"I'm not only getting through ... give us better data, better understanding.":
  Jerry Linenger, "Letters to My Son-March 13, 1997: Good report on you from Mommy," S-M CD-ROM.

"Compared to what they endured ... I float effortlessly.":
  Jerry Linenger, "Letters to My Son-March 8, 1997: Space Station is a five star hotel," S-M CD-ROM.

"Last night it got really, really, really dark ... without the ventilators, only stillness.":
  Jerry Linenger, "Letters to My Son-March 20, 1997: All alone in a dark room," S-M CD-ROM.

Page 95
"[Mir] is a small place ... no difficult problem with that up here.":
  "NASA-4/Mir-23 Mission Status Report Weekly Update-April 11, 1997," S-M CD-ROM.

"The ethylene glycol caused the concern ... we need to fix our problems.":
  "NASA-4/Mir-23 Mission Status Report Weekly Update-April 11, 1997," S-M CD-ROM.

"Not really ... I was trained to work on those systems and assist the crew where I could.":
  "NASA-4/Mir-23 Mission Status Report Weekly Update-April 11, 1997," S-M CD-ROM.

Page 96
"It's basically a long telescoping tube ... on the Docking module.":
  "NASA-4/Mir-23 Mission Status Reports-Week of April 25, 1997," S-M CD-ROM.

Page 97
"I flew to the window ... the Progress must have missed hitting the Base Block.":
  Linenger, Off the Planet, 169.

"There's two sets of difficulties ... about as much difficulty as you can imagine.":
  "NASA-4/Mir-23 Mission Status Report Weekly Update-May 9, 1997," S-M CD-ROM.

Profile: Jerry M. Linenger

"Biographical Data: J. M. Linenger, M.D., M.S.S.M., M.P.H., Ph.D. (Captain, Medical Corps, USN)," NASA Johnson Space Center Homepage, Online, <http://www.jsc.nasa.gov/Bios/htmlbios/linenger.html> (Last Updated September 1997; Accessed 7 October 1999).

Linenger Shuttle-Mir oral history.

It's all down hill from here

Jerry Linenger, "Letters to My Son-April 30, 1997: It's all down hill from here," S-M CD-ROM.

Failures to Communicate

Culbertson Shuttle-Mir oral history.

Marshburn Shuttle-Mir oral history.

Wetherbee Shuttle-Mir oral history.

Direct Quotes
Page 100
"Communications were so bad that I had to give up on them.":
  Personal conversation with Linenger, Morgan author file.

"We had very limited Com ... we didn't have it.":
  Marshburn Shuttle-Mir oral history.

"disconnect between the teams ... without any voice communication.":
  Marshburn Shuttle-Mir oral history.

"You come AOS ... they'd wait 'til the next time.":
  Culbertson Shuttle-Mir oral history.

"impetuosity about getting things done.":
  Culbertson Shuttle-Mir oral history.

"Titov one day said to me ... not just be test pilots all the time.":
  Wetherbee Shuttle-Mir oral history.

Meanwhile on Earth, NASA-4

"NASA-4 and NASA-5/Mir-23 Mission Status Reports: May 2, 1997-May 30, 1997," S-M CD-ROM.

Sang Shuttle-Mir oral history.

Direct Quotes
  Unless otherwise noted, direct quotes taken from the Sang Shuttle-Mir oral history.

Page 101
"So, have you had problems ... because they really would feel like it was not a big deal.":
  Marshburn Shuttle-Mir oral history.

"When we have ... questions ... all the critical systems will function very well.":
  Van Laak Shuttle-Mir oral history.

"There are a few things ... and [its] life is extremely long.":
  Culbertson Shuttle-Mir oral history.

"I think the Soyuz is really the biggest insurance ticket ... The fire is the worst case that you can imagine, I think.":
  "American Astronaut Foale to do repair work aboard Mir," Florida Today Homepage, Online, <http://www.flatoday.com> (Article Dated 13 May 1997; Accessed 5 April 2000).

STS-84: Delivery and Pick Up

Precourt Shuttle-Mir oral history.

"STS-84 (84)," Kennedy Space Center Homepage, Online, <http://science.ksc.nasa.gov/shuttle/missions/sts-84/mission-sts-84.html> (Last Updated 31 August 2000; Accessed September 2000).

"STS-84 Fact Sheet: STS-84/Atlantis, Sixth Shuttle-Mir Docking," KSC Release No. 69-97, May 1997, KSC Fact Sheets Homepage, Online, <http://www-pao.ksc.nasa.gov/ksc-pao/release/1997/69-97.htm> (Release Dated May 1997; Accessed December 2000).

"STS-84 Press Kit," Kennedy Space Center Homepage, Online, <http://science.ksc.nasa.gov/shuttle/missions/sts-84/sts-84-press-kit.txt> (Last Updated August 2000; Accessed 29 March 2000).

Direct Quotes
Page 103
"Your NASA-5 crew member would like to report ... It's good to be back on U.S. soil.":
  "STS-84 Space Shuttle Atlantis, May 15-24, 1997: Mir hand off and Biorack mark mission," IS-1997-05-001.084JSC, NASA Johnson Space Center Homepage, Online, <http://vesuvius.jsc.nasa.gov/pao/factsheets/missions/084/9705001084.html> (Last Updated May 1997; Accessed 3 March 2001).

"So on -84 ... a really, really memorable experience.":
  Precourt Shuttle-Mir oral history.

NASA-5 Mike Foale: Collision and Recovery

Colin Foale, Waystation to the Stars: The Story of Mir, Michael, and Me (North Pomfret, VT: Trafalgar Square Publishing, 2000).

Foale Shuttle-Mir oral history, 16 June 1998.

Foale Shuttle-Mir oral history, 7 July 1998.

Foale Shuttle-Mir oral history, 31 July 1998.

"Mir Mortals," BBC Homepage.

"NASA-5/Mir-23 and Mir-24 Mission Status Reports: May 16, 1997-October 10, 1997," S-M CD-ROM.

Direct Quotes
  Unless otherwise noted, direct quotes taken from the Foale Shuttle-Mir oral histories.

Page 105
"Scott is too tall; Wendy is too short; but Mike is just right.":
  Personal conversation with Rhonda Foale, Morgan author file.

Page 106
"To be sure, that the Mir looked and even smelled better ... spent cleaning up.":
  Linenger, Off The Planet, 215.

Page 107

"how he would don the respirator ... and activate the fire extinguisher.":
Linenger, *Off The Planet*, 218.

"Jerry and I talked for a long time ... It will change.'":
"Mir Mortals," BBC Homepage.

"I certainly enjoy their company ... chatting and talking.":
"NASA-5/Mir-23 Mission Status Report Weekly Update-June 13, 1997: Interview with Mike Foale," S-M CD-ROM.

Page 108

"I feel like I'm living in a garage,":
"Weekly Update-June 13, 1997: Interview with Mike Foale."

"just on general maintenance tasks ... I like doing that stuff.":
"Mir-23: Foale's Home Away from Home." NASA Human Spaceflight Homepage. Online, <http://spaceflight.nasa.gov/history/shuttle-mir/mir23/status/week13/foale.html> (Last Updated 30 May 1997; Accessed 4 April 2000).

Page 109

"it was difficult to make out the station ... looked very similar to the clouds below it.":
"Mir Mortals," BBC Homepage.

"Michael, get in the escape ship! ... the whole station shook.":
"Mir Mortals," BBC Homepage.

"The decompression alarm system ... the station started to spin.":
"Mir Mortals," BBC Homepage.

Page 110

"We watched the polar lights and the stars in complete silence.":
"Mir Mortals," BBC Homepage.

Page 111

"Oh, my feelings! ... I realized instantly that I'd made a mistake.":
"Mir Mortals," BBC Homepage.

Page 112

"No one who has been there thinks of Mir simply as a pile of metal ... I felt it was like a living creature.":
"Mir Mortals," BBC Homepage.

Page 114

"[Of] course, I am getting excited ... less electric power.":
"NASA-5/Mir-24 Mission Status Report Weekly Update-September 19, 1997: Status Report from Mike Foale," S-M CD-ROM.

Profile: C. Michael Foale

"Biographical Data: C. Michael Foale (Ph.D.)," NASA Johnson Space Center Homepage. Online, <http://www.jsc.nasa.gov/Bios/htmlbios/foale.html> (Last Updated May 2000; Accessed October 2000).

Foale Shuttle-Mir oral history, 16 June 1998.

Foale Shuttle-Mir oral history, 7 July 1998.

Foale Shuttle-Mir oral history, 31 July 1998.

Meanwhile on Earth, NASA-5

Sarah Enticknap, "Status Briefing, July 31, 1997," NASA Human Spaceflight Homepage. Online, <http://spaceflight.nasa.gov/history/shuttle-mir/mir23/status/week22/pressconf.html> (Last Updated 1 August 1997; Accessed 17 January 2000).

Meredith, "Hearing on the Mir Safety," NASA Headquarters Homepage.

"NASA-5/Mir-23 and Mir-24 Mission Status Reports: May 16, 1997-October 10, 1997," S-M CD-ROM.

"NASA Office of Inspector General Letter to Congress Concerning Shuttle-Mir Program, August 29, 1997," NASA Headquarters Homepage. Online, <http://www.hq.nasa.gov/office/oig/hq/inspections/mir/> (Last Updated 12 September 1997; Accessed 21 January 2000).

"Panels Give Astronaut a "GO" for Launch to Mir," News Release 95-3, NASA Spacelink Homepage, Online, <http://spacelink.nasa.gov/NASA.News/NASA.News.Releases/Previous.News.Releases/97.News.Releases/97-09.News.Releases/97-09-25.NASA.Gives.Go.for.Launch.to.Mir> (Release Dated 25 September 1997; Accessed 18 October 2000).

Zimmerman Shuttle-Mir oral history.

Direct Quotes

Unless otherwise noted, direct quotes taken from the Zimmerman Shuttle-Mir oral history.

Page 116

"There has been sufficient evidence ... who can say when you have learned enough?":
Meredith, "Hearing on the Mir Safety," NASA Headquarters Homepage.

"Not only is the Mir station deemed to be a satisfactory life support platform ... the safety of our American astronauts.":
"Panels Give Astronaut a "GO" for Launch to Mir," News Release 95-3, NASA Spacelink Homepage.

"It put our leadership ... we really can show what good partners we will be.":
Wolf Shuttle-Mir oral history.

"Making a change in crew assignments ... during this upcoming mission.":
Enticknap, "Status Briefing, July 31, 1997," NASA Human Spaceflight Homepage.

"Wendy was still a critical part ... it's a misnomer to think Wendy wasn't on that mission.":
Wolf Shuttle-Mir oral history.

Needed On Mir

"Mir-23 Special Update-June 26, 1997: Audio-only Communication with Mike Foale," NASA Human Spaceflight Homepage, Online, <http://spaceflight.nasa.gov/history/shuttle-mir/mir23/status/special/foale.html> (Last Updated 26 June 1997; Accessed 3 March 2001).

STS-86: *Loaded with Experience*

Lawrence Shuttle-Mir oral history.

Statement of Chairman F. James Sensenbrenner, Jr., U.S. House of Representatives Committee on Science, Washington, DC, 25 June 1997.

"STS-86 (87)," Kennedy Space Center Homepage, Online, <http://science.ksc.nasa.gov/shuttle/missions/sts-86/mission-sts-86.html> (Last Updated 31 August 2000; Accessed September 2000).

"STS-86 Fact Sheet: STS-84/*Atlantis*, Seventh Shuttle-Mir Docking," KSC Release No. 155-97, KSC Fact Sheets Homepage, Online, <http://www-pao.ksc.nasa.gov/kscpao/release/1997/155-97.htm> (Release Dated September 1997; Accessed December 2000).

"STS-86 Press Kit," Kennedy Space Center Homepage, Online, <http://science.ksc.nasa.gov/shuttle/missions/sts-86/sts-86-press-kit.txt> (Last Updated September 1997; Accessed 15 March 2000).

Titov Shuttle-Mir oral history.

Wetherbee Shuttle-Mir oral history.

Direct Quotes
Page 118
"without a potentially life-threatening situation.":
Sensenbrenner statement.

Page 119
"Anatoly told us ... water that we bring up for them to use.":
Wetherbee Shuttle-Mir oral history.

Joint Effort

Nield and Vorobiev, ed., *Phase 1 Joint Report.*

Flynn Shuttle-Mir oral history.

NASA-6 David Wolf: *Recommitment to Mir*

Cable Network News, Inc., "Repairing Mir: Robot camera on Mir abandoned after malfunction," CNN Homepage, Online, <http://www.cnn.com> (Article Dated 17 December 1997; Accessed 27 August 1999).

"Mir finally hears about John Glenn's upcoming flight." *Florida Today* Homepage, Online, <http://www.flatoday.com> (Article Dated 26 January 1998; Accessed 4 April 2000).

"NASA-6/Mir-24 Mission Status Reports: September 12, 1997-January 23, 1998," S-M CD-ROM.

Wolf, "Letters Home," S-M CD-ROM.

Wolf Shuttle-Mir oral history.

Direct Quotes
Unless otherwise noted, direct quotes taken from the Wolf Shuttle-Mir oral history.

Page 122
"The Space Shuttle *Atlantis*' hatch shut ... against the dimly moonlit Earth.":
Dave Wolf, "Letters Home-November 14, 1997: Little things mean a lot up here," S-M CD-ROM.

"Now, be careful down there ... You take care of yourselves, and I'll be okay up here.":
Wetherbee Shuttle-Mir oral history.

Page 123
"The central command post ... bring us back to Earth.":
Dave Wolf, "Letters Home-October 31, 1997: My mother would be proud of me," S-M CD-ROM.

"It was almost eerie ... Supplies had arrived.":
Dave Wolf, "Letters Home-October 21, 1997: Progress supply ship docking," S-M CD-ROM.

Page 124
"quite good ... cleaner than when I came.":
"Mir 24 Weekly Update-October 17, 1997: David Wolf Reports from the Mir," NASA Human Spaceflight Homepage, Online, <http://spaceflight.nasa.gov/history/shuttle-mir/mir24/status/week10/wolf.html> (Last Updated 31 January 1998; Accessed 16 May 2000).

Page 125
"It's important ... It's what puts the people in charge.":
"NASA-6/Mir-24 Mission Status Report Weekly Update-November 7, 1997: Interview with David Wolf," S-M CD-ROM.

Page 126
"My job was issuing ... spacesuit carbon dioxide scrubbers.":
Wolf, "Letters Home-November 14, 1997: Little things mean a lot up here," S-M CD-ROM.

Page 127
"I love space ... But I get better at it every day.":
"NASA-6/Mir-24 Mission Status Report Weekly Update-November 7, 1997: Interview with David Wolf," S-M CD-ROM.

Page 130
"fixer-upper ... change the plan during the mission.":
"NASA-6/Mir-24 Mission Status Report Weekly Update-November 7, 1997: Interview with David Wolf," S-M CD-ROM.

"Ghostly outlines ... Set like a gem stone into the Earth's crust.":
Wolf, "Letters Home-November 14, 1997: Little things mean a lot up here," S-M CD-ROM.

Page 131
"After about a month ... with the Australians first.":
"NASA-6/Mir-24 Mission Status Report Weekly Update-December 12, 1997: Press Conference with the Mir-24 Crew," S-M CD-ROM.

Page 132
"The whole team ... sounds of drilling and wrenching subside, he emerges.":
Dave Wolf, "Letters Home-December 22, 1997: The Great Blackcurrant Jelly Juice Spill," S-M CD-ROM.

Profile: David A. Wolf

"Biographical Data: David A. Wolf (M.D.)," NASA Johnson Space Center Homepage, Online, <http://www.jsc.nasa.gov/Bios/htmlbios/wolf.html> (Last Updated October 2000; Accessed 15 November 2000).

Wolf Shuttle-Mir oral history.

After Weighing the Risks

"Press Briefing-September 25, 1997: Getting Settled in on Mir," NASA Human Spaceflight Homepage, Online, <http://spaceflight.nasa.gov/history/shuttle-mir/mir24/status/week7/goldin.html> (Last Updated 26 September 1997; Accessed April 2000).

Getting Settled in on Mir: A letter home from David Wolf

Dave Wolf, "Letters Home-October 6, 1997: Getting Settled in on Mir," S-M CD-ROM.

Meanwhile on Earth, NASA-6

"Culbertson and Parazynski lend expertise to space station crews," News Release 99-108, NASA Spacelink Homepage, <http://spacelink.nasa.gov/NASA.News/NASA.News.Releases/Previous.News.Releases/99.News.Releases/99-09.News.Releases/99-09-24.Space.Station.Crew.Expertise> (Release Dated 24 September 1999; Accessed 13 September 2000).

"NASA-6/Mir-24 Mission Status Reports: October 10, 1997-November 28, 1997," S-M CD-ROM.

Sharipov Shuttle-Mir oral history.

"U.S., Russia Name International Space Station Crews," News Release 97-269, NASA Spacelink Homepage, <http://spacelink.nasa.gov/NASA.News/NASA.News.Releases/Previous.News.Releases/97.News.Releases/97-11.News.Releases/97-11-17.U.S.-Russia.Names.ISS.Crews> (Release Dated 17 November 1997; Accessed 13 September 2000).

Direct Quotes
Page 134
"high point ... But we were all there together.":
Patti Moore Shuttle-Mir oral history.

Who Went to Mir and Who Didn't

Nield and Vorobiev, ed., Phase 1 Joint Report.

Direct Quotes
Page 135
"all volunteer force.":
"Mission Highlights STS-89," IS-1998-01-001.089JSC, NASA Johnson Space Center Homepage, Online, <http://vesuvius.jsc.nasa.gov/pao/factsheets/missions/089/Hilit89.pdf> (Last Updated 10 March 2000; Accessed 3 March 2001).

STS-89: *Last NASA Astronaut to Mir*

Dunbar Shuttle-Mir oral history.

"*Endeavour* arrives at Mir to pick up homesick astronaut," *Florida Today* Homepage.

"STS-89 (89)," Kennedy Space Center Homepage, Online, <http://science.ksc.nasa.gov/shuttle/missions/sts-89/mission-sts-89.html> (Last Updated 31 August 2000; Accessed September 2000).

"STS-89 Fact Sheet: STS-89/*Endeavor*, Eighth Shuttle-Mir Docking," KSC Release No. 220-97, KSC Fact Sheets Homepage, Online, <http://www-pao.ksc.nasa.gov/kscpao/release/1997/220-97.htm> (Release Dated December 1997; Accessed December 2000).

"STS-89 Press Kit," Kennedy Space Center Homepage, Online, <http://science.ksc.nasa.gov/shuttle/missions/sts-89/sts-89-press-kit.txt> (Last Updated January 1998; Accessed 29 March 2000).

Direct Quotes
Page 137
"We're just discussing the fact ... grown ... like a glove.":
"*Endeavour* brings fresh astronaut researcher to space station Mir," *Florida Today* Homepage, Online, <http://www.flatoday.com> (Article Dated 25 January; Accessed April 2000).

"When we opened the hatch ... I was on the Shuttle side.":
Dunbar Shuttle-Mir oral history.

NASA-7 Andy Thomas: *Smoother Sailing*

"NASA-7/Mir-24 and Mir-25 Mission Status Reports: September 12, 1997-January 23, 1998," S-M CD-ROM.

Thomas Shuttle-Mir oral history.

Chris van den Berg, "MIRnews Information," Association de Muurkrant Homepage. Online, <http://www.infohuis.nl/muurkrant/mirmain.html> (Last Updated 29 December 1999; Accessed April 2000).

Direct Quotes
Unless otherwise noted, direct quotes taken from the Thomas Shuttle-Mir oral history.

Page 140
"The weather had been questionable ... know what was happening.":
Andy Thomas, "Letters from the Outpost-February 1998: The Flight Begins," S-M CD-ROM.

"We could see the station ... more spacious by comparison.":
Andy Thomas, "Letters from the Outpost-March 1998: Rendezvous and Docking with Space Station Mir," S-M CD-ROM.

Page 142
"Their Soyuz appeared ... protruding solar panels.":
Thomas, "Letters from the Outpost-March 1998: Rendezvous and Docking with Space Station Mir," S-M CD-ROM.

"It was strange to see them all again here in orbit.":
Andy Thomas, "Letters from the Outpost-April 1998: The First Days on Mir," S-M CD-ROM.

Page 143
"when we had a lot of people aboard and it was very crowded.":
"STS-91 *Discovery*/Mir Crew Interview," NASA Human Spaceflight Homepage. Online, <http://spaceflight.nasa.gov/shuttle/archives/sts-91/crewint.html> (Last Updated 7 June 1998; Accessed 19 June 2000).

"The mood ... is excellent ... they do not complain about their modest housing.":
Chris van den Berg, "MIRnews 9 February 1998 (406)," Association de Muurkrant Homepage. Online, <http://www.infohuis.nl/muurkrant/mirmain.html> (Last Updated 29 December 1999; Accessed April 2000).

"it then became a lot easier ... in some ways.":
"STS-91 *Discovery*/Mir Crew Interview," NASA Human Spaceflight Homepage.

"Because we spend a lot of time ... perfectly natural thing.":
"NASA-7/Mir-25 Mission Status Report Weekly Update-February 13, 1998: Andy Thomas Talks About Life on Mir," S-M CD-ROM.

Page 146
"He did this in reasonable and certainly comprehensible Russian.":
Chris van den Berg, "MIRnews 4 March 1998 (411)," Association de Muurkrant Homepage. Online, <http://www.infohuis.nl/muurkrant/mirmain.html> (Last Updated 29 December 1999; Accessed April 2000).

"excellent German ... it was clear that Musabayev had picked up a lot of the German language ... in his native country":
Chris van den Berg, "MIRnews 28 April 1998 (420)," Association de Muurkrant Homepage. Online, <http://www.infohuis.nl/muurkrant/mirmain.html> (Last Updated 29 December 1999; Accessed April 2000).

"gave him more energy.":
Chris van den Berg, "MIRnews 2 January 1998 (401)," Association de Muurkrant Homepage. Online, <http://www.infohuis.nl/muurkrant/mirmain.html> (Last Updated 29 December 1999; Accessed April 2000).

"You need to be able to psychologically remove yourself ... get away from things.":
"NASA-7/Mir-25 Mission Status Report Weekly Update-March 20, 1998," S-M CD-ROM.

"seamless, a beautiful piece of work ... executed the work flawlessly.":
"NASA-7/Mir-25 Mission Status Report Weekly Update-March 20, 1998," S-M CD-ROM.

Page 147
"You see texture ... it depends on the season.":
"NASA-7/Mir-25 Mission Status Report Weekly Update-April 10, 1998: A Conversation with Andy Thomas." S-M CD-ROM.

"I wasn't too surprised ... the best part about being back.":
"NASA-7/Mir-25 Mission Status Report Weekly Update-May 1, 1998: Interview with Andy Thomas," S-M CD-ROM.

Profile: *Andrew S. W. Thomas*

"Biographical Data: Andrew S. W. Thomas (Ph.D.)," NASA Johnson Space Center Homepage. Online, <http://www.jsc.nasa.gov/Bios/htmlbios/thomas-a.html> (Last Updated May 2000; Accessed 13 June 2000).

Thomas Shuttle-Mir oral history.

A View from Space

Andy Thomas, "Letters from the Outpost-May 1998: A Typical Day on Space Station Mir," S-M CD-ROM.

Earth Observations: More than mere beauty

"Astronaut Photographs from Space Shed Light on Changes on Earth," Release: J00-37. NASA Johnson Space Center Homepage. Online, <http://www.jsc.nasa.gov/pao/media/rel/2000/J00-37.html> (Release Dated 7 June 2000; Accessed 19 June 2000).

STS-91: *Closing Out Shuttle-Mir*

"STS-91 (91)," Kennedy Space Center Homepage. Online, <http://science.ksc.nasa.gov/shuttle/missions/sts-91/mission-sts-91.html> (Last Updated 31 August 2000; Accessed September 2000).

"STS-91 Fact Sheet: STS-91/*Discovery*, 9th and last Shuttle-Mir Docking," KSC Release No. 56-98, KSC Fact Sheets Homepage. Online, <http://www-pao.ksc.nasa.gov/kscpao/release/1998/56-98.htm> (Release Dated May 1998; Accessed December 2000).

"STS-91 Press Kit," Kennedy Space Center Homepage. Online, <http://www.shuttle.nasa.gov/shuttle/archives/sts-91/presskit/dsodto.html> (Last Updated 7 March 2000; Accessed March 2000).

Direct Quotes
Page 155
"as a point of light out on the horizon, like a bright star. ... It was a great moment.":
Thomas Shuttle-Mir oral history.

"Charlie, this place is in bad shape ... Maybe I'll just stay ... unsafe.":
Precourt Shuttle-Mir oral history.

"frat house":
Foale Shuttle-Mir oral history, 16 June 1998.

"Well, you know ... Maybe I'll just stay.":
Precourt Shuttle-Mir oral history.

"the condition of the inside of the space station was better ... nice and dry.":
Precourt Shuttle-Mir oral history.

"Charlie, take this wrench ... for all the work that we have ahead of us.":
"Shuttle Departs From Mir as a New Era Begins" *New York Times* Homepage. Online, <http://www.nyt.com/> (Article Dated 9 June 1998; Accessed 30 August 1999).

Looking Back, Looking Forward: *Conidering Shuttle-Mir*

Daniel J. Boorstin, *The Discoverers* (New York: Random House Vintage Books, 1985), 190-97.

Capps Shuttle-Mir oral history.

Frank L. Culbertson, Jr., "What's in a Name?," Paper Submitted to 10th Congress of The Association of Space Explorer, 3 October 1996, S-M CD-ROM.

Culbertson Shuttle-Mir oral history.

Phase 1 Review News Conference, 15 August 1998, Morgan author file.

Rumerman, comp., *U.S. Human Spaceflight, A Record of Achievement 1961-1998.*

Direct Quotes
Unless otherwise noted, direct quotes taken from the Culbertson Shuttle-Mir oral history.

Page 158
"Two households … ancient grudge.":
Romeo and Juliet, prologue, lines 1-3.

"The Shuttle-Mir program … good training in crisis management.":
"Value of Mir Experience Debated," *Florida Today* Homepage, Online. <http://www.flatoday.com> (Article Dated 31 May 1998; Accessed 19 September 2000).

Page 159
"Suddenly, we brought a team together … And we were a program.":
Capps Shuttle-Mir oral history.

"We had no idea how much we would learn … handle the unexpected.":
"NASA-7/Mir-25 Mission Status Report Weekly Update-May 25, 1998: Interview with Frank Culbertson," S-M CD-ROM.

"What I hope the American public can glean … is not capable of doing.":
Precourt Shuttle-Mir oral history.

"it's the people you fly with … you'll have a great flight.":
Lucid Shuttle-Mir oral history.

Shuttle-Mir Accomplishments
McDonald, *Mir Mission Chronicle.*

Nield and Vorobiev, ed., *Phase 1 Joint Report.*

Phase 1 Review News Conference, 15 August 1998, Morgan author file.

Comparing the Increments
Blaha Shuttle-Mir oral history.

Foale Shuttle-Mir oral history, 16 June 1998.

Foale Shuttle-Mir oral history, 7 July 1998.

Foale Shuttle-Mir oral history, 31 July 1998.

Lucid Shuttle-Mir oral history.

Thagard Shuttle-Mir oral history.

Thomas Shuttle-Mir oral history.

Wolf Shuttle-Mir oral history.

All direct quotes taken from the Blaha Shuttle-Mir oral history.

International Space Station-Phase 2
"Space Station Assembly," NASA Human Spaceflight Homepage, Online, <http://spaceflight.nasa.gov/station/assembly/index.html> (Last Updated 21 February 2001; Accessed 3 March 2001).

Mir Space Station
"4 U.S. Envoys Ordered to Leave Russia, Additional Expulsions Likely In Move Seen As Retaliation," Washingtonpost.com Homepage, Online, <http://www.washingtonpost.com> (Article Dated 24 March 2001; Accessed 10 April 2001).

"Accommodation of a Soyuz TM as an Assured Crew Return Vehicle," Research and Technology Highlights and NASA Langley Highlights Homepage, Online, <http://larcpubs.larc.nasa.gov/randt/1993/RandT/SectionH/H4.html> (Last Updated 25 July 2000; Accessed 23 April 2001).

"After 86,331 Orbits, Mir Space Station's 15 Years In Space Ends (Feb 19, 1986-2001)," Visual Satellite Observer's Homepage, Online, <http://www.satellite.eu.org/sat/vsohp/mir.html> (Last Updated 28 March 2001; Accessed 23 April 2001).

Robert A. Braeunig, "The Mir Complex," Rocket and Space Technology Homepage, Online, <http://users.commkey.net/Braeunig/space/specs/mir.htm> (Last Updated 1997; Accessed 24 April 2001).

"The Country's Leadership Must React Adequately to the Scuttling of the Mir Orbiter," *Pravda* Homepage, Online, <http://english.pravda.ru/main/2001/03/24/3141.html> (Article Dated 24 March 2001; Accessed March 25, 2001).

"Heads up for Mir reentry," The Yomiuri Shimbun, 4 March 2001, Morgan author file.

"Hugh Williams: The stunning fall of the Mir," CNN.com Homepage, Online, <http://www.cnn.com/2001/TECH/space/03/23/williams.debrief/index.html> (Article Dated 23 March 2001; Accessed 26 April 2001).

McDonald, *Mir Mission Chronicle.*

"Mir burns, slow for descent," Houston Chronicle Homepage, Online, <http://www.chron.com/content/interactive/space/missions/> (Article Dated 22 March 2001; Accessed 26 April 2001).

"Mir Components: Docking Module," NASA Human Spaceflight Homepage, Online, <http://spaceflight.nasa.gov/history/shuttle-mir/ops/mir/dm.html> (Last Updated 7 August 1998; Accessed 12 April 2001).

"Mir Docking Module: Orbiter Docking System (ODS)," NASA Human Spaceflight Homepage, Online, <http://spaceflight.nasa.gov/shuttle/archives/sts-74/orbit/payloads/dm/ods.html> (Last Updated 7 November 1995; Accessed 12 April 2001).

"Mir EVA activities," SeeSat-L Archives Homepage.

"Mir flashes over South Pacific islands in final, fiery moments," Houston Chronicle Homepage, Online, <http://www.chron.com/content/interactive/space/missions/> (Article Dated 23 March 2001; Accessed 26 April 2001).

Shuttle-Mir Timeline

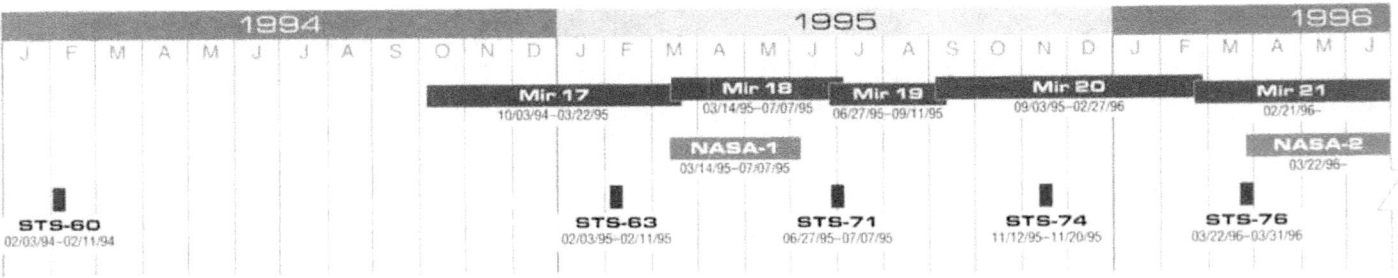

"Mir Orbital Complex Content," Energia Homepage, Online, <http://www.energia.ru/english/energia/mir/mir-structure.html> (Last Updated 30 April 2001; Accessed 2 May 2001).

"Mir Overview," Boeing Homepage, Online, <http://www.boeing.com/defense-space/space/liftoff/sts91/mirov.html> (Last Updated 18 July 2000; Accessed 24 April 2001).

"Mir Space Station," Visual Satellite Observer's Homepage, Online, <http://www.satellite.eu.org/sat/vsohp/mir1.html> (Last Updated 28 March 2001; Accessed 23 April 2001).

"Mir Space Station, Crews and Launch Vehicles," The Astronaut Connection ... Homepage, Online, <http://www.nauts.com/vehicles/80s/mir/index.html> (Last Updated 15 February 2000; Accessed 24 April 2001).

"Mir Space Station Deorbit," U.S. Department of State Homepage, Online, <http://www.state.gov/r/pa/prs/ps/2001/index.cfm?docid=1035> (Release Dated 2 March 2001; Accessed 10 April 2001).

"Most Long-duration Manned Space Flights," Energia Homepage, Online, <http://www.energia.ru/english/energia/mir/mir-long.html> (Last Updated 30 April 2001; Accessed 24 April 2001).

Portree, *Mir Hardware Heritage*.

"Russia to compensate damage if Mir fragments fall down," Itar-Tass (Moscow) News Agency, 3 March 2001, Morgan author file.

"Russian Debt May Deplete Space Funds," Space.com Homepage, Online, <http://space.com/news/spaceagencies/russia_space_funds_010305.html> (Article Dated 5 March 2001; Accessed 10 April 2001).

"Russian PM Orders Demise of Aged Mir" Reuters Homepage, Online, <http://news.excite.com/news/r/010105/07/science-space-mir-dc> (Article Dated 5 January 2001; Accessed 1 April 2001).

Asif A. Siddiqi, "Mission Data for all Missions to Mir," Russian Aerospace Guide Homepage, Online, <http://www.mcs.net/~rusaerog/asifs/Mir.pdf> (Last Updated 30 March 1998; Accessed 30 April 2001).

"Soyuz TM," Federation of American Scientists Space Policy Project Homepage, Online, <http://www.fas.org/spp/guide/russia/piloted/soyuz.htm> (Accessed 23 April 2001).

"Spacecraft: Manned: Mir: Chronology," RussianSpaceWeb.com Homepage, Online, <http://www.russianspaceweb.com/mir_chronology.html>, (Last Updated 27 April 2001; Accessed 27 April 2001).

Patrick E. Tyler, "Mir Space Station Sizzles to Ending Over Pacific," New York Times Homepage, Online, <http://ea.nytimes.com/cgi-bin/email> (Article Dated 23 March 2001; Accessed 19 April 2001).

"Valery Polyakov, Deputy Director, Institute of Biomedical Problems," Institute for Biomedical Problems Homepage, Online, <http://www.ibmp.rssi.ru/webpages/engl/Articles/artic4_e.html> (Last Updated 21 July 1996; Accessed 24 April 2001).

Mark Wade, ed., "Buran," Encyclopedia Astronautica Homepage, Online, <http://www.friends-partners.org/mwade/craft/buran.htm> (Last Updated 12 March 2001; Accessed 10 April 2001).

Mark Wade, ed., "Mir-Shuttle Docking Module," Encyclopedia Astronautica Homepage, Online, <http://www.rocketry.com/mwade/craft/mirodule.htm> (Last Updated 22 December 2000; Accessed 24 April 2001).

Mark Wade, ed., "Soyuz," Encyclopedia Astronautica Homepage, Online, <http://www.friends-partners.org/mwade/project/soyuz.htm> (Last Updated 12 March 2001; Accessed 23 April 2001).

Mark Wade, ed., "Soyuz!," Encyclopedia Astronautica Homepage, Online, <http://www.friends-partners.org/mwade/articles/soyuz.htm> (Last Updated 12 March 2001; Accessed 23 April 2001).

Joanne Welsh to Rebecca Wright, 26 April 2001, email, Morgan author file.

Zimmerman Shuttle-Mir oral history.

Direct Quotes
Page 163
"six school buses ... debris strikes over the years.":
Linenger, *Off The Planet*, 78.

"a bit like a frat house ... better looked after.":
Foale Shuttle-Mir oral history, 16 June 1998.

"Looking into the station ... seeing familiar surroundings.":
Linenger, *Off The Planet*, 95.

Page 164
"It was very wonderful, a wonderful view.":
Titov Shuttle-Mir oral history.

"We don't have a 100-percent safety guarantee.":
"Russia to compensate damage if Mir fragments fall down," Itar-Tass (Moscow) News Agency.

"I am especially sad ... intellectual potential.":
Tyler, "Mir Space Station Sizzles to Ending Over Pacific" New York Times Homepage.

Page 165
"provide the psychological impetus ... 'going in a proper direction.'":
Precourt Shuttle-Mir oral history.

"It's a shame ... a great deal of hope.":
Semyachkin Shuttle-Mir oral history.

Soyuz in Pravda
Portree, *Mir Hardware Heritage*, 3.

Mir Expeditions
McDonald, *Mir Mission Chronicle*.

Portree, *Mir Hardware Heritage*.

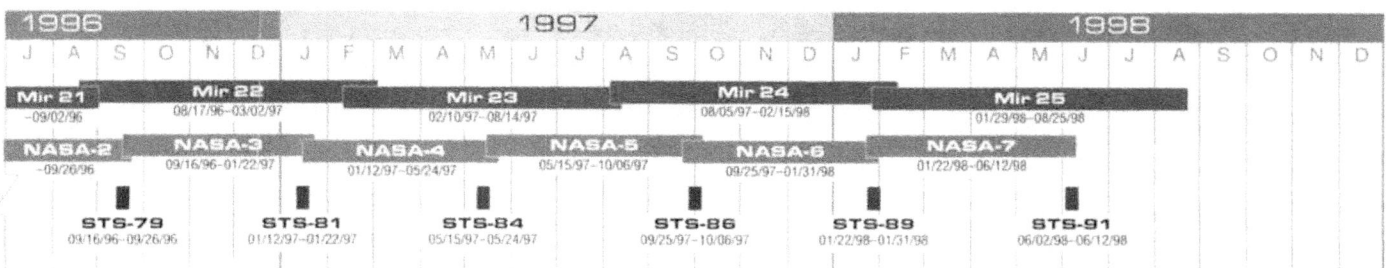

Imagery

Images printed in this publication are from the NASA archives. The photo identification number of each image is printed below with corresponding page number. Exceptions to this statement are noted. NASA acknowledges the following for providing imagery used in this publication: the Russian Space Agency, Analytical Graphics, Inc., Michael R. Barratt, L.D. Stevenson, Don Bourque, *Space News*, and AP World Wide Photos.

Front cover: Composite of Shuttle-Mir images from the NASA archive. Table of Contents: **p. iv**, S95—13884; **p. vi**, KSC-98EC-0227; **p. viii**, STS076-713-036; **p. ix**, KSC-97EC-1245; **p. x**, STS071(S)072.

History's Highest Stage: **p. 2** left, 70-H-1075; **p. 2** right, AS11-40-5878; **p. 3**, S74-24913; **p. 6**, S90-38573; **p. 7**, S98-02782.

STS-60: **p. 8**, STS060-31-028; **p. 9** left, STS060-09-015; **p. 9** right STS060-12-018.

Meanwhile on Earth: **p. 11**, S95-05246; **p. 12**, KSC-97EC-1249; **p. 13** (composite), NM18-310-005, NM21-401-020.

STS-63: **p. 14**, STS063-06-004; **p. 15**, STS063-712-072.

NASA-1: **p. 16**, S94-41077; **p. 18** top, S95-07311; **p. 18** bottom, S95-07315; **p. 19**, S95-07293; **p. 20**, S95-07325; **p. 21** top, S95-07323; **p. 21** bottom, S95-07333; **p. 22**, S78-35299; **p. 23**, STS063-708-088; **p. 24**, NM18-304-017; **p. 25** top, NM18-301-033; **p. 25** bottom, NM18-303-004; **p. 26** top, NM18-305-029; **p. 26** bottom, NM18-303-032; **p. 27**, NM18-305-008; **p. 28**, NM18-305-026; **p. 29** top, NM18-309-026; **p. 29** bottom, STS071-112-004; **p. 30**, JSC2001e16597; **p. 31** top, JSC2001e16598; **p. 31** left, JSC2001e16596; **p. 31** right, JSC2001e16599; **p. 31** bottom, JSC2001e16595.

STS-71: **p. 32**, STS071-122-013; **p. 33** top, STS071-117-020; **p. 33** bottom, STS071-319-005; **p. 35**, STS071(S)074.

STS-74: **p. 36**, STS074-318-005; **p. 37**, KSC-95EC-1604.

Training and Operations: **p. 38** top, S94-41093; **p. 38** bottom left, S95-08563; **p. 38** bottom right, S95-08567; **p. 39** top left, S94-41064; **p. 39** top middle, S95-03306; **p. 39** top right, S94-35074; **p. 39** right, S98-05705; **p. 39** right, S96-17476; **p. 39** bottom, S86-31548; **p. 40** top left, S97-06567; **p. 40** top right, S97-11030; **p. 40** bottom, S95-04320; **p. 41** top, S98-15840; **p. 41** bottom, S97-11026; **p. 42**, S97-12217; **p. 43** top, S96-12746; **p. 43** bottom, 97e00864; **p. 44** top, JSC2000e11559; **p. 44** bottom, NM21-402-013; **p. 45**, NASA6-349-022; **p. 46**, S99-15928; **p. 47**, S94-46620; **p. 48**, STS079S167; **p. 49** top, S94-46594; **p. 49** bottom, S94-41088; **p. 50**, STS079(S)150.

STS-76: **p. 52**, STS076-371-002; **p. 53**, STS076-730-000G.

NASA-2: **p. 54**, S95-21463; **p. 56**, STS076-345-025; **p. 57**, STS076-713-083; **p. 58**, NM21-386-024; **p. 59**, S97-06343; **p. 60**, NM21-382-010; **p. 61** top, NM21-388-007; **p. 61** bottom, NM22-427-012; **p. 62**, KSC-96EC-1040; **p. 63** left, NM22-420-024; **p. 63** right, NM21-381-034; **p. 64**, STS079-332-018; **p. 65**, JSC2000e11530; **p. 66**, S94-28588; **p. 67**, STS076-736-017.

STS-79: **p. 68**, STS079-349-022; **p. 69** top left, STS079-355-017; **p. 69** top right, STS079-309-022; **p. 69** middle, STS079-350-001; **p. 69** bottom, STS079(S)099.

NASA-3: **p. 70**, KSC-96EC-1007; **p. 72** top, S96-10559; **p. 72** bottom, STS079-331-021; **p. 73**, S88-45293; **p. 74** left, STS079-334-015; **p. 74** right, NM22-28-005; **p. 75**, STS079-817-034; **p. 76**, NM22-416-017; **p. 77** top, STS079-349-023; **p. 77** bottom, NM22-23-014; **p. 78**, STS081-306-017; **p. 79**, KSC-97EC-0188; **p. 80** left, STS081-336-001; **p. 80** top right, STS089-333-018; **p. 80** bottom right, NM23-108-025; **p. 81** top left, STS060-28-021; **p. 81** right, STS084-326-036; **p. 81** bottom, STS084-303-004; **p. 83**, S94-35075.

STS-81: **p. 84**, STS081-369-003; **p. 85** top, NM23-77-005; **p. 85** bottom, STS081-312-010.

NASA-4: **p. 86**, NM23-53-023; **p. 88** top, STS081-372-035 ; **p. 88** STS083-372-035; **p. 89**, S95-10762; **p. 90** top, NM23-74-020; **p. 90** bottom left, NM23-104-021; **p. 90** bottom right, NM23-111-026; **p. 91** bottom, NM23-64-026; **p. 91** top, STS084-303-029; **p. 92** top, STS084-319-034; **p. 92** middle, STS086-405-034; **p. 92** bottom, NM23-37-004; **p. 93**, STS081-358-007; **p. 94** top, STS084-389-024; **p. 94** bottom, NM21-449-005; **p. 95** bottom, STS086-405-024; **p. 95** top, NM23-56-019; **p. 96**, NM23-36-003; **p. 98**, NM23-48-002; **p. 99** top, NM23-91-025; **p. 99** bottom left, NM23-45-038; **p. 99** right, NM23-48-032; **p. 100**, NM23-54-030.

STS-84: **p. 102**, STS084-380-019; **p. 103** top left, STS084-379-034; **p. 103** top right, STS084-730-002; **p. 103** bottom, STS084-377-026.

NASA-5: **p. 104**, S96-14349; **p. 106**, STS084-310-025; **p. 107** top, S87-45890; **p. 107** bottom, STS084-371-021; **p. 108** top left, NM23-47-37; **p. 108** top right, S84e5069; **p. 108** bottom, STS086-369-002; **p. 109** top, STS086-364-028; **p. 109** bottom, STS084-305-018; **p. 110** top to bottom, STS086-387-014, STS086-388-001, S89e5189, STS086-373-015; **p. 111**, S86e5781; **p. 112** top, NASA5-326-007; **p. 112** bottom, S86e5400; **p. 113**, NASA5-331-029; **p. 116**, photo by Steve Elfers, courtesy of *Space News*; **p. 117**, STS086-720-100.

STS-86: **p. 118**, STS086-371-004; **p. 119**, STS086-404-009.

NASA-6: **p. 120**, NASA6-316-018; **p. 122** top, S97-11022; **p. 122** bottom, STS086-342-036; **p. 123** top, NASA6-306-032; **p. 123** bottom, NASA6-321-016; **p. 124** top to bottom, STS74-324-027, STS079-336-014, STS86-301-011; **p. 125**, S90-45136; **p. 126** top, NASA6-335-014; **p. 126** bottom, NASA6-340-017; **p. 127**, NASA6-341-033; **p. 129**, STS086-720-055; **p. 130**, NASA6-319-022; **p. 131**, STS076-714-017; **p. 132**, NASA6-325-022; **p. 133**, NASA6-337-022; **p. 134**, NASA6-344-008; **p. 135**, STS086-350-022.

STS-89: **p. 136**, STS089-391-004; **p. 137** top, STS089-390-008; **p. 137** bottom, STS089-335-001.

NASA-7: **p. 138**, STS091-379-018; **p. 140** top, STS089-397-010; **p. 140** bottom, STS089-368-002; **p. 141**, STS089-346-007; **p. 142** top, S89e5236; **p. 142** bottom, NASA7-301-008; **p. 143**, S92-49268; **p. 144** top, STS089-330-026; **p. 144** bottom, NASA7-312-031; **p. 145**, NASA7-715-36E; **p. 146**, STS091-361-037; **p. 147**, NASA7-307-001; **p. 148**, NASA7-715-068; **p. 149**, STS091-711-028; **p. 150** top, 51F-36-059; **p. 150** bottom left, NM21-735-062; **p. 150** bottom right, NASA6-709-080; **p. 151** top, NM21-766-065; **p. 151** bottom, 51B-166-001; **p. 152** top, STS071-746-076; **p. 152** bottom left, STS074-736-047; **p. 152** bottom right, STS081-720-037; **p. 153** top left, NM21-752-048; **p. 153** top right, NM21-763-093; **p. 153** bottom left, STS076-723-094; **p. 153** bottom right, NM21-771-076.

STS-91: **p. 154**, STS091-383-004; **p. 155** bottom left, STS091-370-035; **p. 155** top right, STS091-367-013.

Looking Back, Looking Forward: **p. 156**, S98-13530; **p. 161**, STS097-704-074.

Mir Space Station: **p. 162**, STS086-710-007; **p. 165**, computer-generated images courtesy Analytical Graphics, Inc., **p. 166**, STS086-334-024; **p. 167** top, STS074-334-016; **p. 167** bottom, STS079-324-014; **p. 168**, S95-22141; **p. 169**, STS081-351-018; **p. 170**, S79e5219; **p. 171**, STS063-28-024; **p. 172** top, STS076-341-023; **p. 172** bottom, STS091-336-021; **p. 173** top left, STS084-305-021; **p. 173** bottom left, STS074-302-023; **p. 173** top right, NM23-56-016; **p. 173** bottom right, STS071-105-021; **p. 174**, S79e5124; **p. 175** top, S93-41263; **p. 175** middle, S93-41271; **p. 175** bottom, S93-41213; **p. 176**, STS063-711-080; **p. 177** top, AP World Wide Photos; **p. 177** bottom, STS071-117-034; **p. 178** top, STS074-318-003; **p. 178** middle, NM21-395-024; **p. 178** bottom, NM22-421-011; **p. 179** top, NM23-111-026; **p. 179** middle, STS086-342-036; **p. 179** bottom, STS091-363-017.

Back cover: S98-13530.

Index